J. Peinke   J. Parisi   O. E. Rössler
R. Stoop ————————————————

Encounter with Chaos

Self-Organized Hierarchical Complexity in Semiconductor Experiments

With 182 Figures

Springer-Verlag

Berlin Heidelberg New York
London Paris Tokyo
Hong Kong Barcelona
Budapest

Dr. *Joachim Peinke*
Centre de Recherche sur les très Basses Températures, C.N.R.S.,
BP 166, F-38042 Grenoble Cedex 9, France

Priv.-Doz. Dr. *Jürgen Parisi*
Physics Institute, University of Zurich, Schönberggasse 9,
CH-8001 Zurich, Switzerland

Prof. Dr. *Otto E. Rössler*
Institute for Physical and Theoretical Chemistry, University of
Tübingen, Morgenstelle 8, W-7400 Tübingen, Fed. Rep. of Germany

Dr. *Ruedi Stoop*
Physics Institute, University of Zurich, Schönberggasse 9,
CH-8001 Zurich, Switzerland

ISBN 3-540-55845-4 Springer-Verlag Berlin Heidelberg New York
ISBN 0-387-55845-4 Springer-Verlag New York Berlin Heidelberg

Library of Congress Cataloging-in-Publication Data. Peinke, J. (Joachim), 1956– Encounter with Chaos: Self-organized Hierarchical Complexity in Semiconductor Experiments/J. Peinke ... [et al.]. p. cm. Includes bibliographical references and index. ISBN 3-540-55845-4 (Berlin: alk. paper). — ISBN 0-387-55845-4 (New York: alk. paper) 1. Semiconductors. 2. Nonlinear theories. QC611.P42 1992 537.6'22 — dc20 92-25600

Typesetting: Macmillan India Ltd., Bangalore-25

54/3140/SPS-543210 – Printed on acid-free paper

Preface

Our life is a highly nonlinear process. It starts with birth and ends with death; in between there are a lot of ups and downs. Quite often, we believe that stable and steady situations, probably easy to capture by linearization, are paradisiacal, but already after a short period of everyday routine we usually become bored and seek change, that is, nonlinearities. If we reflect for a while, we notice that our life and our perceptions are mainly determined by nonlinear phenomena, for example, events occurring suddenly and unexpectedly.

One may be surprised by how long scientists tried to explain our world by models based on a linear ansatz. Due to the lack of typical nonlinear patterns, although everybody experienced nonlinearities, nobody could classify them and, thus, study them further. The discoveries of the last few decades have finally provided access to the world of nonlinear phenomena and have initiated a unique inter-disciplinary field of research: nonlinear science. In contrast to the general tendency of science to become more branched out and specialized as the result of any progress, nonlinear science has brought together many different disciplines. This has been motivated not only by the immense importance of nonlinearities for science, but also by the wonderful simplicity of the concepts. Models like the logistic map can be easily understood by high school students and have brought revolutionary new insights into our scientific under-standing.

At this point, we do not want to withhold Jim Yorke's important choice when he coined the word "chaos" for the special phenomena arising from nonlinearity. This word appears to have been the best advertising slogan for such research and, indeed, has exerted an almost magical attractive power. Here, one should become cautious. On the one hand, chaos has a preconceived meaning for our daily experiences. On the other, chaos characterizes particular phenomena of nonlinear models. Of course, the two meanings seem to be related, but it is not at all evident that they are identical. Even though we are familiar with some features of nonlinear models, it would be a mistake to lull ourselves into a false sense of security, thinking that we know what chaos is. There may be further inter-esting things which we have denoted as chaos in our daily lives (and, thus, in science) and yet we are unable to model with our present scientific understanding.

Nevertheless, the present book is devoted to the world of nonlinear science. We do not intend to review all the excellent experimental work done in this field, but we want to take the reader on a journey into the nonlinear world of a tiny semiconductor crystal. Taking this system as an example, we attempt to show what can be explained using these new methods of investigation. The multiplicity of nonlinear phenomena observed experimentally was one major motivation to write this book, an introduction to the basic concepts of nonlinear dynamics. Furthermore, we shall be pleased if we succeed in stimulating readers to embark on their own "nonlinear studies".

It is fortuned that we are able to treat the nonlinear dynamical behavior of a nonequilibrium semiconductor system, originally known as electronic device which exhibits a highly reproducible physics. No other physical system has lately shaped our life in such a comprehensive way. Surprisingly, it was the semiconductor and its precision which opened the door on the phenomenon of chaos. That these findings represent the rule rather than being the product of exotic working conditions will be outlined in the present book. We start with the fundamental transport properties of semiconductor physics that lead to the nonlinear dynamics. Emphasis is laid on the example of low-temperature impact ionization breakdown in extrinsic germanium. The main part of the book concerns the hierarchical classification of experimental results according to their degree of spatio-temporal complexity. Finally, we give a well-founded mathematical background to the different probabilistic and dynamical characterization methods applied. Based on a unified thermodynamical formalism, the theoretical description is intended to provide a more or less comprehensive understanding of the distinct scalings extracted experimentally.

Tübingen and Zurich, February 1992

Joachim Peinke
Jürgen Parisi
Otto E. Rössler
Ruedi Stoop

Contents

Acknowledgements

The present book would not exist without the initiating impetus given by Wolfgang Metzler. The authors greatly benefited from the collective theoretical and experimental work with Reinhold Braig, Hans Brauchli, Wilfried Clauß, Minh Duong-van, Hans-Peter Herzel, Fack Hudson, Achim Kittel, Michael Klein, Oscar E. Lanford III, Martin Lehr, Klaus Michael Mayer, Andreas Mühlbach, Uwe Rau, Reinhard Richter, Brigitte Röhricht, Eckehard Schöll, Heinrich Seifert, and Tamas Tel. The authors are indebted to Marie-Luise Fenske and Hans-Günther Wener for assistance during the preparation of the manuscript. The experiments were performed at the Physical Institute of the University of Tübingen, where Rudolf P. Huebener gave us his kind support. The authors profited from an unbureaucratic and generous financial grant from the Volkswagen Foundation. Last but not least, the effective cooperation with Helmut Lotsch from Springer-Verlag is gratefully acknowledged.

1 Introductory Remarks

In analogy with an old anarchist slogan, chaos is not the absence of order, but it represents a higher form of it (Fig. 1.1). Nowadays, chaos has become popular. Scientists even started to develop their own understanding of chaos by playing on computers with simple algorithms like, for example, the logistic map. New insights into the foundations of science and nature were obtained. There are chaotic deterministic systems whose determinism cannot be experienced in principle. The question whether God plays dice or not may be seen in a new light. Yet, besides these deep philosophical implications, chaos has opened a door for scientists to create lovely pictures of fractals on a computer screen. The subjective opinion of their beauty is founded on the harmony exhibited, which probably corresponds to that of nature. Without overstating, one might claim that the field of nonlinear dynamics and chaos is in keeping with today's *Zeitgeist*, where not only is unconventionality becoming conventional, but where it is also more fashionable to play with computers than to play in nature.

Chaos, as a phenomenon of nonlinear dynamics, in its historical view represents nothing but a product of our computerized world. Up to now, nearly all major progress has been based upon computer work, although chaos had been seen much earlier in experimental systems without being characterizable at that time. Once one has become familiar with chaos, which can be achieved simply by playing with elementary algorithms on a computer, it is at once easy to rediscover chaos everywhere in nature. However, the essential advantage of computer simulations lies in the high precision provided by the advanced semiconductor chip technology and the artificial singularization of distinct phenomena. In contrast to that, nature usually reflects an accumulation of nonlinearities which first have to be disentangled. Nevertheless, there is a lot of good experimental work proving the existence of chaos in all conceivable kinds of different dynamical systems, such as the dripping faucet, the laser, road traffic, and the X-ray luminosity of neutron stars (merely to mention a few examples).

Working experimentally on chaos in physics, one has often quite soon the problem of falling between two stools. The one stands for nonlinear science, the other for system-inherent theory. On the one hand, with respect to nonlinear science, there exists the superiority of the computer simulations. Proving evidence of a known nonlinear phenomenon in an experimental system often gets criticized contemptuously as deserving no general interest. On the other hand, it is nearly impossible to achieve the standards of accordance between experiment and theory that have been established for linear problems, i.e.,

Fig. 1.1. Basin of attraction for the perturbed complex logistic map $z_{n+1} = z_n^2 + C + 0.3\,\text{Re}\{z_n\}$ with $C = -0.03 + 0.75i$ (from [1.1])

equilibrium situations. This drawback directly results from the nonlinearities involved, usually to some extent guiding the physical system in question towards a "new world". The system becomes enslaved by universal nonlinearities. A generic bifurcation like, for example, the Hopf bifurcation, always looks the same no matter what the investigated system is. But it is not always easy or even possible to deduce such bifurcation from first principles underlying the theoretical, physical understanding of the system. Particularly for these macroscopic effects of dissipative multicomponent systems giving rise to spontaneous structure formation, it sometimes becomes hopelessly difficult to explain the observation by means of conventional physical models. Then, one has to take over the methods of nonlinear science. We point out that such forms of paradigm change are common in physics. Quite often, the main part of solving a problem has been done once the appropriate simplification has been found (for instance, in the case of the quasiparticle concept). In some respects, one might regard nonlinear dynamics as a comparable simplifying concept, even though nowadays it is not accepted by everybody.

The experimentalist may handle the discrepancy between the system-dependent theory of the underlying physics and the system-independent model of abstract nonlinear dynamics in different ways. To gain best agreement, one should find appropriate working conditions. In this connection, externally periodically driven systems seem to behave well. In order to explain many different nonlinear phenomena, the system investigated has to display damped or undamped periodic oscillations which obey the theoretical model. Such kinds of systems have provided the most accurate experimental results during recent years (note, e.g., the universal quasiperiodic transition to chaos at the golden mean). But even in periodically forced systems, unexpected surprises may arise. Application of two different experimental set-ups can lead to different dynamical responses, although much effort has been invested to prepare the same arrangements. Unavoidable changes in the nonlinear elements are capable of causing a totally different dynamics. Of course, such apparently erratic behavior indicates one of the characteristic features of nonlinear systems. Well, then the question, what sense a system-specific physical model can have if it is not able to describe two similar experimental situations, must be posed.

Another approach results from investigating systems with the maximum degree of self-organization by keeping all working conditions as constant as possible. Here we have in mind the class of dissipative systems where self-organized structures may emerge as a consequence of a nonequilibrium state. These dissipative patterns can develop in space and/or in time. Accordingly, they introduce new length and time scales. Upon looking at these systems, one usually breaks new ground. So far, we are aware of only poor knowledge on that score. Experience taken from equilibrium situations of the system in question does not help much. Some basic concepts of self-organization, symmetry, structure formation, interaction of subsystems, and their hierarchical configuration are known from the pioneering work by Prigogine and Haken. But still there exists no universal overall knowledge about these systems.

Each subclass of dissipative systems exhibits its own nonequilibrium world. At this stage, research becomes particularly exciting, since one has the impression of taking part in solving the age-long problem of how nature operates and how life may be created. Pretty often, involvement of typical nonlinear phenomena can be recognized, providing an actual chance to approach the system considered. In our opinion, it is always more important to find access to a complex system behavior by an abstract theory of nonlinear dynamics than to get stuck by the traditional system-inherent ansatz. We therefore believe that a lot of hitherto unsolved nonlinear problems – not only in physics – would be unravelled to some extent by changing the method of looking at the phenomenon accordingly.

With the purpose of offering the reader a certain intuition for the nonlinear world of chaos, we outline in the following the simplest type of chaotic flow that is possible in three dimensions. To understand the nature of chaos, one may either read *Anaxagoras* (456 B.C. [1.2]) or study a mechanical mixing system, the rotating taffy puller, which captures the essence of his model [1.3]. Two synchronously moving pairs of arms in Fig. 1.2a serve to automatically stretch

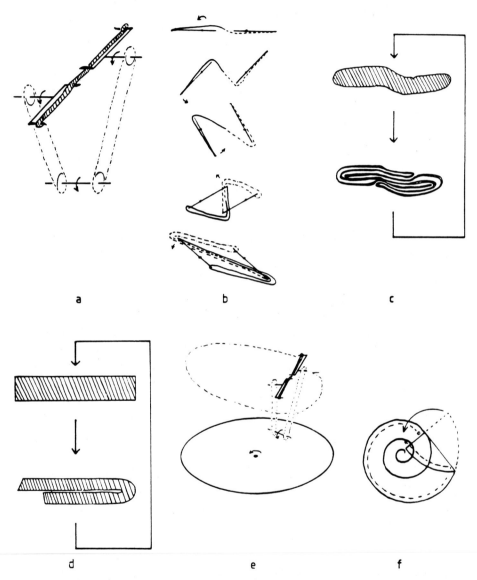

a   b   c

d   e   f

Fig. 1.2a–f. Scheme of the operating principle of a mechanical chaos machine: **a** taffy puller; **b** subsequent positions of the hands over half a cycle; **c** overall two-dimensional map that would apply if the whole process were strictly two-dimensional (think of two confining glass plates, one in front of and one behind the taffy); **d** simplified overall map; **e** rotating taffy puller; **f** zero-thickness version of a simplified rotating taffy puller ("paper model"). The dashed arrows in **e** and **f** each show the path in the three-dimensional space of a single point ("light bulb") (from [1.3])

and fold a piece of taffy – that sticky, expansible, sweet material from which caramel candy is made. Figure 1.2b shows a number of subsequent "shots" viewed from the side. Figure 1.2c illustrates the underlying law, and Fig. 1.2d, an equivalent, simplified two-dimensional map. Figure 1.2e puts the taffy puller on a platform rotated in synchronization with it ("lazy susan"), whereas Fig. 1.2f presents an essentially equivalent, but "infinitely thin", flow caused by an expanding, rotating, and ever again folded-back motion in a sheet of paper.

The pictures are self-explanatory. Stretching plus folding implies physical mixing, because neighboring points (think of two raisins suspended in the taffy) diverge exponentially, while previously distant points are brought close together. If the original piece of taffy consists of labeled (black or radioactive) molecules in the right-hand portion and nonlabeled ones in the left-hand portion, it takes only about 24 rotations (or less than half a minute if there is one rotation per second) until a "molecular sandwich" of black and white molecules has been achieved [if the original height is 1 cm and $(1/2)^{24}$ cm is the diameter of a molecule].

The rotating taffy puller of Fig. 1.2e forms a dynamical system in the sense of mathematics [1.4]. If a tiny little light bulb is assumed to be present inside the taffy and a dark room is assumed, then the visible light path, i.e., the bulb's trajectory, forms an invertible flow (in the mathematical sense). This flow happens to be governed (in the simplest case) by the more or less walking-stick-shaped two-dimensional diffeomorphism of Fig. 1.2d. The "paper flow" of Fig. 1.2f again traces out such a line path, but with the additional simplifying assumption that the thickness of the original taffy is zero.

These pictures serve to show that in three dimensions it is very easy to generate an inextricably complicated "tangle" of hairlines ("spaghetti") that, nevertheless, is locally simple (everywhere parallel) and everywhere invertible and, moreover, governed by a very simple overall map. Such a disciplined tangle describes the simplest type of chaos.

It would be surprising if this simple spatial principle ("three-dimensional blender") did not readily arise spontaneously in very simple equations and systems.

The principle of Fig. 1.2e was apparently already known to *Anaxagoras*. In his conservative cosmology, everything existed in a state of perfect mixture for an infinity of time. Then the sole immiscible substance, mind, at one point in space and time (nowadays called an initial condition) started a recurrent motion. In Greek, the technical term invented by Anaxagoras was *perichoresis* – "running around". Anaxagoras thus invoked the main ingredients of the above machinery to generate a process of "unmixing" – since he needed to explain the emergence of simple ingredients out of chaos. The machine of Fig. 1.2e was, of course, introduced by *Rössler* [1.3] with the opposite goal in mind – to explain the emergence of chaos out of simple things. The two views differ only in the direction of time.

The word chaos, incidentally, originally had the meaning "emptiness" ("yawning"). Since the 4th century B.C., however, that is, since the time of Anaxagoras, its meaning has changed towards denoting mixture and turmoil.

The paper flow of Fig. 1.2f suggests that it should be very simple to set up an equation with the same behavior. Try the following. Take a two-variable linear oscillator with an unstable focus, like

$$\ddot{x} = -x + 0.15\dot{x} \tag{1.1}$$

or, equivalently,

$$\dot{x} = -y \, , $$
$$ \tag{1.2}$$
$$\dot{y} = x + 0.15y \, .$$

Then, add a third variable, z, that tends towards a value close to zero whenever x is less than a certain threshold value, but rises autonomously as long as x is larger. A simple example reads

$$\dot{z} = 0.2 + z(x - 10) \, , \tag{1.3}$$

where the value 10 gives the threshold. What is still lacking is a folding mechanism. Until this moment, as long as z is large, the "pivot" around which the linear suboscillator is rotating is just the same as before (uncoupled case). If the pivot is instead assumed to be displaced downwards along the y-axis during the elevation of z, however, the trajectory when touching down again will no longer reach the same point as if there had been no intervening elevation, but will be displaced somewhat downwards and towards the right in proportion to its former height. This can be achieved by introducing a feedback from z to x – by subtracting z on the right-hand side of the first line, for example.

The result is the set of three-variable equations [1.5]

$$\dot{x} = -y - z \, , $$
$$\dot{y} = x + 0.15y \, , \tag{1.4}$$
$$\dot{z} = 0.2 + z(x - 10) \, .$$

A numerical simulation of this flow can be found in Fig. 1.3. One sees that the equation indeed functions as expected. One also notices that an actual cross section through the flow cannot be noninvertible like an ideal paper flow, but must be invertible, since the right-hand sides of (1.4) are all analytical functions, meaning that the existence and uniqueness theorems for ordinary differential equations [1.6] are fulfilled. A cross section through the flow of Fig. 1.3 therefore has roughly the form of the map shown in Fig. 1.2d, the walking-stick map.

The simplest algebraic equation generating such a map, incidentally, reads

$$a_{n+1} = \gamma a_n(1 - a_n) - b_n \, , $$
$$ \tag{1.5}$$
$$b_{n+1} = (\delta b_n - \varepsilon)(1 - 2a_n) \, ,$$

which is easy to put into a computer. An appropriate set of parameters is $\gamma = 3.8$, $\delta = 0.4$, and $\varepsilon = 0.02$. A rectangular box that is never left under iteration

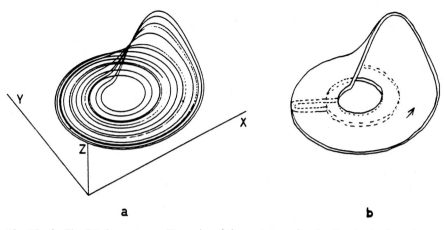

Fig. 1.3a, b. The Rössler system. **a** Dynamics of the prototype chaotic attractor in three-dimensional phase space obtained from (1.4). **b** Schematic presentation of the prototype attractor in three-dimensional phase space accentuating the walking-stick form of the cross section through the flow (from [1.3])

can be defined by the axis lengths 0.04 . . . 0.98 for a and -0.0294 . . . 0.0306 for b, respectively (see [1.7] for pictures). Equation (1.5) is related to the well-known *Hénon* map [1.8], but contains an additional quadratic term ($2\delta a_n b_n$, second line) which is necessary if the "taffy" is indeed to be folded realistically. [Setting δ equal to zero in (1.5), so that the latter becomes equivalent to an orientation-preserving version of Hénon's map, incidentally, also produces realistic foldings – but only if instead of the first iterate the second one is looked at. This second iterate is, however, again described by an equation with two quadratic terms on its right-hand side.] Therefore, it is a matter of taste which of the two maps to choose for one's walking stick. While it is possible to study chaos purely on the basis of such maps (whereby many interesting results have already been found [1.9]), this implies a concentration on detail that interferes with the present purpose of classifying whole flows rather than their internal structure.

There remains just one major "little detail" concerning folded-over maps that should be mentioned here. Already while observing the taffy puller of Fig. 1.2a in a display window of a candy store (like the one in a shopping mall of Salt Lake City), one sees that at some places little nonexpanding "pockets" form in the taffy. This occurs in the "knee" regions and is the consequence of a "lack of stretching" there. At least one "contracting pocket" almost always exists in the folded-over cross section formed by a simple volume-contracting flow like that of Fig. 1.3. Therefore, such flows almost always contain one or several periodic attractors – usually of very high periodicity – buried in the attracting chaotic regime. The existence of these pockets, nonetheless, does not interfere with the fine structure of the overall attracting object (the "red line" with zero area, but uncountably many times the length of the original rectangle [1.10]) that is formed as a limiting cross section.

In other words, the flow of Fig. 1.3 and a rotating taffy puller (as in Fig. 1.2e, but with a simplified folding mechanism) are very close indeed if the fact that the area of a cross section is not preserved by the realistic equation is taken into account. In terms of the original taffy picture, this means that the taffy must contain a liquid matrix that may be "squeezed out" in part during the stretching that precedes the folding over. If one wants to stick to the picture, this implies that the lost volume has to be made up for by new taffy. The latter is to be automatically poured in after every round in such a way that the original cross section (think of a rectangular pan) is again filled out completely. The newly added material (which also has the function of remoisturizing the remnant of the last iteration) is conveniently pictured as having a different color. If the original cross section was red, for example, the new cross section may be all white except for the walking-stick-shaped red inset. At the next iterate, only the "walking stick within the walking stick" will be red, and so forth. Hence, the attracting "red line" in the limit [1.11]. Computer graphics can be very helpful in verifying that this picture, derived from dough processing, indeed describes what occurs inside the map of (1.5), for example [1.3, 7].

Problems

1.1   Demonstrate numerically the sensitive dependence on initial conditions of (1.5) using the parameter values given in the text. Calculate a_n and b_n ($n = 1, \ldots, 20$) for three different initial conditions ($a_0 = 0.8$, $b_0 = 0$; $a_0 = 0.80001$, $b_0 = 0$; $a_0 = 0.8$, $b_0 = 0.00001$). Determine the value of n for which the relative errors become larger than 10%. Calculate a_{50} and b_{50} for the initial condition $a_0 = 0.8$, $b_0 = 0$ with different precision in the routine employed (in the case of a FORTRAN program, use, for example, one time REAL and the next time DOUBLE PRECISION variables). Again determine the value of n for which the relative errors become larger than 10%.

1.2   Determine numerically (on a computer with graphic utilities) the structure of Fig. 1.1. Transform the corresponding equation (given in the legend) into real variables ($z \rightarrow x + iy$). Change the initial conditions in the range $-1.5 \le x_0 \le 1.5$, $-1.5 \le y_0 \le 1.5$ using, for example, 100 steps. Calculate x_n and y_n ($n = 1, \ldots, 50$). If under this iteration the condition $x_n^2 + y_n^2 > 20$ is fulfilled for a value of n, stop the procedure and start with the next initial condition. Mark all points black that do not satisfy the above condition (within the first 50 iterations). Choose a small part of the structure obtained and calculate the magnification. Finally, repeat the numerical procedure for different values of the constant C. The final structure is called the basin of attraction, the significance of which will become evident in Chap. 3.

2   Semiconductor Physics

In this chapter we present a comprehensive outline of a model semiconductor system (as an appropriate example of chaos in experiment) considered from the viewpoint of the underlying physics. Section 2.1 is devoted to fundamental aspects in terms of the nonlinear semiconductor transport properties. Section 2.2 gives a brief survey of recent chaos experiments performed on a variety of different semiconductor systems. Sections 2.3 and 2.4 deal with the systematic physical characterization and the most essential results of a particular semiconductor experiment, respectively. Here we focus on spatio-temporal instabilities of the electric charge transport during low-temperature impact ionization breakdown in extrinsic germanium.

2.1   Fundamentals of Nonlinear Dynamics

This section reports on the various types of semiconductor instabilities, the macroscopic understanding of which can be achieved by treating the charge carriers as solid state plasma and investigating their collective response. Independent of the particular ansatz, there always exists a strongly nonlinear region of negative differential conductivity due to the nonequilibrium phase transition involved. Finally, possible electric transport mechanisms of hot carrier dynamics are pointed out.

2.1.1   Historical Remarks

Certain effects of electric conductivity that are currently included under the generic term "nonlinear dynamics" were specified formerly as semiconductor instabilities. These consisted of both spatial and temporal transport phenomena, such as the self-generated formation of current filaments or electric field domains and of current or voltage oscillations, respectively. In what follows, we summarize some important aspects that have resulted from previous research on semiconductor instabilities.

The first experimental observation of spontaneous current oscillations in a semiconductor was published in the late 1950s by *Ivanov* and *Ryvkin* [2.1]. The term "spontaneous" emphasizes that these oscillations were obtained under

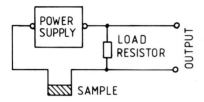

Fig. 2.1. Schematic illustration of the basic oscillator (from [2.2])

constant external conditions. The circuit shown schematically in Fig. 2.1 and explicitly defined as a new type of semiconductor oscillator by *Larrabee* and *Steele* [2.2] represents the prototype for all subsequent experimental activity. Here a constant voltage source is applied to the series combination of the semiconducting sample and a load resistor. The voltage drop at the load resistor can be regarded as a response of the system to the constant bias voltage. This standard experimental set-up was used in numerous studies of spontaneous semiconductor oscillations. In addition to the electric field, the different sample materials investigated were sometimes supplied with several other – temporally constant – control parameters, e.g., external illumination or an external magnetic field. Although the ambient temperature range of some experiments extended up to room temperature, the majority of the measurements has been performed at liquid helium temperatures.

Particularly during the 1960s, both experimental and theoretical research activities on semiconductor instabilities seem to have reached a peak in their intensity, as can be concluded from the huge number of papers written in that decade. In fact, most authors of later review articles have stated that it is practically impossible to provide a complete list of all relevant publications on this topic. We, therefore, do not intend to give a detailed synopsis of semiconductor instabilities. Yet, it is emphasized that a large part of this work can be found in the *Proceedings of the International Conferences on Semiconductor Physics* of that decade [2.3]. Further access to these developments may be feasible via the more recent summaries given in the monographs by *Bonch-Bruevich* et al. [2.4], *Pozhela* [2.5], and *Schöll* [2.6]. According to the multitude of diverse types of instabilities reported, a rough classification in terms of typical characteristic features will be attempted.

2.1.2 Plasma Ansatz

It is important to note that a major breakthrough in the macroscopic understanding of semiconductor instabilities could be achieved by introducing the well-known plasma concept into solid state physics. In this model, the collective response behavior of the charge carriers is investigated globally. In analogy with the gas plasma, the solid state plasma can be defined as a neutral concentration of charges embracing electrons, holes, and ionized impurities. Therefore, one distinguishes between an intrinsic and an extrinsic carrier plasma, the former containing an equal number of electrons and holes, while the latter represents either an electron plasma or a hole plasma. In the intrinsic plasma, both types of carriers are mobile, whereas the extrinsic plasma consists of only one type of mobile charge carriers. So, instabilities which can be described by the plasma theory are called plasma instabilities. A detailed discussion on this topic can be found elsewhere [2.5].

Different types of plasma instabilities are discernible due to their distinct dependence on various control parameters. In particular, the influence of an external magnetic field on an instability often plays an important role. For instance, the application of longitudinal magnetic fields that are oriented parallel to the electric field leads to characteristic dependences of the plasma frequency. Certain instabilities, like the helicon instability, can be created only in the presence of a longitudinal magnetic field.

For some other instabilities, the plasma type according to the characterization given above is important. For example, Alfvén waves are expected exclusively in an intrinsic plasma. If we consider a two-stream instability, the required interaction between two different particle beams running in opposite directions can happen in the case of an electron and hole current. The pinch effect represents a spatial instability where the plasma contracts to a filament structure due to the self-induced magnetic field of the mobile carriers. Such a phenomenon presupposes definite criteria for a minimum current density.

With respect to a further characterization of semiconductor instabilities, the way in which a plasma is generated remains interesting. Now the sample material and the experimental set-up become important. On the one hand, plasma states in a semiconductor can be created by external excitations, for example, by optical irradiation. On the other hand, a semiconductor plasma may also arise from intrinsic properties, such as the autocatalytic process of avalanche breakdown. Here the carrier multiplication is accomplished by a strong interaction between bound and mobile charge carriers, the latter having gained kinetic energy from the applied electric field. Depending upon the magnitude of the bias voltage and the ambient temperature, an avalanche breakdown can produce either an intrinsic or an extrinsic plasma. For an inhomogeneous semiconductor material, e.g., pin diodes, injection of charge carriers through the contact areas may also give rise to a plasma.

The above brief discussion of semiconductor instabilities has shown that their causality is more or less predetermined by external conditions. Hence, we conclude that the intrinsic characteristic features of a distinct semiconducting material or device play different roles in different kinds of plasma instabilities.

2.1.3 Negative Differential Conductivity

Besides the question of whether a semiconductor instability can be more appropriately described by the global macroscopic approach of plasma physics or the more fundamental ansatz of a system-immanent microscopic theory, the

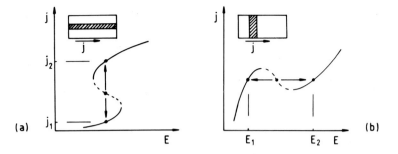

Fig. 2.2a, b. Schematic current density vs. electric field characteristics with S-shaped **a** and N-shaped **b** negative differential conductivity providing instability regimes $j_1 < j < j_2$ and $E_1 < E < E_2$, respectively. The insets illustrate a filament formation (**a**) and a domain formation (**b**) of two phases 1 and 2

most essential problem remaining is whether there exists negative differential conductivity (ndc) in connection with the instability. In general, instabilities are easily explained in terms of ndc [2.4–7].

From the viewpoint of nonlinear theory, ndc in the simplest case corresponds to an unstable solution of the system equation. In solid state physics, one distinguishes between two kinds of ndc that lead to two differently shaped current density (j) versus electric field (E) characteristics, as shown in Fig. 2.2. Commonly, it is argued that an S-shaped j–E characteristic favors the formation of current filaments (Fig. 2.2a), while an N-shaped j–E characteristic gives rise to the formation of electric field domains (Fig. 2.2b). These arguments can, in an oversimplified way, be summarized as follows. A system forced to an unstable state of ndc splits up spatially into two different regions each of which is governed by stable electric variables (indicated by arrows in Fig. 2.2). Thus, only the spatially averaged electric variables possess unstable values. It is emphasized that, as an alternative to spatial instabilities, ndc systems quite often display temporal instabilities. An argument similar to that used above for the space domain now holds for the time domain. Instead of an average over space, a time average has to be considered.

So far, we have seen that the existence of ndc is essential for explaining semiconductor instabilities. The difficulty lies in the experimental verification of ndc since cases in which the ndc leads to negative differential conductance (NDC) in the macroscopic current–voltage (I–V) characteristic are rare. Note that usually in experiments the I–V characteristic can be measured only as a spatial and temporal average. If the I–V characteristic contains an S-shaped or N-shaped region of NDC, it is obvious that the nonaveraged microscopic j–E characteristic must have the correspondingly shaped ndc. But, on the other hand, the experimental observation of an instability if the I–V characteristic displays positive differential conductance (PDC) can be explained in terms of an underlying j–E characteristic having positive differential conductivity (pdc) and/or ndc.

In what follows, the latter point is illustrated with the help of two examples. Let us start with the first case, assuming an ndc $j-E$ characteristic, which may be caused, for example, by some microscopic transport properties. If we further consider a spatial inhomogeneity as a consequence of the ndc, the space-averaged $I-V$ characteristic can easily be shown to display any form ranging from NDC to PDC [2.8]. The main idea is to characterize the spatial inhomogeneity by two different specific resistivities, each corresponding to one stable branch of the underlying $j-E$ characteristic (Fig. 2.2). An essential cause of the existence of NDC in the $I-V$ characteristic lies in the dependence of the growth of the spatial inhomogeneity on a distinct control parameter (e.g., I or V). It is easily demonstrated that for spatial inhomogeneities changing only by degrees as a function of the control parameter an $I-V$ characteristic with nothing else but PDC can be obtained, even though the underlying ndc gives rise to a spatial instability. Such a situation may be verified experimentally by looking at the gradual growth behavior of a current filament in a PDC region of the $I-V$ characteristic [2.9]. In a similar way, it is obvious that the presence of spontaneous current and/or voltage oscillations due to ndc may lead to NDC as well as PDC in the time-averaged $I-V$ characteristic.

A second example of the main difficulty in deciding whether or not the experimental situation investigated exhibits ndc causing an instability is given by *Pozhela* [2.5]. Here a case is pointed out where, on the one hand, no ndc can be extracted immediately from an $I-V$ characteristic having only PDC obtained by direct current and direct voltage measurements. But, on the other hand, a frequency-dependent ndc can be concluded from time-resolved current and voltage measurements exhibiting NDC in a certain frequency range of the oscillatory behavior. This kind of dynamical ndc may explain the existence of undamped spontaneous oscillations with a distinct frequency.

To conclude, there exists a complex interplay between the possible existence of ndc on the microscopic level and the concrete observation of NDC or PDC in the macroscopic $I-V$ characteristic. Obviously, NDC always requires ndc, while PDC can result from both ndc and pdc. Hence, one recognizes the general difficulty in connecting the appearance of instabilities (in the case of ndc) with the particular shape of the measured $I-V$ characteristic.

2.1.4   Transport Mechanisms

The appearance of instabilities in the electric current flow of a semiconductor is generally associated with characteristic nonlinearities of system-immanent transport properties. As a matter of fact, the differential electric conductivity defined as

$$\sigma_{ij}^{d} = \partial j_i / \partial E_j$$
$$= \partial (qn\mu E_i)/\partial E_j \tag{2.1}$$

often depends sensitively upon the electric field applied instead of being constant. The indices i and j denote one of the three possible space coordinates, j_i and E_j denote the current density and the electric field in the i- and j-directions, respectively, while q is the elementary charge, n, the number density, and μ, the mobility of the charge carriers. From (2.1) we see that the possible electric field dependence of the quantities $n(E)$ and $\mu(E)$ can give rise to ndc.

Charge carriers which change their characteristic properties in the presence of an electric field are usually referred to as "hot carriers" if their nonequilibrium energy distribution can be appropriately described by an effective carrier temperature. In a metal, these nonlinear effects are difficult to observe. Here the carrier concentration is simply too high, and the excessive power dissipation leads to destruction of the sample before the electric field values necessary for appreciable deviations from Ohm's law are reached. However, in a semi-conductor, because of the low carrier concentration, the nonlinear regime and the case of extreme nonequilibrium can be established at levels of the current density which are quite tolerable. Of course, the electric field necessary for the generation of hot carriers is expected to decrease with decreasing sample temperature due to the reduction of the carrier–phonon interaction. For a thermodynamical description in terms of nonequilibrium phase transitions, we refer to [2.6].

Obviously, spatial and/or temporal instabilities often occur in semiconductors when there is a change in the dominant scattering mechanism due to carrier heating. As an example, the Gunn instability results from hot charge carriers that are scattered into another conductivity band, where they have a lower mobility. With increasing electric field, the semiconductor system undergoes a transition from higher to lower conductivity, giving rise to an N-shaped ndc in the j–E characteristic (which also may lead to an N-shaped NDC in the I–V characteristic). Moreover, the impact ionization avalanche breakdown mentioned above represents a particularly important case yielding a highly non-linear current–voltage relation with potentially S-shaped NDC. The presumed S-shaped ndc in the j–E characteristic is caused by a strong dependence of both the carrier concentration and mobility on the electric field. Here the auto-catalytic nature of the avalanche-like carrier multiplication during breakdown plays an essential role for the nonequilibrium phase transition from a weakly conducting to a highly conducting state.

We point out that there are also some types of instabilities whose origin is not influenced by hot charge carriers. Injection breakdown gives a typical example. Here the mean free path lengths of the injected carriers determine the process ingredients [2.10]. The sample length may, thus, have considerable consequences for the incidence of an instability. In addition to injection breakdown, certain kinds of oscillations can be characterized by their dependence on the sample length. The oscillation frequency is given by the transition time of a perturbation passing through the sample. Think of high-electric-field domains for the case of a Gunn instability. In contrast to this type, one must consider oscillations which are independent of the sample length. These can be generated either homogeneously throughout the whole semiconductor material (global oscillatory behavior) or inhomogeneously at definite sites of the sample (local oscillatory behavior). It is emphasized that a localization may originate from external conditions (as occurs for the case of a p-n junction), but it also may be organized by the instability itself.

Finally, we want to address the potential importance that distinct sorts of impurities present in the semiconductor material investigated can have for the relevant transport mechanism underlying an instability. If there are different impurity levels located within the energy band gap, for avalanche breakdown one expects an $I-V$ characteristic with S-shaped NDC [2.6, 11]. The basic assumption of such a multilevel transport model is that the charge carriers can occupy the conduction band or can be bound to the ground state and the excited states of impurity centers. The different energy levels may lead to the simultaneous occurrence of two or even more different relaxation times in the semiconductor. Analytical conditions for both filamentary and oscillatory instabilities have been derived explicitly by *Schöll* [2.6]. An extended treatment of this topic is also given by *Pozhela* [2.5].

So far, we have reflected on more general aspects of possible nonlinear transport mechanisms responsible for various types of semiconductor instabilities. The understanding of the underlying nonequilibrium phase transitions should afford the basis for an insight into the diverse nonlinear phenomena observed recently in different semiconductor experiments, as is discussed in the following section.

2.2 Recent Experimental Progress

This section gives a brief review on chaos experiments performed on different semiconductor systems. The early research activities on oscillatory instabilities during the 1960s may appropriately be characterized by the statement of *Glicksman* [2.12] that the wide range of semiconductor instabilities can be divided into three groups, namely, the verified (agreement between theory and experiment), the predicted (theory, but no conclusive experiments), and the observed ones (experiments, but no conclusive theory). During this time, considerable progress was made in trying to transfer instabilities from the last to the first category with help of the plasma ansatz. But there still remained a lot of unanswered questions concerning these instabilities. For example, abruptly changing shapes of the oscillatory modes, or unexpected frequency behavior, or even unreproducible oscillations often led to very confusing attempts at characterization.

These unsolved problems somehow formed a basis for new research activities. The crucial impetus was received from recent advances in the theory of nonlinear dynamical systems and deterministic chaos. Clearly, these highly exciting new developments have added a new dimension to investigations of the complex behavior in semiconductors and have motivated many new measurements. During the past years, it has become distinctly clear that semiconductors and their electronic conduction behavior represent an interesting and highly productive example of the experimental investigation of nonlinear dynamics and deterministic chaos. Thus, as an experimental system which is easy to handle, semiconductors play an important role in nonlinear science, just as they did previously in the field of plasma physics. On the other hand, nonlinear science can explain the complex temporal behavior of semiconductors, formerly characterized as states with some kind of noise, in terms of simple deterministic chaotic mechanisms.

In what follows, we summarize the highlights of these developments. It is interesting to note that the dominant part of the semiconductor instabilities studied experimentally so far with respect to possible chaotic behavior is associated with the phenomenon of avalanche breakdown. Apparently, the autocatalytic nature of this process plays an important role, similar to the situation in chemical reaction systems, which have also been studied extensively. Avalanche breakdown of shallow impurities (with an ionization energy of only a few meV) has turned out to be a particularly promising research subject. To investigate such systems, the experiments must be performed in the low-temperature regime. Here the substantial reduction of the thermal noise at low temperatures represents an additional distinct advantage. Indeed, such low-temperature experiments allowed the observation of extremely pure oscillatory states. Low-temperature experiments have been performed with a wide range of materials: n-Si [2.13], n-InSb [2.14], n-GaAs [2.15, 16], and p-Ge [2.17–19]. Even though all these studies concerned the regime of avalanche breakdown of shallow impurities, for some systems further special experimental conditions were required for the observation of chaotic oscillations. Two groups reported the observation of chaotic oscillations only in conjunction with an applied magnetic field [2.14, 16]. Two other groups apparently required additional optical excitations which had to be temporally periodic (driven system) in some cases [2.15, 17].

Experiments studying avalanche breakdown of deep impurities (with an ionization energy in the range of about 100 meV) are not restricted to the low-temperature regime. One group [2.20] reported the observation of chaotic oscillations at room temperature in Cr-compensated GaAs, without any further additional experimental conditions.

In addition to these experiments in which the chaotic oscillations are associated with electronic breakdown mechanisms, in semi-insulating GaAs chaotic behavior has been found at room temperature due to domain instabilities [2.21, 22]. Two other groups reported the observation of chaotic behavior due to plasma instabilities. One [2.23] studied the chaotic behavior in a p^+nn^+-Ge sample at 77 K, which can be explained by helicon instabilities in an electron–hole plasma in an applied longitudinal magnetic field. The other [2.24] reported on the chaotic behavior in an electron plasma in Ni-compensated n-Ge at 77 K. Here the nonlinear dynamics has been explained in terms of the interaction between the plasma and repulsive impurity centers resulting in a dynamical ndc.

The frequency range of the spontaneous oscillations in the semiconductor systems listed above was nearly always in the kilohertz regime. Only the experiments on semi-insulating GaAs [2.21, 22] and the studies by *Aoki* et al. [2.15] yielded oscillation frequencies of only a few hertz. We note that the possible frequency range of semiconductor instabilities extends from millihertz up to gigahertz. (In many cases this large extension of the frequency range results from the large variation of the charge carrier concentration characteristic for semiconductors.) It appears likely that the dominance of the kilohertz regime in the reported experimental observations is due to the ease of experimental accessibility and not due to any fundamental reason. Temporal structures with characteristic frequencies in the kilohertz range are particularly suitable for detailed temporal analysis without overextending the necessary time period. It is this highly favorable frequency range which distinguishes semiconductor systems as objects for detailed experiments on nonlinear dynamics. On the other hand, for hydrodynamical experiments and in chemical reactions, the characteristic frequencies often extend below 1 Hz, resulting in extremely long periods necessary for data acquisition.

From these arguments, we are not surprised to find that semiconductor experiments have contributed significantly to the field of nonlinear dynamics and chaos. In particular, semiconductor experiments were able to confirm many theoretical predictions in this field. Initially, the experimental work concentrated on the verification of the different major universal routes to chaos predicted theoretically [2.25], for example, the intermittent switching between different oscillation modes (Pomeau–Manneville scenario), the period-doubling cascade (Feigenbaum–Grossmann scenario), and the quasiperiodic route to chaos (Ruelle–Takens–Newhouse scenario). Moreover, the transition from low-dimensional ordinary chaos to higher-dimensional hyperchaos (Rössler scenario [2.26]) has been realized experimentally. More recently, the main experimental interest shifted to two new developments in nonlinear dynamics. First, modelling of quasiperiodicity by the circle-map formalism has opened up a new approach to the analysis of dynamical states. A method for studying multifractal scaling properties based on the thermodynamics of strange sets (spectrum of scaling indices $f(\alpha)$ [2.27]) and inherent to chaos has been developed. Second, the combined behavior of temporal and spatial structures generated in nonlinear dynamical systems has moved towards the center of attention [2.28]. In terms of semiconductor systems, spatially resolved observations of the oscillatory behavior have been possible by means of localized electron-beam perturbation [2.29, 30] and proper configurations of ohmic contact probes [2.18, 23, 29].

The theoretical and experimental advances in the field of deterministic chaos of the last ten years have provided a new perspective for the understanding of noise (irregular temporal behavior) and turbulence (irregular temporal and spatial behavior) [2.31]. Here the concepts of deterministic chaos represent an important supplement of the concept of stochasticity, which previously had dominated the subjects of noise and turbulence. The experimental studies of semiconductors discussed in this section have contributed significantly to our present understanding of nonlinear behavior, instabilities, and deterministic chaos. In particular, the high precision of the recent semiconductor experiments yielded impressive confirmation of detailed theoretical predictions. The electronic transport in semiconductors represents an experimental system which is extremely well suited for observations by means of advanced electronic measuring techniques. Here the direct coupling between the electronic phenomena as the objects of study and the measurement principles yields a distinct advantage, in addition to the favorable frequency range of most of the observed phenomena. Due to these features, the electronic transport in semiconductors seems set to remain an important testing ground for nonlinear dynamics in the future.

On the other hand, results that are obtained with the help of system-independent characterization methods emphasizing the underlying nonlinear dynamics may give some further insight into the physics of semiconductors. For instance, the fractal dimensionality evaluated numerically for a distinct time series of an instability provides information about the minimum number of independent variables (i.e., the actively participating degrees of freedom) necessary to formulate the system-immanent equations of motion. Note that, even for an apparently noisy state governed by low-dimensional chaotic dynamics, it is possible to describe the electric properties of a semiconductor (involving at least 10^{10} charge carriers) by no more than a few system variables in total. Certainly, this kind of nonlinear collective treatment represents one of the most exciting ideas setting the tone of this research field. It can be seen to be closely connected with the plasma ansatz discussed already in Sect. 2.1.

2.3 Model Experimental System

The following two sections concentrate on a thorough treatment of one particular semiconductor experiment, the instabilities of which are characterized from the viewpoint of the underlying physics. This representative semiconductor system consists of single-crystalline p-doped germanium electrically driven into low-temperature avalanche breakdown via impurity impact ionization [2.18, 29]. The huge diversity of complex spatio-temporal behavior observed will provide the basis for the detailed analysis in terms of nonlinear dynamics theory to be presented in Chap. 3.

The present section is devoted to a systematic description of the samples and circuitry used.

2.3.1   Material Characterization

The semiconductor material investigated was p-Ge with an impurity concentration in the range of 10^{14} acceptors per cm^3 and the compensation ratio definitely smaller than 10^{-2}, thus, being neither of ultrahigh purity nor highly doped. On the one hand, the doping is so high such that the electric transport properties are clearly dominated by the acceptor impurities, or the holes, respectively. On the other hand, the doping is weak enough such that overlapping of the localized acceptor states can be excluded, i.e., unwanted hopping or impurity band conduction mechanisms [2.32, 33] do not play a significant role. The experiments reported here were performed with two different charges of extrinsic germanium material, the specification of which is summarized in Table 2.1. We emphasize that, for both materials, no remarkable difference in the experimental results could be recognized up to now. Thus, we conclude that the global system behavior, especially the nonlinear dynamics, reported in Chaps. 2 and 3 is typical for such p-Ge material.

To specify in detail the acceptor material in question, we present in Fig. 2.3 a measurement of a typical photoconductivity spectrum. Comparing this spectrum with the spectrum of the energy levels of all possible acceptor materials [2.34] allows one to identify different acceptor materials present in the semiconductor system. Here we obtained for one representative sample three different shallow acceptors, namely, indium, aluminum, and gallium, each contributing at least 10% to the total impurity concentration. For each impurity different energy levels were detected.

These materials provided the basis for preparing samples with different geometrical shapes. The typical size was in the range of a few millimeters. Usually, the surfaces of the samples were successively polished and etched, in order to obtain an ideal surface structure. No significant influence on the basic nonlinearities due to different surface treatment could be observed, even if the etching procedure was totally omitted. Thus, we conclude that the basic nonlinear effects do not depend on the surface structure of the sample used.

As the next preparation step, ohmic contacts were put on the sample. For that, we applied two different techniques, namely, alloying and ion implantation. As an alloying material, we utilized aluminum, indium, or a mixture of gallium

Table 2.1. Characteristic properties of the extrinsic semiconductor material investigated

Material	Single-crystalline p-Ge	
Source of supply	Darien Magnetics, USA	Hoboken, The Netherlands
Crystal orientation	(111)	(100)
Specific resistivity (300 K)	$\sim 10\,\Omega\,cm$	$16.8\text{--}19.2\,\Omega\,cm$
Acceptor concentration	$2 \times 10^{14}\,cm^{-3}$	$1.7 \times 10^{14}\,cm^{-3}$
Acceptor material	Unknown	In, Al, Ga
Donor concentration	$< 6 \times 10^{12}\,cm^{-3}$	$< 10^{12}\,cm^{-3}$

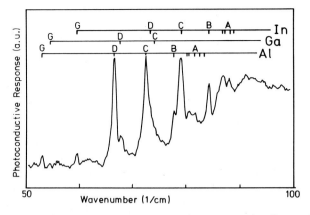

Fig. 2.3. Photoconductivity spectrum of a representative Ge sample obtained from the measured photocurrent signal as a function of the wavenumber of the exciting irradiation. Further details are given in [2.34]

and indium. For obtaining ion-implanted contacts, an ion beam of boron with an acceleration energy of 60 keV and a surface density of 5×10^{14} cm^{-2} was used. Again, the application of different contact materials did not change the basic results. From this, we conclude that the nonlinearities in our semiconductor system originate from a bulk effect and are not contact-induced. This statement is further confirmed by experimental observations of localized oscillation centers in the bulk of the semiconductor, as reported in Sect. 2.4.4.

In addition to the influence of different contact materials, the influence of different contact geometries was systematically investigated. Contact shapes ranging from capacitor-like parallel planes to spike-like planar forms were arranged on the crystal surface. Here we found that different geometries of the facing contacts resulted in different shapes of the current–voltage characteristic. But such a phenomenon can be easily understood, as already discussed in Sect. 2.1.3. Regardless of these changes in the current–voltage characteristics, the basic nonlinear effects could still be detected.

Not only the geometry, but also the distance between the contacts was varied. For relatively large distances in the range of a few millimeters, differently prepared contacts always led to more or less the same results. For distances in the range of a few 100 μm and smaller, only samples with ion-implanted contacts showed reasonable results. We emphasize that we could observe dynamical instabilities (described in the following sections) even in samples with the smallest contact separation of 10 μm.

The geometry of one particular example (with semiconductor material from Darien Magnetics) including the shape of the ohmic contacts is shown in Fig. 2.4. We have chosen to present this particular sample, because the majority of the experimental results given in this book were obtained on that sample. The special feature of this sample derives from the inner contacts B and C, which can

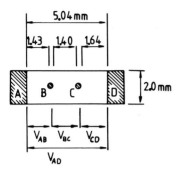

Fig. 2.4. Scheme of a particular sample geometry. The ohmic contacts are indicated by the hatched areas labelled A, B, C, D. The thickness of the sample is 250 μm

be used as local potential probes. The corresponding partial voltages are labelled accordingly.

2.3.2 Experimental Set-up

The investigation of the electric transport properties at low temperatures of the p-Ge material introduced in Sect. 2.3.1 was performed with two different experimental arrangements. One set-up served for recording current–voltage characteristics and spontaneous oscillations, another for detecting spatially inhomogeneous current distributions.

The experimental set-up used in the first case is shown schematically in Fig. 2.5. Here the sample was in direct contact with the liquid helium bath. By the help of vapor pressure regulation, the temperature could be varied continuously between 4.2 K and 1.7 K. The accuracy was better than 1 mK at 4.2 K and about 2 mK at 1.7 K. Since infrared irradiation may affect the semiconductor system at these temperatures, the sample configuration was surrounded by a copper shield, which still allowed the liquid helium to reach the interior. Of course, this shielding was also in direct contact with the liquid helium bath. To provide the ohmic contacts of the sample with an electric field, a d.c. bias voltage V_0 was applied to a series combination of the sample and the load resistor R_L. In addition, a d.c. magnetic field B oriented parallel or perpendicular to the broad sample surfaces could also be applied by a superconducting solenoid surrounding the semiconductor sample.

So far, we have introduced the essential external control parameters, namely, the bath temperature T_b, the magnetic field B, the bias voltage V_0, and the load resistor R_L. The attainable precision of these parameters was in the range of 10^{-5}. As a response of the system to these external parameters, typically, the voltage drop along the load resistor is taken. But also the voltage drop over the sample or even smaller parts of the sample were investigated, as shown later on. Two different measurements were performed: first, the time-averaged and, second, the time-resolved system behavior via current and/or voltage signals. Time-averaged signals served for obtaining, e.g., the current–voltage

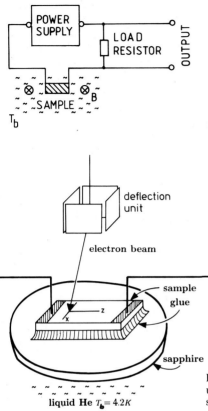

Fig. 2.5. Scheme of the experimental arrangement used for conventional low-temperature measurements without external irradiation

Fig. 2.6. Scheme of the experimental arrangement used for low-temperature scanning electron microscopy

characteristic of a sample. The time behavior of the system response, mainly under time-constant external conditions, provided information on spontaneous oscillations or rising noise. These measurements required the most stabilized external conditions. For the electric power supply, thus, batteries were used as far as possible. We even had to pay attention that the voltage output of the battery remained as constant as could be. Usually, such high standards were met by employing commercial car batteries.

As the second experimental set-up applied for two-dimensional imaging of stationary and dynamical current structures, we equipped a scanning electron microscope with a liquid helium cryostage, the scheme of which is sketched in Fig. 2.6. Here the samples were glued onto 1 mm thick sapphire discs using Stycast cement for good thermal contact. During the experiments, the bottom of the sapphire substrates was in direct contact with the liquid helium bath kept at 4.2 K or below, whereas the top side of the samples was exposed to the vacuum of the microscope column and could be scanned directly with the electron beam. The electric measuring configuration corresponds to the one shown in Fig. 2.5. By operating the germanium samples in the impact ionization breakdown regime, we recorded the beam-induced change of the electric conductance as a function of the coordinates of the beam focus on the specimen surface. In particular, we observed the beam-induced current change in the voltage-biased samples via an 1 Ω load resistor. With the additional help of a beam-blanking unit, it was possible to chop the electron beam periodically in time. Thus, time-correlated responses of the system to very small perturbations of the electron beam could be detected sensitively via conventional lock-in techniques. A more detailed discussion of the operating principle of low-temperature scanning electron microscopy can be found elsewhere [2.35].

Before we turn to the presentation of some typical results obtained from the semiconductor experiment discussed above, it should be pointed out that an inherent attribute of nonlinear systems is their extremely high sensitivity to smallest changes of the relevant control parameters applied. Thus, good reproducibility of specific nonlinear effects could be observed only in measurements performed on one sample. In this case, for instance, distinct oscillatory temporal and/or filamentary spatial structures were found repeatedly even after time periods of several years, if the control parameters had been realized with the high precision required. Also small changes in the sample under consideration (e.g., renewed preparation of ohmic contacts or some material split-off near the crystal edges) did not affect the reproducibility of instabilities once observed. Comparing results from different samples, however, only an overall agreement in the basic nonlinear behavior was recognized, i.e., impact ionization breakdown associated with complex spatio-temporal current flow could generally be observed.

2.4 Experimental Results

This section briefly reviews the global static and dynamical current transport behavior of the example experimental system characterized in the preceding section. The basic nonlinear phenomena are discussed in terms of the underlying semiconductor physics, taking into account the influence of different experimental control parameters.

2.4.1 Static Current–Voltage Characteristics

The experimental arrangement sketched in Fig. 2.5 served for measuring the current–voltage characteristics. Here both the voltage drop V along the sample and the current I as the voltage drop along the precision load resistor (R_L = 100 Ω) were taken in their space and time average. Figure 2.7 shows a typical $I-V$ characteristic. For small voltages applied, the semiconductor remains highly insulating. In this regime, resistance values larger than 100 GΩ could be measured. On increasing the voltage bias, we found a distinct threshold value V_b where the sample conductance changed abruptly. Just at this point, the resulting

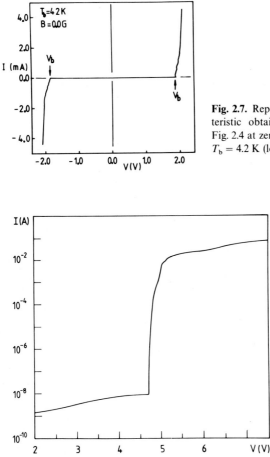

Fig. 2.7. Representative current–voltage characteristic obtained from the sample sketched in Fig. 2.4 at zero magnetic field and the temperature $T_b = 4.2$ K (load resistance $R_L = 100\,\Omega$)

Fig. 2.8. Current–voltage characteristic obtained from a distinct sample of 11 mm contact distance at an applied transverse magnetic field $B = 30$ mT and the temperature $T_b = 4.2$ K (load resistance $R_L = 1\,\Omega$)

current flow drastically increases by several orders of magnitude (typically, from a few nA up to a few mA for extremely small incremental changes in voltage, as can be clearly seen from the semilogarithmic plot of the I–V characteristic in Fig. 2.8). Accordingly, this critical current rise is referred to as breakdown event. Depending on the effective length of the sample, i.e., the distance between the ohmic contacts, the apparent breakdown voltage V_b corresponds to a critical electric field of about 5 V/cm, called the breakdown electric field. Our experimental finding that different samples always had the same characteristic value (provided the contact distance was not smaller than 100 μm [2.36]) gives some hints to the physical origin of the breakdown phenomenon, as follows.

At liquid helium temperature, most of the mobile charge carriers are frozen at the impurity centers. This can easily be understood from an energy argument. Note that the binding energy of the charge carriers to the impurities is about 10 meV (see also Fig. 2.3), whereas the thermal energy at 4.2 K (liquid helium under normal pressure) corresponds to 0.35 meV. So, a bulk semiconductor material offers high resistivity when the applied electric field is small. The conductivity of the sample under these conditions arises from the few mobile carriers which may have been generated either thermally or by stray radiation. On increasing the applied electric field, these mobile carriers gain more and more in kinetic energy (hot carriers if this new energy distribution can be characterized by a fictive temperature $T^* > T_b$), which may at times become sufficient to ionize the neutral impurity centers upon impact (impurity impact ionization). At the critical electric field, avalanche multiplication of mobile charge carriers leads to a reversible nondestructive breakdown effect (avalanche breakdown). This process persists until nearly all impurities are ionized, as can be seen from the carrier density (n) versus voltage (V) characteristic in Fig. 2.9a. Low-temperature avalanche breakdown in extrinsic germanium is, thus, produced through impact ionization of shallow impurities by mobile hot carriers heated up by the applied electric field. The mobility of the conduction carriers is controlled by additional scattering processes with ionized impurities and acoustic phonons of the lattice vibrations. In particular, conductivity and Hall effect measurements have shown that throughout the breakdown region the mobility and drift velocity display a pronounced nonlinearity which sensitively depends upon the density of the mobile charge carriers. The characteristic electric field dependences of these intrinsic system quantities, namely, the carrier mobility (μ) and drift velocity (v) vs. voltage (V) characteristics, shown in Fig. 2.9b and c, respectively, are discussed elsewhere [2.37] in detail.

Just beyond the breakdown event, i.e., at voltages slightly above V_b, the global behavior of the integral $I-V$ characteristic (Figs. 2.7, 8) exhibits a strongly nonlinear curvature with sometimes S-shaped NDC (in the case of current-controlled measurements done with a large load resistance) or hysteresis (in the case of voltage-controlled measurements done with almost zero load resistance), depending upon the load resistor applied and, thus, the corresponding load line that becomes effective. This nonlinear part of the $I-V$ curve changes into a nearly linear one at higher voltages in the post-breakdown regime. Clearly, the extrapolation of this linear part does not meet the origin. It is moreover emphasized that further nonlinearities can occur at even higher values of the voltage applied [2.38]. In such relatively high dissipative conductivity states one only has to pay attention that the liquid helium bath continues to be capable of sufficiently cooling the sample investigated. In our experience, the electric power dissipated in the sample per unit surface must not exceed the critical value of about 0.3 W/cm^2. Otherwise, unwanted boiling of the helium is to be expected [2.39]. Following these arguments, the sample geometry used in our experiments was always chosen to have a reasonable surface-to-volume ratio.

Let us return to the existence of NDC in the system under consideration. As already discussed in Sect. 2.1.3, this experimental fact represents the central point in our intention to characterize the system behavior. Figure 2.10 gives various $I-V$ characteristics recorded via different local potential probes that are placed along the same sample (cf. Fig. 2.4). Here the bias voltage was always

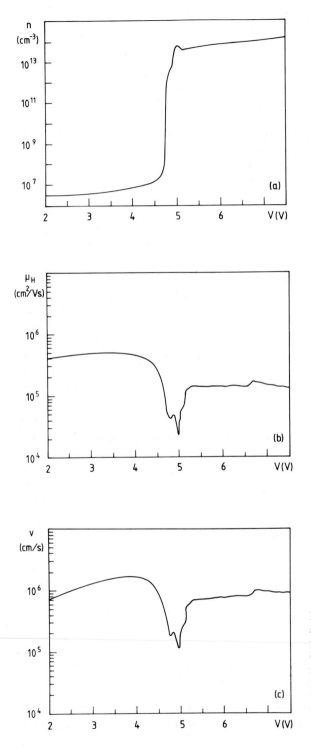

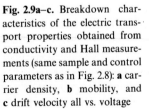

Fig. 2.9a–c. Breakdown characteristics of the electric transport properties obtained from conductivity and Hall measurements (same sample and control parameters as in Fig. 2.8): **a** carrier density, **b** mobility, and **c** drift velocity all vs. voltage

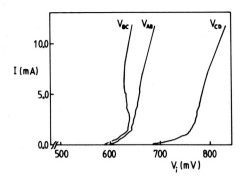

Fig. 2.10. Current–voltage characteristics obtained from different parts of the sample sketched in Fig. 2.4 at zero magnetic field and the temperature $T_b = 4.2$ K (load resistance $R_L = 100\ \Omega$)

connected to the outer ohmic contacts A and D (i.e., V_{AD} represents the total voltage drop over the sample), and the partial voltage drops V_{AB}, V_{BC}, and V_{CD} were measured with respect to the inner contacts B and C, accordingly. The current in turn flowing through all three sample parts was found from the voltage drop at the load resistor, as described in Sect. 2.3.2. Specifically, the $I-V_{BC}$ characteristic (the left-hand side curve in Fig. 2.10) displays a pronounced part of NDC in the breakdown region. Compared to that extent, only tiny parts of NDC are indicated in the remaining two characteristics.

In order to check whether there exists a spatially inhomogeneous distribution of the electric field along the sample during breakdown, we looked at the particular critical electric field values corresponding to the local characteristics of Fig. 2.10. Verification of almost identical breakdown fields in the only small range between 4.15 V/cm and 4.21 V/cm gives rise to the assumption that the electric field extends homogeneously in the direction of the current flow. Therefore, it is obvious that the ohmic contacts do not play any dominant role for the breakdown process considered. In fact, the nonlinear system behavior potentially governed by an NDC characteristic mainly results from a volume effect of the bulk material, while surface- or contact-induced phenomena can be ignored (see also the arguments given by *Mönch* [2.40]). In our opinion, this experimental finding is of fundamental importance, since it clearly supports the philosophy intended for the present book that the nonlinear dynamics of a real-world macroscopic system obeys the basic idea of spontaneous self-organization underlying the universal approach of synergetics [2.41] and information thermodynamics [2.42].

The specific shape of the nonlinear parts in the $I-V$ characteristics sensitively depends on the external control parameters, such as the transverse magnetic field, the temperature, and the electromagnetic irradiation. The influence of the last two parameters arises from the coupling of an additional energy source to the material. Such extra energy influx diminishes the freeze-out effect of the mobile charge carriers to the impurities. Experimentally, an overall tendency towards the formation of PDC parts in the $I-V$ characteristic could be observed as expected [2.43]. In the presence of an external magnetic field applied perpendicular to the direction of the electric field, the mobile charge carriers are deflected due to the Lorentz force. This results in a corresponding shift of the $I-V$ characteristic, including an increase of the critical electric field at which breakdown sets in. The increase of the breakdown field with increasing magnetic field is associated with an increase in the magnetoresistance. This effect may be ascribed to a diminution of mobility for the conduction carriers. Here we assume that the charge carriers gain the kinetic energy necessary to ionize in one mean free path and that the effect of the transverse magnetic field is to diminish the projection of the mean free path in the direction of the electric field (cooling effect of the magnetic field on the hot carriers). To obtain the same energy from the electric field in one mean free path, a higher electric field is then necessary. At low temperatures where the carrier mobility, and hence the Lorentz force, is high, the magnetoresistive effect is already evident at small applied magnetic fields (in the range of some 10^{-4} T) [2.37]. In particular, the striking carrier multiplication effect taking place during avalanche breakdown as a consequence of the autocatalytic process of impurity impact ionization leads to a further strengthening of the magnetic field dependence [2.44]. We point out that the structural change in the form of the $I-V$ characteristic developing under a variation of the above control parameters will be discussed in Sect. 3.2 from the viewpoint of catastrophe theory.

Finally, another important parameter that affects the shape of the nonlinear part in the $I-V$ characteristic is given by the geometric form of the ohmic contacts applied. For a systematic investigation, contact areas ranging from capacitor-like parallel planes to spike-like planar forms were arranged on the crystal surface. On the one hand, ideally parallel planes of the facing contacts give rise to large hysteretic or NDC regions in the characteristic. On the other hand, spike-like shaped contacts are more likely to favor nonhysteretic or PDC curvatures in the breakdown regime. These experimental findings can be well understood if we assume the growth behavior of a current filament forming during avalanche breakdown to be a sensitive function of the prevailing contact shape [2.8].

2.4.2 Temporal Instabilities

Up to now, the $I-V$ characteristics have been discussed only as time-averaged measurements. With the help of time-resolved experiments, we revealed self-generated temporal instabilities closely linked to the nonlinear breakdown regime. In particular, spontaneous current oscillations via the load resistor and/or spontaneous voltage oscillations over the total length as well as over distinct parts of the sample were detected. These undriven temporal oscillations which per se appear at constant applied control parameters (bias voltage, load resistance, magnetic field, temperature) can be roughly divided into three different types [2.45].

First, switching oscillations due to the stochastical switching of the current between two different conducting states were found in the parameter range near the breakdown threshold, where the S-shaped nonlinearity of the $I-V$ characteristic just vanishes as a consequence of sufficiently high applied magnetic fields (corresponding to a just vanishing breakdown hysteresis at voltage-controlled operation). At this point, the curvature of the characteristic displays an infinite slope. So far, the existence range of the switching oscillations has been reached under the following conditions, namely, by changing the load resistance (in order to adjust the slope of the load line to that of the NDC part in the $I-V$ characteristic), by increasing the bath temperature beyond 4.2 K (in order to attain critical phase transition behavior of the underlying nonlinear physics), or by applying a transverse magnetic field of sufficient strength. For the last case, a distinct time-resolved current structure $I(t)$ of highly rectangular pulses with constant amplitude is plotted in Fig. 2.11. Under slight increase of the applied bias voltage and fixed magnetic field, the stochastically occurring current pulses show increasing width and density. Remarkable features of these oscillations are the very slow time scale (typically, a few seconds) and the relatively large amplitude (typically, a few mA). Hence, the switching oscillations could be measured directly with an $x-y$ recorder when starting from distinct levels of the $I-V$ characteristic. In the present case, we observed a switching between a weakly conducting pre-breakdown state and a highly conducting immediate breakdown state. This sort of current instability will be taken up again in Sect. 3.2 for an interpretation on the basis of the fold topology of a cusp catastrophe. From the point of view of semiconductor physics, one might associate this characteristic instability with some kind of random firing of breakdown microplasmas, as is outlined elsewhere [2.46] in detail.

Second, relaxation oscillations attributed to stochastical firing of individual avalanche breakdown bursts were found already in the weakly conducting pre-breakdown state. Corresponding to the initial rising part of the time-averaged

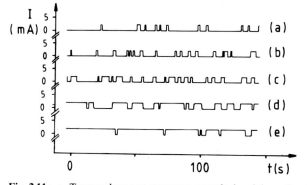

Fig. 2.11a–e. Temporal current structures near the breakdown threshold obtained from the sample sketched in Fig. 2.4 at different bias voltages $V_0 = 2.378$ V **a**, $V_0 = 2.379$ V **b**, $V_0 = 2.380$ V **c**, $V_0 = 2.381$ V **d**, $V_0 = 2.382$ V **e**, and the following constant parameters: transverse magnetic field $B = 25$ mT and temperature $T_b = 4.2$ K (load resistance $R_L = 1\,\Omega$)

$I-V$ characteristic shown in Fig. 2.12, the time-resolved current profiles $I(t)$ traced in Fig. 2.13 reveal increasing firing density and magnitude of the break-down bursts with increasing applied bias voltage. These relaxation oscillations typically occur on a time scale of a few milliseconds with amplitudes in the range of a few hundred μA. Their specific shape is given by an abrupt rise and a much slower subsequent relaxation of the current. This type of oscillatory behavior has recently been demonstrated to sensitively depend on the electric circuit employed [2.47]. Here the mutual interplay between the system-immanent NDC associated with intrinsic relaxation times, on the one hand, and the external circuitry, on the other, determines the shape of the relaxation oscil-lations. An extended theoretical treatment of circuit-induced instabilities with respect to nonlinear dynamics can be found elsewhere [2.48]. Nevertheless, we point out that this relaxation-type oscillatory behavior is clearly dominated by intrinsic transport properties. It can, thus, be interpreted as sporadically firing current filaments that form prior to the stationary state of stable filamentary conduction in the post-breakdown regime [2.46].

Third, spontaneous current and voltage oscillations superimposed upon the steady d.c. signals of typically a few mA and a few hundred mV, respectively, were found in the strongly nonlinear breakdown region, immediately after the relaxation oscillations have vanished. The relative amplitude was about 10^{-3}, and the frequency varied in the range of 0.1–10 kHz. Their existence range is characterized by an already initiated, but not fully developed, breakdown process. However, these oscillations disappear in the linear post-breakdown region where spatially homogeneous current flow already takes place. Figure 2.14 gives typical time traces of spontaneous current oscillations. Most

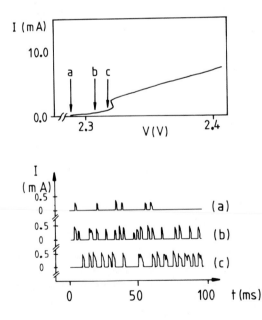

Fig 2.12. Current–voltage characteristic obtained from the sample sketched in Fig. 2.4 at the applied transverse mag-netic field $B = 20$ mT and the temper-ature $T_b = 4.2$ K (load resistance $R_L = 100\,\Omega$)

Fig. 2.13a–c. Temporal current struc-tures in the pre-breakdown regime obtained at different bias voltages $V_0 = 2.280$ V **a**, $V_0 = 2.310$ V **b**, $V_0 = 2.320$ V **c** corresponding to the markers of Fig. 2.12 (same sample and control parameters)

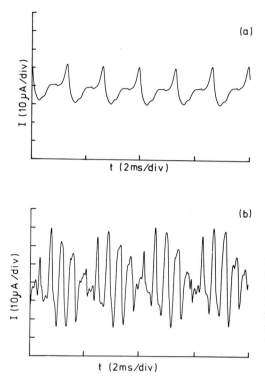

Fig. 2.14a, b. Temporal current structures in the nonlinear breakdown regime obtained from the sample sketched in Fig. 2.4 at different transverse magnetic fields $B = 0.53$ mT **a**, $B = 1.91$ mT **b** and the following constant parameters: bias voltage $V_0 = 2.330$ V and temperature $T_b = 2.1$ K (load resistance $R_L = 100\,\Omega$)

characteristically, their shape (including amplitude and frequency) is extremely sensitive to very small changes of the control parameters. Parts (a) and (b) display a periodic and a quasiperiodic state of the underlying system attractor, respectively, obtained under slight variation of the transverse applied magnetic field. In particular, the wide variety of self-generated oscillatory behavior embracing the universal transitions to low-dimensional chaos favors this type of oscillation as an appropriate testing ground for the generic theory of nonlinear dynamics. While these more interdisciplinary aspects will be reserved for Chap. 3, we concentrate in the following on a detailed treatment of spontaneous oscillations in the light of the system-immanent theory of semiconductor physics, the ingredients of which are outlined in Chap. 2.

From the material characterization of the present experimental system reported in Sect. 2.3.1 it becomes clear that holes remain as the only possible relevant charge carriers. Therefore, the formation of a hole plasma (i.e., an extrinsic plasma) during electric breakdown can be concluded. Since the measurements were performed with two different material specifications (Table 2.1) and, nevertheless, gave about the same results, no exotic material parameter is expected to cause the instabilities. More likely, the oscillation mechanism should rather be affected, in principle, by the overall properties of the original material used. It definitely does not originate from pure contact effects, because neither different geometries nor different materials of the ohmic contacts have any essential influence on the occurrence of the spontaneous oscillations. Also, instabilities have never been localized in the close vicinity of the contact areas, as will be demonstrated in Sect. 2.4.4. That the oscillations are not induced by the external circuitry was verified by systematic changes in the experimental set-up. For example, a capacitance connected in parallel to the sample did not affect the frequency of an oscillatory state. All experimental facts support the notion that the nonlinear dynamics in our semiconductor system undoubtedly results from a bulk effect.

To explain the origin of the spontaneous oscillations, it is of central importance that these oscillations are closely related to the nonlinear part of the $I-V$ characteristic and, moreover, that this nonlinearity on the macroscopic level can be explained by an underlying microscopic ndc. That the nonlinearity on the macroscopic level derives from negative differential conductivity on the microscopic level is in accordance with the experimental evidence of spatially inhomogeneous current structures developing in the nonlinear part of the characteristic (Sect. 2.4.3). Therefore, the spontaneous oscillations can be attributed to ndc inherent to the breakdown mechanism.

An explanatory model approach based on semiconductor physics assumes that the mobile charge carriers undergo generation–recombination processes with respect to the ground and excited states of impurity centers [2.6]. The idea that such a multilevel transport model is applicable to our p-Ge system is based on the photoconductivity spectrum presented in Fig. 2.4. Theoretically, involving at least two impurity levels, analytical conditions can be explicitly deduced for the appearance of an NDC characteristic as well as for both filamentary and oscillatory instabilities [2.6, 11, 49]. In the framework of a hot electron transport theory taking into account impact ionization of impurity centers, the moment expansion of the semiclassical Boltzmann equation yields a set of coupled hydrodynamical balance equations for the mean carrier density, the mean momentum, and the mean energy per carrier. They constitute, together with Maxwell's equations for the electric field, a macroscopic nonlinear dynamical system. The time scale on which these variables change is determined by the generation–recombination lifetime, the momentum relaxation time, the energy relaxation time, and the dielectric relaxation time, respectively. Since the macroscopic dynamics is governed by the slow variables, which of the above variables must be considered as relevant dynamical transport variables depends upon the relations between these time scales. The fast variables can be eliminated adiabatically from the dynamics [2.41].

So far, theoretical modelling of impact-ionization-induced instabilities observed experimentally has been attempted in various simplified versions that are capable of predicting local or global bifurcations to spatio-temporal pattern formation. These can be classified according to the different time scales involved. Moreover, the experimental evidence of spatially localized oscillation centers in the boundary region of current filaments (reported in Sect. 2.4.4) supports the central idea of "breathing" filament dynamics that originates from the possible interaction between longitudinal relaxation instabilities and transverse filamentary instabilities [2.49]. Nevertheless, our semiconductor-theoretical understanding of these phenomena is still lagging behind the experiments. A full understanding would require the derivation of a macroscopic nonlinear dynamical system exhibiting such behavior from the microscopic level of elementary transport processes. This is not yet available; there exist only model approaches to identify the relevant physical mechanisms at different levels of semiconductor transport theory.

Finally, two open questions remain which cannot be explained on the basis of such physical models. The first concerns the relatively low frequency range of the spontaneous oscillations observed (compared to the characteristic time scales of semiconductor breakdown). The second problem arises from the fact that it is practically impossible to give a detailed description of the wide variety of nonlinear effects intrinsic to the present semiconductor system, especially, the complex chaotic behavior together with its sensitive dependence on distinct control parameters.

In order to attain an independent approach to the slow macroscopic system dynamics of these breakdown instabilities, we have evaluated the typical relaxation times from the response behavior to external pulsed excitations as follows. Instead of measuring the conventional d.c. $I-V$ characteristic, we looked at the time dependence of the corresponding a.c. plot, obtained by applying triangular bias voltage pulses with varying repetition rate and recording the resulting time response of the current as a function of the sample voltage. With the pulses applied, the bias voltage V_0 periodically increases from 0 V to about 2 V (pulse height) over a time interval of 2 ms (pulse width) and then abruptly decreases to the initial zero voltage level within a few μs. Figure 2.15 displays an example of two different a.c. $I-V$ curves, measured by applying distinct sequences of voltage pulses having time distances of more and less than

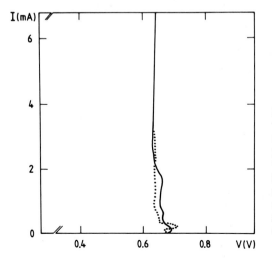

Fig. 2.15. Time-dependent current–voltage characteristics obtained from the sample sketched in Fig. 2.4 by applying triangular bias voltage pulses with different repetition rates 70 Hz (*solid curve*) and 340 Hz (*dotted curve*) at the following constant parameters: transverse magnetic field $B = 1.49$ mT and temperature $T_b = 2.1$ K (load resistance $R_L = 1.1$ kΩ)

10 ms (solid curve and dotted curve, respectively). We emphasize that these findings can be reproduced utilizing a time series of double pulses, the distance of which is varied accordingly. Then the current response to the first voltage pulse always gives the solid curve, while the second pulse can produce the dotted $I-V$ curve, provided the distance between the two pulses is smaller than 10 ms. The graphs of Fig. 2.15 have been averaged over 100 pulse cycles, but recorded simply during the rising branch of the bias voltage, i.e., for increasing current.

Obviously, the structure of the time-dependent $I-V$ characteristic changes appreciably in the highly nonlinear breakdown regime of negative differential resistance, just where the current instabilities develop. The system response behavior is, thus, governed by a certain "memory" capability, still remembering the preceding excitation if the temporal distance between the voltage pulses applied is not too long. The extracted time constant of about 10 ms clearly supports the relatively low-frequency scale observed experimentally for the relevant relaxation-type system dynamics [2.18, 29, 30, 36, 38, 44, 45, 47]. Moreover, we have also performed combined d.c. and a.c. $I-V$ measurements under a variation of the experimental parameters in the following ranges: load resistor $R_L = 1\,\Omega$–$100\,k\Omega$, magnetic field $B = 0$–$50\,mT$, and temperature $T = 1.7$–$4.2\,K$. Here the bias voltage V_0 consisted of sinusoidal or rectangular pulses superimposed upon the nonzero, constant, d.c. level. Depending on the parameter region investigated, characteristic relaxation times have been evaluated from the temporal profiles of the current response ranging from 0.1 ms up to 10 ms and, thus, covering the whole frequency span observed experimentally. Most importantly, an overall tendency of slowing down the time constant could be recognized on approaching the phase transition located in the strongly nonlinear breakdown region (see also Fig. 2.15). A thorough treatment of these pulse measurements is given elsewhere [2.50].

The physical origin of the complex nonlinear dynamics observed in the present semiconductor system is not yet fully understood theoretically. As already indicated above, first model attempts involve the autocatalytic process of impact ionization of one or more impurity levels coupled either with dielectric relaxation of the electric field or with energy relaxation of the hot charge carriers. The central idea of "breathing" current filaments together with long-range coupling of spatially separated oscillation centers via energy exchange may roughly explain some filamentary and oscillatory behavior of our samples. But, on the other hand, a detailed description of the rich variety of spatio-temporal nonlinear dynamics cannot be expected on the basis of such physical models. Like in turbulence, it seems to be impossible to describe the global system behavior starting from first principles. Furthermore, it is well known that complex nonlinear behavior can be modelled with an astonishingly high precision by universal ad hoc models.

The still open question which arises is: To what extent can the physical mechanism discussed contribute to the understanding of the relatively slow macroscopic time scale observed experimentally for the present semiconductor system? Certainly, the possible increase of the carrier lifetime in the lowest excited bound state of the hydrogen-like multilevel energy spectrum of p-Ge [2.34] may affect the dynamics of the breakdown instabilities. The spatial identification of spontaneous oscillations in the boundary region of the current filaments (Sect. 2.4.4) gives a further hint for critical slowing-down behavior. Note that these privileged sites of the experimental oscillators are just very close to the sample locations where phase transitions between different conducting states occur (see the definition of "current filaments" in Sect. 2.4.3). Nevertheless, we feel – in the sense of *Haken* [2.41, 42] – that all traditional disciplines in physics which are concerned with the macroscopic behavior of multicomponent systems require new ideas and concepts based on the interdisciplinary synergetic approach, in order to cope with self-organizing systems. In the first step, we have derived a phenomenological reaction–diffusion model from the generic Rashevsky–Turing theory of morphogenesis [2.51], which, so far, looks highly promising.

2.4.3 Spatial Structures

In addition to the appearance of spontaneous current and voltage oscillations as reported in Sect. 2.4.2, the strongly nonlinear breakdown regime of the time- as well as space-averaged $I-V$ characteristic is further associated with the self-sustained development of spatially inhomogeneous current flow in the formerly homogeneous bulk semiconductor. These structures usually evolve in the form of current filaments representing highly conducting channels embedded in a much more weakly conducting semiconductor medium [2.6]. The complex spatial behavior can be globally visualized by means of two-dimensional imaging of the current filament structures via low-temperature scanning electron microscopy [2.35]. These measurements are performed with the experimental arrangement sketched in Fig. 2.6. As already outlined in Sect. 2.3.2, the sample is locally perturbed by the electron beam, and a proper response signal is recorded as a function of the coordinates (x, y) of the beam focus. We register the change in the electric conductivity, i.e., the beam-induced current change $\delta I(x, y)$ in the voltage-biased sample, when the electron beam is scanned over a current filament (Fig. 2.16).

Qualitatively, the imaging process resulting from the interaction of the electron beam with the semiconductor sample can be understood as follows. Taking the present p-Ge system as an example and assuming a typical beam energy of 26 keV, the beam injects about 10^4 electron–hole pairs per incident electron in the generation volume of a few micrometer diameter. If the beam is directed onto a nonconducting region where no filamentary current flow can occur, the injected hot carriers may locally induce avalanche breakdown by impurity impact ionization, resulting in a significant current increment $\delta I(x, y)$ in the voltage-biased sample. On the other hand, if the beam is focused on a highly conducting region with filamentary current flow where most of the shallow impurities are already ionized, no significant current increment is obtained. In this way, the boundaries of a current filament can be detected.

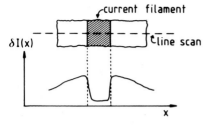

Fig. 2.16. Origin of the response signal used to image current filaments in a semiconductor via low-temperature scanning electron microscopy. *Top*: Part of the semiconductor sample as seen from the top side with a current filament in the vertical direction marked by the hatched section in the center. The horizontal dashed line indicates a line scan of the electron beam. *Bottom*: Beam-induced current change vs. the beam coordinate for the line scan shown at the top

The present experiments were performed using a beam voltage of 26 kV and a beam current in the range of 5–100 pA. This corresponds to an absorbed beam power of 0.1–2 μW, which is small compared to the Joule heat dissipated in the sample during avalanche breakdown (typically, in the milliwatt range). The temperature increase at the position of the beam spot was estimated to be less than 1 mK, assuming that the beam energy is dissipated in the generation volume of about 4 μm diameter [2.9]. In order to obtain the necessary signal detection sensitivity, the beam current was chopped, typically, at 5 kHz, and the response signal was detected using a lock-in technique. Clearly, for undisturbed recording of the intrinsic filamentary current flow, the chopping frequency must be sufficiently low such that the sample is able to respond properly. Since we have measured relaxation times of the electron-beam induced signal of about 10 μs, the modulation frequency used appears adequate. Therefore, if the lock-in output is taken to control the brightness on the video screen of the scanning electron microscope, current filaments appear as dark channels within bright regions indicating their boundaries. The spatial resolution limit of approximately 10 μm attainable for stationary current images has been determined from individual line scans where the amplitude of the beam-induced response signal is directly recorded as a function of the beam position perpendicular to the direction of the current flow (Fig. 2.16).

Figure 2.17 shows two-dimensional images of typical current filament patterns spontaneously forming in the nonlinear breakdown regime of the I–V characteristic. The scanned portion of the sample surfaces represents exactly the space between the parallel ohmic contacts located beyond the top and the bottom of the images. The bright regions correspond to a large response signal $\delta I(x, y)$, indicating the highly sensitive boundary regions between different conducting phases. The enveloped dark regions mark the current filament channels extending along the y-direction. The photographs in Fig. 2.17a and b display self-sustained stationary patterns of singlefilamentary and multi-filamentary shape, respectively. Note that the present imaging method detects only the filament portions sufficiently close to the irradiated sample surface.

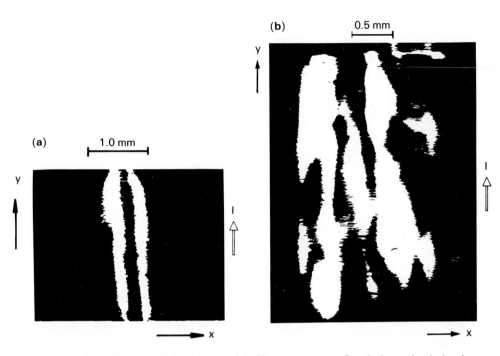

Fig. 2.17a, b. Brightness-modulated images of the filamentary current flow during avalanche break-down obtained by low-temperature scanning electron microscopy from distinct samples with 2.5 mm **a** and 3.4 mm **b** contact distance at the different bias voltages $V_0 = 1.35$ V **a** and $V_0 = 2.00$ V **b**, and the following constant parameters: zero magnetic field and temperature $T_b = 4.2$ K (load resistance $R_L = 1\ \Omega$). The dark bands within the bright regions correspond to the filament channels extending along the y-direction

As reported elsewhere [2.9, 52] in detail, nucleation and growth of fila-mentary current patterns in the nonlinear breakdown regime are often accom-panied by abrupt changes between different stable filament configurations via noisy current instabilities. Here a common phenomenon is that there exists a minimum as well as a maximum value of the filament current. On the one hand, with decreasing applied electric field, the filament size cannot decrease continuously to zero, but abruptly decays to zero at a distinct minimum diameter. On the other hand, with increasing applied electric field, a continuous growth of the existing filaments takes place only up to a distinct maximum size, beyond which an additional filament is nucleated. The ensuing multifilamentary current flow becomes more and more homogeneous if the semiconductor system further approaches its linear post-breakdown region at higher electric fields. On this way, the sample current usually exhibits pronounced jumps as a conse-quence of the abrupt nucleation (or annihilation) of individual filaments.

A comprehensive understanding of these experimental findings can be attained by a phenomenological theoretical description based on the electric power balance in the present nonequilibrium system [2.53]. The existence of a minimum diameter for current filaments is obvious from simple arguments balancing the power gain of moving hot charge carriers in the volume of the cylindrical filaments and the power losses across the filament boundaries, thus giving rise to some critical surface-to-volume ratio. The maximum filament diameter was found to be given by the competing contributions of the power losses generated by the inhomogeneous current flow in the ohmic contact layers and those across the filament boundaries. In particular, minimization of the power sum of the contact dissipation and the filament surface dissipation could directly explain particular bifurcation patterns observed experimentally. Here the current has partially split up into different filament branches. An example of such a complex multifilamentary current structure is given in Fig. 2.17b.

2.4.4   Spatio-Temporal Behavior

So far, the imaging method of low-temperature scanning electron microscopy has been applied to investigate the stationary filament structure of the electric current flow in a semiconductor. Moreover, this technique can serve also for simultaneously visualizing the dynamical part of the spatially inhomogeneous current distribution. Hereto, we have combined both space- and time-resolved measurements as described in the sequel.

First, we looked at the temporal response of the sample current to the local electron-beam induced perturbation as a function of the beam coordinate [2.9, 52]. The results presented in Fig. 2.18 suggest that the electron-beam

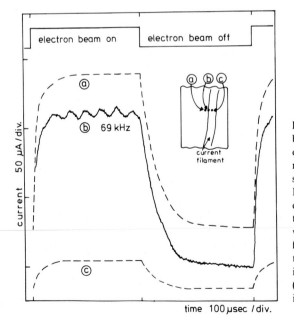

Fig. 2.18. Time dependence of the beam-induced current response obtained by low-temperature scanning electron microscopy for the same experimental situation as in Fig. 2.17a. The trace at the top indicates the chopped operation of the electron beam. The traces (a–c) were obtained when the beam was focused on the corresponding locations at the sample surface marked in the inset: (a) outside the filament, (b) on the filament boundary, (c) inside the filament

irradiation can stimulate oscillations if the beam is focused on the boundary region of a current filament. The beam was turned on and off periodically with a 50% duty cycle as shown by the upper trace. The three temporal recordings of the response signal $\delta I(t)$ were obtained by shifting the beam focus from location (a) to location (c) across the filament boundary as indicated schematically in the inset. The beam-induced current measured via a standard boxcar-averaging technique decreased as expected, i.e., recordings (a) and (c) with the location of the beam focus outside and inside the current filament region exhibit a relatively large and small signal amplitude, respectively. When the beam was focused on location (b) of the filament boundary, superimposed stable current oscillations could clearly be observed as shown on the temporal trace (b). Beam-induced oscillations of the response signal $\delta I(t)$ appeared in the frequency range 10–120 kHz, depending upon the location of the beam focus along the filament boundary and upon the beam intensity. Note that in the experimental situation of Fig. 2.18 no spontaneous current oscillations were observed without the beam irradiation of the filament boundary. Clearly, the temporal profile of the beam-induced sample response supports our notion that it is the boundary region of the spatial current patterns where the spontaneous oscillations, and hence the chaotic behavior of the electric current, develop during avalanche breakdown.

The second way to study the spatio-temporal dynamics of the present semiconductor system arises from a resonance imaging method [2.9, 30], the principle of which is as follows. If the sample investigated already reveals spontaneous oscillations without electron-beam irradiation, the local regions where these self-generated oscillations take place can be imaged two-dimensionally by resonant interaction with the applied electron beam. One only has to modulate the beam just at the intrinsic frequency of the spontaneous oscillations observed in the unirradiated sample and to scan over the specimen surface, in order to pick up the regions oscillating spontaneously. Note that a distinct resonance signal arises in the temporal current response if the focus of the properly modulated electron beam approaches the location of the oscillation center. The present technique, thus, provides spatially resolved analysis of the nonlinear system dynamics.

Figure 2.19 shows two-dimensional images of both stationary and dynamical current structures in the nonlinear breakdown region, recorded successively at different beam-modulation frequencies. The bias voltage was applied to the triangular ohmic contacts, the tips of which are indicated by the full markers (the open triangle means a nonactive contact). The surface of the intervening sample part was scanned by the electron beam (horizontal traces), and the beam-induced current change $\delta I(x, y)$ was plotted vertically (y-modulation). The well-established stationary current filament pattern (Fig. 2.19a) extending between the facing active contacts can be recognized through pronounced response signals originating from its enveloping boundaries, where the local conductivity of the sample is most sensitive to the beam-injected perturbation. Accordingly, as in the brightness-modulated images of Fig. 2.17, the electron beam was chopped at a frequency of 5 kHz. The conspicuous dynamical current structure

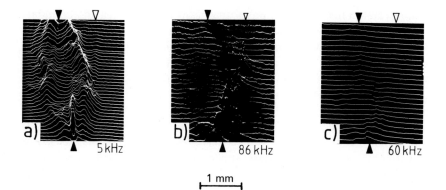

Fig. 2.19a–c. *Y*-modulated images of the spatio-temporal current flow during avalanche break-down obtained by low-temperature scanning electron microscopy from a distinct sample with 3.2 mm contact distance with different beam-modulation frequencies 5 kHz **a**, 86 kHz **b**, and 60 kHz **c** at the following constant parameters: bias voltage $V_0 = 1.4450$ V, zero magnetic field, and temperature $T_b = 4.2$ K (load resistance $R_L = 1 \Omega$). **a** Standard image of a stationary current filament recorded at low electron-beam modulation; **b** resonance image of spontaneously oscillating current filament regions recorded at resonant electron-beam modulation; **c** off-resonance image recorded at nonresonant electron-beam modulation. Resonant response signals arise in the right boundary of the current filament extending in the vertical direction

(Fig. 2.19b) taken at resonant electron beam excitation clearly identifies spontaneously oscillating sample regions on the right-hand side boundary of the current filament oriented in the vertical direction (Fig. 2.19a). Of course, the 86 kHz beam modulation frequency used for the resonance image is far too high in view of the typical sample response times necessary to obtain the standard images of stationary current patterns. In comparison, the off-resonance case (Fig. 2.19c) due to nonresonant electron beam excitation displays neither a stationary filament nor an oscillatory resonance structure, as expected. Clearly, the modulation frequency used (60 kHz) is too far from the intrinsic frequency of the spontaneous current oscillation (86 kHz) to receive any resonant response signal.

From these experiments (including some results not shown here [2.9, 18, 29, 30, 38, 47, 52]), we conclude that the origin of different oscillatory modes lies in the simultaneous existence of spatially separated oscillation centers located close to distinct boundary regions of the current filaments. More generally, the spatial discrimination of temporal current flow dynamics in the vicinity of different conducting phases yields a powerful tool for gaining deeper insight into the mutual interplay between spatial and temporal current structures developing during semiconductor breakdown. Here, the electron beam was demonstrated to act as an exemplary control parameter which can be manipulated both spatially and temporally.

Problems

2.1 Discuss the temperature dependence of the charge carrier concentration of a semiconductor in the intrinsic and extrinsic ranges and compare with that of a metal. (See [2.33].)

2.2 Describe qualitatively the connection between the existence of negative differential conductivity and the formation of semiconductor plasma instabilities. (See [2.8].) What can we learn from the particular shape of the measured current–voltage characteristic?

2.3 Start from the well-known semiclassical balance equations for the mean density, energy, and momentum of the charge carriers and look at the relevant time scales of the characteristic relaxation times involved. (See [2.6].) What are the necessary prerequisites for generating so-called hot charge carriers? Derive the critical electric field sufficient to initiate the autocatalytic process of avalanche-like carrier multiplication at the onset of impact ionization breakdown. Discuss the effects leading to spatio-temporal transport instabilities via phase transition analogies. What are the key obstacles preventing an overall understanding of the responsible semiconductor-physics mechanism?

2.4 Calculate the macroscopic current–voltage $(I-V)$ characteristic starting from a linearized microscopic current density versus electric field $(j-E)$ characteristic. Use the following ansatz: $j = \sigma_1 E$ $(0 < E < E_0)$, $j = \sigma_2 E$ $(E > E_0 - \Delta E)$ with $\sigma_1 < \sigma_2$ and $0 < \Delta E/E_0 < 1$. Assume that in the unstable region $\sigma_1 E_0 < j < \sigma_2 (E_0 - \Delta E)$, $E_0 - \Delta E < E < E_0$ two phases having conductivities σ_1 and σ_2 coexist in the form of a filamentary current density structure. The phases σ_1 and σ_2 have cross sections A_1 and A_2, respectively, where $A_1 + A_2$ gives the total cross section A of the semiconductor sample. Assume further that $A_1 = A(1 - x)$ and $A_2 = Ax$ with the parameter x varying linearly with the current density j between 0 and 1 in the instability region. Derive the condition for which negative differential resistance appears in the $I-V$ characteristic.

3 Nonlinear Dynamics

This chapter focuses on the methods for analyzing time-dependent phenomena. The main emphasis is laid on the characterization of nonlinear dynamical system behavior. We do not intend to give another more or less complete treatment of what is known at present about nonlinear systems in general, but to classify a representative natural system – low-temperature impact ionization semiconductor breakdown – in the light of nonlinear dynamics. Guided by the results of this experimental situation, we introduce various characterization methods at such a level that any reader should be able to understand the basic arguments without requiring too much knowledge in advance. For a more comprehensive – particularly with respect to theory – introduction to the field of nonlinear dynamics, we refer to other books [3.1–11].

Since semiconductor breakdown belongs to the subclass of dissipative nonlinear systems, our considerations are restricted to the corresponding transport phenomena. The huge variety of conservative or Hamiltonian systems is thereby excluded. Moreover, we abstain from the treatment of partial differential equations, even though they are capable of describing the mutual interplay between spatial and temporal degrees of freedom in a system and, thus, representing the "mother" of nonlinear dynamics theory (e.g., the well-known Navier–Stokes equation). Note that solutions of partial differential equations are arbitrarily complicated in general and, so far, confined to rather exceptional cases. Just the reduction to ordinary differential equations without any spatial components was responsible for initiating research into nonlinear dynamical systems [3.12]. Most importantly, the treatment of low-dimensional ordinary differential equations with only a small number of state variables involved can still provide an adequate understanding of a widespread diversity of different temporal phenomena, particularly if the consequences of nonlinearities are considered.

For a possible classification of such dissipative ordinary differential equations, we build up a hierarchical order principle according to the degree of their complexity, i.e., with respect to the relevant number of actively participating degrees of freedom. Accordingly, the present chapter is organized as follows. After an introduction to the fundamental aspects of the theory of nonlinear dynamics (Sect. 3.1), we start in Sect. 3.2 with nonlinear ordinary differential equations of one variable capable of describing only temporally constant phenomena. In this sense of hierarchy, the simplest temporally varying state, the periodic oscillation, is treated in Sect. 3.3. The higher periodic oscillations follow in Sect. 3.4. Nonperiodic chaotic oscillations with their higher-order analogs are discussed in Sect. 3.5. Finally, Sect. 3.6 is devoted to the interwoven relationship between the temporal and spatial phenomena of the nonlinear dynamics considered. Guided by the results of a particular semiconductor experiment, the ingredients of the system-immanent physics of which are outlined in Chap. 2, we aim to span the gap from order to chaos.

Besides this logical ordering of the different sections with respect to each other, each individual section is also built up internally in a similar way. Thus, we always begin with a summary of the theoretical basis necessary to describe the prevailing dynamical state. Particular emphasis is thereby laid on the characterization methods. We then deal with the specific dependence of the dynamical state upon distinct external parameters (control parameters), i.e., we look at its bifurcation behavior. Finally, the current experimental results are characterized in terms of these system-independent methods of nonlinear dynamics theory. It is important that theoretically obtained features of non-linear dynamical systems are verified by real-world experiments, in order to demonstrate their universal validity.

3.1 Basic Ideas and Definitions

In this section we introduce some fundamental concepts used in the theory of nonlinear dynamics. They are also illustrated with the help of examples. For a more detailed treatment, we refer to the monograph by *Guckenheimer* and *Holmes* [3.2].

A dynamical state is characterized by the temporal behavior of a quantity z (Fig. 3.1). In a theoretical model z can be thought of as a variable, whereas for an experimental situation z is a measurable quantity, i.e., an observable. A basic assumption is that the time dependence of z can be described by a differential equation. As already mentioned above, we restrict ourselves to ordinary differential equations of the form

$$\dot{x} = f_\mu(x) .$$ (3.1)

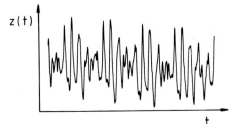

$z(t)$

t

Fig. 3.1. Typical time trace of a quantity z

Here x is an n-dimensional variable ($\dot{x}$, its time derivative), μ, a value of an m-dimensional control parameter space, and f, a continuous n-dimensional function. f should be continuous or, more precisely, should fulfill the Lipschitz condition, in order to warrant uniqueness of the solution. Since we claimed to consider only dissipative systems, any volume element in variable space (phase space) must contract under the time development of (3.1). This holds for the condition $\int dx \, \mathrm{div} f < 0$. The function f_μ generates the flow $\phi_\mu(x, t)$ in such a way that for each initial condition x_0 its temporal behavior is determined by (3.1). From ϕ_μ we obtain the solution $x(x_0, t)$ for any given x_0. Thus, ϕ defines the solution curve which is often termed the orbit or trajectory of the differential equation starting from x_0. The quantity z mentioned above is, therefore, a function of $\phi_\mu(x_0, t)$. A trajectory can be uniquely figured in the phase space defined by the components of the variable (Fig. 3.2). Such a presentation, together with its two-dimensional projection, is called a phase portrait.

Next we define some particular objects in phase space. x_E means a fixed point or an equilibrium point if $f_\mu(x_E) = 0$. This condition guarantees that x_E does not change in time any more because the time derivative of x_E is zero. Depending on the temporal evolution of its neighboring points, x_E is denoted as a stable, an asymptotically stable, or an unstable fixed point. In the case of a stable fixed point, there exists – in the sense of Lyapunov – for any neighborhood U of x_E in phase space a smaller neighborhood U_1 of x_E with $U_1 \subset U$ such that all trajectories starting in U_1 remain in U for $t > 0$. An asymptotically stable fixed point means that all solutions $x(t)$ starting in U_1 approach x_E asymptotically for $t \to \infty$. A fixed point is unstable if it is not stable. The characteristic features of these different Lyapunov stabilities are summarized schematically in Fig. 3.3.

A more precise partitioning among the different fixed points can be found in Fig. 3.4. According to this representation, a stable fixed point comprises a center

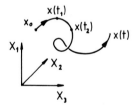

Fig. 3.2. Schematic representation of a trajectory in three-dimensional phase space starting from the initial condition x_0 and passing through the points $x(t_1)$ and $x(t_2)$ at times t_1 and t_2

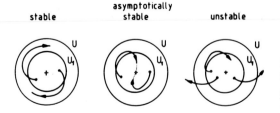

Fig. 3.3. Schematic representation of typical trajectories in phase space around three fixed points of different Lyapunov stability

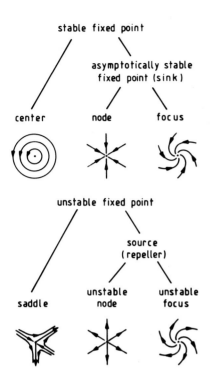

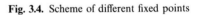

Fig. 3.4. Scheme of different fixed points

or an asymptotically stable fixed point. The latter, which is also called a sink, may be a node or a focus, depending on the way the trajectories approach the fixed point. In the case of an unstable fixed point, one distinguishes between a saddle and a source. The difference lies in the fact that for the latter, which is also called a repeller, all neighboring points diverge in time, whereas for a saddle this holds only for some (usually, for nearly all) neighboring points. A source may be an unstable node or an unstable focus, again depending on the way the trajectories depart in its neighborhood. We emphasize that these definitions of stability are not restricted to the properties of fixed points. They can analogously be applied to any point x in phase space and to the corresponding orbits.

In what follows, these stability terms will be explained with the help of some example models of simple oscillators which represent the basic elements of nonlinear dynamics theory. A more extensive discussion can be found in the monograph by *Thompson* and *Stewart* [3.9].

The differential equation of an undamped linear oscillator is given by

$$\ddot{x} + x = 0 .$$ (3.2a)

As a first-order ordinary differential equation, it can be written as

$$\dot{x}_1 = x_2, \qquad \dot{x}_2 = -x_1 .$$ (3.2b)

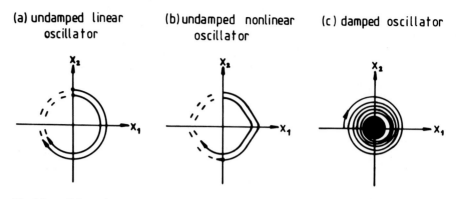

Fig. 3.5a–c. Schematic representation of phase portraits of different oscillators. In cases **a** and **b**, the temporal development of the trajectories starting from two nearby initial conditions is plotted as solid lines

For different initial conditions (x_{10}, x_{20}), oscillations with different amplitudes but equal frequencies are obtained as solutions. As a consequence, the distance between two initial points remains constant in time. This gives an example of a stable orbit (Fig. 3.5).

The differential equation of an undamped nonlinear oscillator is, for instance,

$$\ddot{x} + \sin x = 0 \tag{3.3a}$$

or

$$\dot{x}_1 = x_2, \qquad \dot{x}_2 = -\sin x_1 . \tag{3.3b}$$

Here the solutions for different initial conditions result in oscillations with different amplitudes and different frequencies. As a consequence, initially neighboring points diverge in time. Now we have a sensitive dependence on initial conditions. These trajectories are unstable, more precisely, they are Lyapunov-unstable. If one considers only the orbits in total and does not take into account the temporal development of single points on trajectories in phase space, one finds that the orbits proceed at constant separation. This case is called orbit-stable (Fig. 3.5b).

As the next step, we regard the influence of a small damping term on the oscillators described above. Now the differential equations have the form

$$\dot{x}_1 = x_2, \qquad \dot{x}_2 = -x_1 - \xi x_2 \tag{3.4}$$

and

$$\dot{x}_1 = x_2, \qquad \dot{x}_2 = -\sin x_1 - \xi x_2 . \tag{3.5}$$

For both cases, one finds oscillations with an amplitude decaying in time as the solution. Figure 3.5c illustrates the spiral-type form of the corresponding phase plot. Apparently, the origin plays the role of an asymptotically stable fixed point.

By means of these examples, two further notions can be explained, namely, the attractor and the structural stability. In the case of damped systems, we have seen that the origin attracts trajectories from its neighborhood. That is why this fixed point is called an attractor. For the class of nonlinear dissipative systems, attractors play an important part, whereas they do not exist in nondissipative systems like, e.g., the undamped oscillator. Attractors are embedded in a neighborhood U such that all trajectories starting from any point of U will approach the attractor asymptotically (Fig. 3.6). Such a neighborhood is named the basin of the attractor. The course of any trajectory that starts within the basin of an attractor will become more and more similar to the actual structure of the attractor as time goes on. In fact, due to the inevitable noise or finite resolution present, there is no difference between the long-term behavior of a trajectory starting from a point within the basin and that of a trajectory starting from a point of the attractor. The initial part of the trajectory approaching the attractor is called a transient. Accordingly, a trajectory which starts directly on the attractor does not have a transient, it is recurrent.

An attractor can, thus, be defined as a compact set A having the following properties [3.13, 14]: (1) A has a shrinking neighborhood such that for almost all initial conditions the limit set of the orbit for $t \to \infty$ is A. (2) A is invariant under the flow ϕ. (3) The flow on A is recurrent and, hence, not decomposable. The shape of an attractor may be a point (fixed point) or a closed line. The latter may have a finite length, as for a periodic motion. It is also possible that the line has an infinite length, as is the case for a two-torus, where the attractor line is capable of forming a two-dimensional plane in phase space.

In what follows, we discuss a method to construct an attractor. One does not always have such an easy situation as is the case for the numerical solution of a differential equation where all variables are available in their time behavior. Here the construction of the attractor as a path in variable space or phase space represents no more than a trivial task. However, there are more problems in experiments, where quite often only the temporal course of a few observables (not always necessarily in keeping with the relevant system variables) can be measured. For this case, an important method is available for constructing the attractor out of a one-dimensional time signal $z(t)$ [3.15]. A d-dimensional phase

Fig. 3.6. Scheme of the phase portrait of an attractor. Trajectories starting from initial conditions located inside the basin (*white region*) approach the attractor asymptotically. Trajectories starting from initial conditions located outside the basin (*hatched region*) head elsewhere

space is reconstructed by a time transformation of the signal $z(t)$ to $z(t + n\tau)$, where τ represents an arbitrarily chosen time constant and $n = 1, 2, \ldots, d - 1$. Now, the signals $z(t)$ and $z(t + n\tau)$ are stated to reflect independent variables. Their phase portrait, often called the Takens–Crutchfield construction, then coincides with the real graph.

In order to illustrate the idea of this construction method, let $z(t)$ be a harmonic periodic signal, e.g., $z(t) = \sin \omega t$. It is a well-known fact that the corresponding attractor has the form of a circle, which can be obtained by plotting $\sin \omega t$ vs. $\cos \omega t$. For the construction of the attractor, we take an arbitrary time constant τ to get a second signal $z(t + \tau) = \sin \omega(t + \tau)$. With the help of the identity $\sin \omega(t + \tau) = \sin \omega\tau \cos \omega t + \cos \omega\tau \sin \omega t$, one recognizes that the second signal contains the $\cos \omega t$ term. The phase portrait $z(t)$ vs. $z(t + \tau)$, thus, gives at least a deformed circle – an ellipse. Only in the particular case of $\omega\tau = k\pi$ $(k \in \mathbb{Z})$, i.e., $\tau = kT/2$, where T denotes the period of the signal $z(t)$, does one end up with nothing but a straight line. However, the probability of this case is of measure zero for an arbitrarily chosen τ. Note that, when analyzing a dynamical state via the attractor, it is only its principal shape (e.g., point-like or circle-like) which represents the important information. This can be easily extracted from the construction method above. It will quite often be successfully applied to experimental conditions to be dealt with later on.

We turn next to the second term, the structural stability [3.16]. The example mentioned above and illustrated in Fig. 3.5 demonstrates the role of the origin changes. In the case of an undamped motion, the origin is a stable fixed point or a center. It becomes an asymptotically stable fixed point due to the presence of an arbitrarily small damping. Note that the appearance of the asymptotically stable fixed point influences the structural course of the trajectories in its neighborhood (cf. the phase portraits in Fig. 3.5). As a consequence, the orbit-stable trajectories of the undamped oscillation vanish in the damped case.

In a more general way, the structural stability can be shown by the influence of a perturbation term $\varepsilon g(x)$ on the structure of the solution of a differential equation,

$$\dot{x} = f_\mu(x) + \varepsilon g(x) . \tag{3.6}$$

Here the coefficient ε gives the magnitude of the perturbation term, whereas $g(x)$ represents an arbitrary function. If there is a function $g(x)$ such that for any value of ε the structure of the attractor (displaying the solution of the ordinary differential equation) is changed in a fundamental way, then we call a system structurally unstable. In the case where the attractor is only changed by a small amount proportional to the magnitude of the perturbation term, we call the system structurally stable. Analogous to the characterization of the Lyapunov instability by a sensitive dependence on the initial conditions of trajectories, the structural instability reflects the sensitive dependence on a perturbation term of a dynamical system.

Consequently, a structurally stable system is robust against small perturbations. Just this property enables one to obtain results that can be reproduced in experiments as well as in computer simulations. Both kinds of investigation are inevitably affected by a certain level of perturbation. Experiments are at least confronted with thermal noise, while computer simulations generally produce rounding errors. If there were no structural stability, one could not obtain any reproducible (i.e., reliable) result and, therefore, not gain any understanding of the system considered. The structural stability plays a fundamental role for our overall comprehension and even our existence. Nevertheless, the same is true for the structural instability. Certainly, it represents the only way to force the system behavior to undergo any form of structural change. In this sense, there is a dialectic relation between structural stability and instability.

Of particular interest are systems capable of displaying stabilities and instabilities in such an ordered way that it becomes possible to understand their behavior. Nonlinear dynamical systems represent typical examples. In physics, controlled variation of external parameters (like pressure, temperature, etc.) may cause transitions from stable to unstable and back to stable system behavior. In mathematics, the situation described above is reflected in the parameter dependence of the structure of a solution. This problem was first investigated by *Jacobi* [3.17] and called "Abzweigung". Afterwards, *Poincaré* [3.18] introduced the term "bifurcation" describing the same phenomenon.

For an ordinary differential equation system, $\dot{x} = f_\mu(x)$, see (3.1), a bifurcation point is a critical value of the control parameter, μ_c, where the system becomes structurally unstable. At this point, vanishingly small perturbations of the differential equations cause qualitative changes in their solutions (and attractors, respectively). With the help of the example oscillators described by (3.2)–(3.5), we have demonstrated that the perturbation term ξx destroys the dynamics of the undamped oscillators. If ξ is taken as a control parameter, then $\xi_c = 0$ represents the bifurcation point. Here the asymptotically stable fixed point, $x_1 = x_2 = 0$, loses its stability and becomes an unstable fixed point (unstable focus) for $\xi < 0$. Note that now the damping is negative, i.e., the system gains energy, giving rise to divergency. Thus, we can define a bifurcation by two characteristic features: (1) The system becomes structurally unstable. (2) The phase portrait (i.e., the attractor) undergoes a geometrical change.

An essential point for the observation of a bifurcation is that the structural instability occurs – in a new sense – in a structurally stable way. Such structural stability of a bifurcation can be expected if one considers the extended control parameter and phase space. Figure 3.7 gives an example of a bifurcation from a fixed point to a limit cycle, illustrated schematically in the corresponding space. There is a family of functions f_μ describing a bifurcation type (also called the generic behavior of f_μ). We denote this bifurcation type as structurally stable if small perturbations do not cause any structural change of the whole bifurcation in the extended control parameter and phase space. Of course, small deviations of the bifurcation point μ_c, for example, may take place under perturbation.

Due to their structural stability, it is possible to group together bifurcations with the same generic behavior. The simplest type of the equation which describes a bifurcation is called its normal form. It is, moreover, possible to

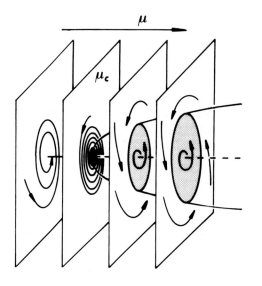

Fig. 3.7. Phase portrait of a Hopf bifurcation in the extended control parameter and phase space. The different planes correspond to the phase spaces for distinct values of the control parameter μ. For $\mu < \mu_c$, there is a stable focus, whereas, for $\mu > \mu_c$, this focus becomes unstable being surrounded by a limit cycle attractor

classify bifurcations with respect to their degree of complexity (in the generic description). Then one obtains an important quantity, namely, the minimum number of control parameters that are necessary to comprehend a structurally stable bifurcation. We denote this number as the codimension of a bifurcation. For the special case of bifurcations of fixed points, the elementary catastrophe theory developed by *Thom* [3.19] provides a complete framework for the classification of all these bifurcation possibilities. However, for dynamical systems with time-dependent solutions, such a powerful classification scheme is still lacking.

The classification of bifurcations yields two essential features. On the one hand, there are countably many bifurcation types characterized by unique attributes. On the other hand, each bifurcation type is completely described by one distinct (the simplest) normal function $g_\mu(x)$. One just has to take into account that this holds only in a close neighborhood of the bifurcation point μ_c in control parameter space and of the attractor in phase space. Thus, it is generally referred to as a local description of a bifurcation. In some sense, the normal function $g_\mu(x)$ can be taken as the structurally appropriate nonlinear expansion of $f_\mu(x)$ around μ_c. The central idea behind all this is no more than finding the simplest expansion form of $f_\mu(x)$ under the condition that the structural bifurcation is still comprehended completely.

Before we continue with the treatment of these aspects in a more formal way, some different types of bifurcations are briefly enumerated. To begin with, one distinguishes between a local and a global bifurcation. The former is characterized by the creation of a new attractor in the direct neighborhood of the attractor which has become unstable. For the latter type, however, it is not sufficient to describe the bifurcation in the immediate vicinity of the attractor becoming unstable. Furthermore, there exists a distinction between a continuous and a discontinuous bifurcation (the latter is called a catastrophe by Zeeman). It depends upon whether or not there is a continuous connection between the attractors on both sides of the bifurcation point in the extended control parameter and phase space. Note that analogous criteria are applied in the case of thermodynamical phase transitions. In this sense, the bifurcation displayed in Fig. 3.7 is local and continuous.

In order to wind up the present introductory section, we formally point out how it is possible, in principle, to reduce any given dynamical system to the normal form necessary to describe its bifurcation behavior. Hereto, let us start with a fixed point x_E of (3.1), $f_\mu(x_E) = 0$. Further, let η be a perturbation of the fixed point. Then, by setting $x = x_E + \eta$, one obtains the differential equation for η

$$\dot\eta = f_\mu(x_E + \eta) \, . \tag{3.7}$$

Assuming η to be small, the Taylor expansion reads

$$\dot\eta = f_\mu(x_E) + f_{\mu,x}(x_E)\eta + \tfrac{1}{2}f_{\mu,xx}(x_E)\eta^2 + \ldots , \tag{3.8}$$

where $f_{\mu,x}(x_E)$ represents the $(n \times n)$ Jacobian matrix $J(f_\mu)$ of the first derivatives of f_μ with respect to x at $x = x_E$ and $f_{\mu,xx}(x_E)$ represents the corresponding matrix of the second derivatives. For small perturbations η, we can approximate with a linear first-order differential equation,

$$\dot\eta = J(f_\mu(x_E))\eta \, , \tag{3.9}$$

giving the temporal development of the perturbation η of the fixed point x_E (linear stability analysis). To solve this linear differential equation, one has to find the spectrum, i.e., the set of eigenvalues, of the Jacobian matrix. Each eigenvalue λ_i is associated with a distinct eigenvector, in the direction of which the temporal development of the perturbation has $\exp(\lambda_i t)$ dependence. For simplicity, we do not deal with cases of degeneracy. The spectrum of eigenvalues can then be divided into three subsets – first, eigenvalues having a negative real part, which describe the asymptotic approximation $\exp[-\operatorname{Re}(\lambda_i)t]$ to the fixed point x_E; second, eigenvalues having a positive real part, which describe the exponential divergence from x_E; and, third, eigenvalues having a zero real part, which describe the critical behavior around x_E. In the case of critical behavior, just the higher-order nonlinear terms of (3.8) become essential. The eigenvectors of each subset of eigenvalues form a distinct subspace. There are stable, unstable, and central subspaces, depending upon the temporal behavior given by the eigenvalues. A fixed point x_E in the neighborhood of which all eigenvalues have a negative real part is asymptotically stable. According to the value of the imaginary part of the eigenvalues, we have a node or a focus. But if, in addition to eigenvalues with a negative real part, there are also some eigenvalues with a positive real part, then x_E is a saddle. Finally, if all the eigenvalues have a positive real part, then x_E represents a repeller, i.e., an unstable node or an unstable focus (Fig. 3.4).

Next we take into account the influence of the variation of the control parameter μ. Now each eigenvalue λ_i becomes a function of μ, hence forming a distinct trace in the complex plane under variation of μ. As a consequence, some eigenvalues may change the sign of their real part on crossing the imaginary axis. In Fig. 3.8 two simple examples are sketched schematically. Part (a) illustrates how one real eigenvalue λ_j becomes positive, whereas the real part of all other eigenvalues remains negative. Just at the point where λ_j is zero, we define the critical value of the control parameter, $\mu = \mu_c$. Then we have the following situation. For $\mu < \mu_c$, $\lambda_j < 0$ and, thus, all eigenvalues have a negative real part, i.e., we have a node. For $\mu > \mu_c$, $\lambda_j > 0$ and, thus, one eigenvalue has a positive real part, the other a negative real part, i.e., we have a saddle. The present example illustrates a transition from a node into a saddle by variation of μ, which is, consequently, denoted as saddle node bifurcation. Part (b) displays how a pair of complex conjugate eigenvalues simultaneously become positive in the real part. This corresponds to a change from a damped to an undamped oscillation, named Hopf bifurcation. Figure 3.7 has already demonstrated such a behavior in the extended control parameter and phase space. With the above examples we intended to show that bifurcation points are characterized by the condition that the real part of the eigenvalues becomes zero. A review of different bifurcations classified by the number of eigenvalues whose real part becomes positive is given in [3.3].

The geometrical description of bifurcations in the extended control parameter and phase space is characterized by three different manifolds, namely, the stable, the unstable, and the center manifold. These generalized subspaces are invariant under flow. They are distinguished by their linearized behavior being stable, unstable, or central. The corresponding eigenvectors represent tangents of the manifolds. It turns out to be important that a bifurcation can be described only on the basis of the center manifold embracing all points in the extended control parameter and phase space which have eigenvalues with zero real part. For example, the Hopf bifurcation is characterized by a two-dimensional center manifold. If we come back to Fig. 3.7, the center manifold then has the form of a straight line in the control parameter regime $\mu \leq \mu_c$, while for $\mu > \mu_c$ it becomes a cone on which there are limit cycle attractors of the periodic oscillations. The technique for reducing any system onto its center manifold is discussed elsewhere in detail [3.2, 20]. This reduction method has been denoted as the enslaving principle by *Haken* [3.3].

With the help of the reduction method, for any system we obtain in the neighborhood of a bifurcation point the representative equation, i.e., the normal

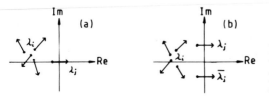

Fig. 3.8. Graphical representation of typical eigenvalues λ_i, λ_j and their paths in the complex plane under a variation of the control parameter

form, of the bifurcation. This normal form is capable of describing the bifurcation behavior of totally different systems. As a consequence, the knowledge of the original system becomes faded. In this sense, we concentrate in the following sections on the discussion of the most important bifurcations explained via the simplest possible equations, in order to find the corresponding classification features of the systems to be looked at. Particularly, we pay attention to the geometrical form of the bifurcations in their extended control parameter and phase space representation. Typical attributes of a bifurcation can then be compared with experimental findings, which enables us to classify the local system behavior.

3.2   Fixed Points

Subsequent to the introduction of some fundamental definitions of nonlinear systems, we now consider the simplest dynamical states, namely, the fixed points. In particular, the development of fixed points is discussed under the influence of changing control parameters. On the one hand, the cases to be presented are intended to provide a foundation for the following sections, where we deal with temporally varying states. On the other hand, the bifurcations of fixed points are applied to different current–voltage characteristics of low-temperature semiconductor impact ionization breakdown at the end of this section.

In what follows, we treat only the variation of fixed points, i.e., the variation of a point x_E of the differential equation (3.1), $\dot{x} = f_\mu(x)$, with $f_\mu(x_E) = 0$, is considered. From the stability analysis in Sect. 3.1, we already know that there are different kinds of fixed points (Fig. 3.4). Here we concentrate on the simplest bifurcations between stable and unstable fixed points. Subsequently, their relationship to catastrophe theory will be demonstrated. Readers interested in a more detailed discussion of this subject are referred to the literature [3.2, 19, 21, 22].

3.2.1   Fundamental Bifurcations

First, we consider bifurcations that are controlled by one single parameter $\mu \in \mathbb{R}$. These are the saddle node bifurcation, the transcritical bifurcation, the pitchfork bifurcation, and the subcritical pitchfork bifurcation. Their characteristic features are summarized in Fig. 3.9. Each bifurcation type is described by the differential equation (3.1). With the help of the first derivative of f_μ (for the more general case of higher dimensions, one has to take the Jacobian matrix), the linear stability of the fixed points, and, thus, their type, can be determined. In Fig. 3.9, $x_E = x_{as}$, $x_E = x_c$, and $x_E = x_{us}$ mean asymptotically stable, central, and unstable fixed points, respectively. The bifurcation diagram represents the

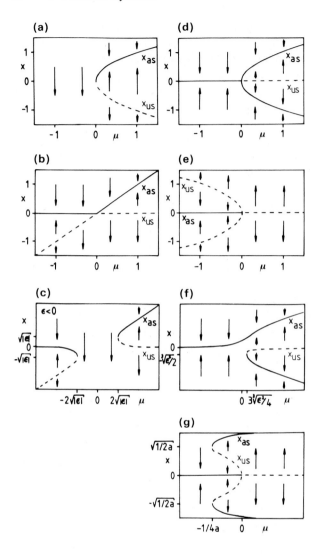

Fig. 3.9a–g. Schematic representation of the fundamental bifurcations of fixed points subject to one control parameter. **a** The saddle node bifurcation obtained from the equation $\dot{x} = \mu - x^2$. Its first derivative reads $Jf_\mu = -2x$. The fixed points are located at $x_E = \pm\sqrt{\mu}$, i.e., the asymptotically stable fixed point is $x_{as} = +\sqrt{\mu}$ ($\mu > 0$), the central fixed point is $x_c = 0$ ($\mu = 0$), and the unstable fixed point is $x_{us} = -\sqrt{\mu}$ ($\mu > 0$). **b** The transcritical bifurcation obtained from the equation $\dot{x} = \mu x - x^2$. Its first derivative reads $Jf_\mu = \mu - 2x$. The fixed points are located at $x_E = 0$ and $x_E = \mu$, i.e., the asymptotically stable fixed points are $x_{as} = 0$ ($\mu < 0$) and $x_{as} = \mu$ ($\mu > 0$), the central fixed point is $x_c = 0$ ($\mu = 0$), and the unstable fixed points are $x_{us} = 0$ ($\mu > 0$) and $x_{us} = \mu$ ($\mu < 0$). **c** The perturbed transcritical bifurcation obtained from the equation $\dot{x} = \mu x - x^2 + \varepsilon$. **d** The pitchfork bifurcation obtained from the equation $\dot{x} = \mu x - x^3$. Its first derivative reads $Jf_\mu = \mu - 3x^2$. The fixed points are located at $x_E = 0$ and $x_E = \pm\sqrt{\mu}$, i.e., the asymptotically stable fixed points are $x_{as} = 0$ ($\mu < 0$) and $x_{as} = \pm\sqrt{\mu}$ ($\mu > 0$), the central fixed point is $x_c = 0$ ($\mu = 0$), and

geometrical form of the bifurcation in the extended control parameter and phase space. Besides the variation of the fixed points as a function of the control parameter, one also recognizes the phase portraits of the fixed points through the arrows, indicating the temporal development of any point in phase space. These phase portraits further provide information on the stability of the fixed points. In the cases of structurally unstable bifurcations, we have added in Fig. 3.9 the corresponding diagrams for one perturbation term.

For the structural stability of a bifurcation one has to consider the perturbed differential equation (3.6), $\dot{x} = f_\mu(x) + \varepsilon g(x)$, where ε represents the small amplitude of the perturbation term and $g(x)$, an arbitrary function. $g(x)$ can be expanded around the bifurcation point (here, always, $\mu = 0$ and $x_E = 0$) in a Taylor series. In this way, the general case of an arbitrary perturbation affecting a bifurcation is investigated by looking at the terms $\varepsilon, \varepsilon x, \varepsilon x^2, \ldots, \varepsilon x^n$. The influence of these perturbation terms on the different kinds of bifurcations develops as follows.

A saddle node bifurcation can be described by the differential equation $\dot{x} = \mu - x^2$. If one adds a constant perturbation term ε, the unperturbed system comes up again with the linear transformation $\tilde{\mu} = \mu + \varepsilon$. Thus, we see that this term only shifts the bifurcation point along the μ-axis by a constant value ε, i.e., there is no structural change of the bifurcation. The same holds for the case of a linear perturbation term εx, which can be eliminated by the linear transformations $\tilde{\mu} = \mu + \varepsilon$ and $\tilde{x} = x + \varepsilon$. Even a quadratic perturbation term εx^2 diminishes after the linear transformation $\tilde{x} = x\sqrt{1 - \varepsilon}$. However, higher-order perturbations (i.e., of order equal to or larger than three) generally lead to structural changes of the system, as illustrated in the bifurcation diagram of Fig. 3.10. Keep in mind that these changes do not affect the structural stability of the bifurcation, because the influence of the perturbation takes place according to $x \propto 1/\varepsilon$ and $\mu \propto 1/\varepsilon^2$. The smaller ε is, the farther away from the

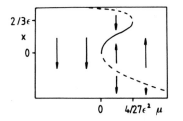

Fig. 3.10. Bifurcation diagram of a saddle node bifurcation subject to a third-order perturbation term (εx^3)

the unstable fixed point is $x_{us} = 0$ $(\mu > 0)$. **e** The subcritical pitchfork bifurcation obtained from the equation $\dot{x} = \mu x + x^3$. Its first derivative reads $Jf_\mu = \mu + 3x^2$. The fixed points are located at $x_E = 0$ and $x_E = \pm\sqrt{-\mu}$, i.e., the asymptotically stable fixed point is $x_{as} = 0$ $(\mu < 0)$, the central fixed point is $x_c = 0$ $(\mu = 0)$, and the unstable fixed points are $x_{us} = 0$ $(\mu > 0)$ and $x_{us} = \pm\sqrt{\mu}$ $(\mu < 0)$. **f** The perturbed pitchfork bifurcation obtained from the equation $\dot{x} = \mu x - x^3 + \varepsilon$. **g** The subcritical pitchfork bifurcation obtained from the equation $\dot{x} = \mu x + x^3 - ax^5$. For further explanation, see text

bifurcation point ($x = 0$, $\mu = 0$) we have a change in the bifurcation diagram. Thus, the local structure of the bifurcation in the vicinity of the bifurcation point remains unchanged. There are similar arguments for perturbation terms of order larger than three.

So far, we have demonstrated the structural stability of the saddle node bifurcation. Note that higher-order perturbations generally lead to the same consequences for all other types of bifurcations. For the analysis of the structural stability it is, therefore, of interest to take into account only perturbation terms having an order smaller than that of the system itself.

A transcritical bifurcation can be described by the differential equation $\dot{x} = \mu x - x^2$. Due to an additive constant perturbation term, the structure of this bifurcation changes, as illustrated in the bifurcation diagram of Fig. 3.9c (for a positive perturbation term). Obviously, the perturbed system no longer reflects the characteristic behavior of the transcritical bifurcation in the neighborhood of the bifurcation point. Instead, there are two facing saddle node bifurcations at $x = \pm \sqrt{\varepsilon}$ and $\mu = \pm 2\sqrt{\varepsilon}$.

A structural stability of the transcritical bifurcation can be attained only if the system obeys the condition that no additive constant perturbation term is possible. Such a condition, for example, might be the initial condition $x = 0$, which always gives rise to a fixed point. These conditions are founded in conservation laws or symmetries of physical systems.

A pitchfork bifurcation and a subcritical pitchfork bifurcation can be described by the differential equation $\dot{x} = \mu x \pm x^3$. Both types of bifurcations are structurally unstable against additive constant perturbation terms, as illustrated in Fig. 3.9f for the case of a perturbed pitchfork bifurcation. Here the typical original bifurcation vanishes, and a stable fixed point together with a separate saddle node bifurcation is left over. Comparing the unperturbed pitchfork bifurcation with the subcritical one, it is obvious that they are analogous with the exception of the exchanged role of their asymptotically stable and unstable fixed points. Both types of bifurcations are structurally stable in systems whose fixed points are located symmetrically with respect to the origin (there exist pairs of positive and negative fixed points). As a consequence of this symmetry condition, perturbation terms capable of changing the structure of the bifurcation are forbidden.

In the case of the subcritical pitchfork bifurcation, we have further shown in Fig. 3.9g the bifurcation diagram for a fifth-order perturbation term. It clearly demonstrates the structural stability of the above equation against perturbations of higher order. This bifurcation diagram will be taken up again in Sect. 3.3.

3.2.2   Catastrophe Theory

Up to now, we have presented bifurcations embracing no more than one variable and one control parameter. These bifurcations are, therefore, assigned to the codimension one. The codimension is equal to the minimum number of control parameters necessary to describe a bifurcation. More precisely, the codimension is defined as the difference between the dimension of the extended control parameter and phase space and the number of equations that are necessary to describe a set of fixed points. From the previous examples, we have found that only the saddle node bifurcation represents a structurally stable codimension-one bifurcation. The other structurally unstable bifurcations can merely be regarded as somehow structurally stable if certain perturbation terms were forbidden (e.g., due to distinct symmetry restrictions).

In what follows, we point out how a structurally unstable bifurcation embedded in a higher-dimensional extended control parameter and phase space may become a structurally stable bifurcation. This procedure is often called the unfolding of an unstable bifurcation. From previous examples, we have learnt that there always exists one characteristic point, namely, the bifurcation point or the center, which sensitively depends upon the perturbation terms. With the help of catastrophe theory, it is possible to systematically construct the differential equations for any bifurcation of fixed points in an extended control parameter and phase space such that it becomes structurally stable. As one peculiarity of the catastrophe theory, the total bifurcation can be described by a single point, the so-called organizing center. The codimension of this center gives the minimum number of control parameters, i.e., the dimension of the control parameter space. The corank of the Jacobian matrix at this point yields the number of eigenvalues with zero real part and, thus, the minimum number of variables, i.e., the dimension of phase space. This procedure is nothing else but the reduction of the system to the center manifold. For more details, see [3.19, 21, 22].

It is worth mentioning that catastrophe theory comprises the evolution of fixed points only for such systems which can be characterized by a potential (in the neighborhood of the critical point). In the case of these systems, investigating the organizing center provides information on the reduction of phase space and the number of control parameters, including the corresponding nonlinear terms of the variables, necessary to describe the structural stability.

As an example, let us consider the pitchfork bifurcation. The organizing center is located at $x = 0$ and $\mu = 0$. It can be described by the differential equation $\dot{x} = x^3$. The corresponding potential reads $V(x) = x^4/4$. We, therefore, have $(dV/dx)_{x=0} = 0$, $(d^2V/dx^2)_{x=0} = 0$, $(d^3V/dx^3)_{x=0} = 0$, and $(d^4V/dx^4)_{x=0} > 0$. The first derivative of the potential gives the dynamics $\dot{x} = dV/dx$ and, thus, the fixed points ($\dot{x} = dV/dx = 0$). From the second derivative of the potential, one obtains the eigenvalues of the linearized problem (here $\lambda = 0$) and, hence, the linear stability of the fixed points (here $x = 0$ is a center). The final nonlinear analysis of the center then requires the next derivatives of the potential. In the present example it is the fourth derivative that determines the stability. The number of higher-order derivatives (of order larger than two) necessary to determine the stability is denoted as the degree of degeneracy that corresponds to the codimension. So the pitchfork bifurcation

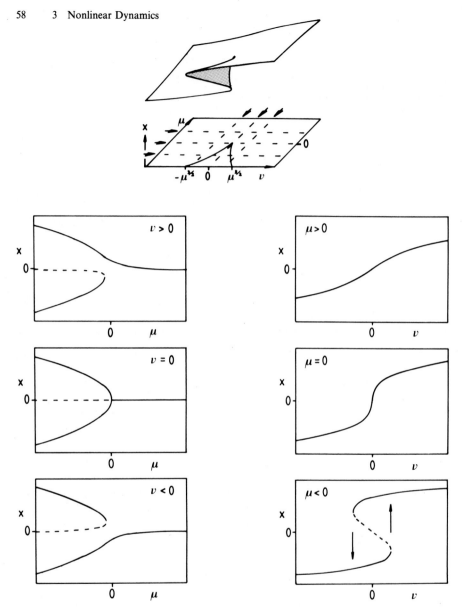

Fig. 3.11. Schematic representation of the cusp catastrophe in the extended control parameter and phase space. The bifurcation diagrams correspond to different cuts through the μ–v control parameter plane (indicated by *arrows*)

has a codimension of two. It is also common to say that the minimum of the potential of the pitchfork bifurcation at $x = 0$ and $\mu = 0$ becomes twofold degenerate. This derives from the fact that here the potential contains a nonlinear term two degrees higher than the simplest quadratic term. The twofold degeneracy can be unfolded by two control parameters μ and v, as follows:

$$V(x) = x^4/4 + \mu x^2/2 + vx .\tag{3.10}$$

This potential and the corresponding dynamical system

$$\dot{x} = dV/dx = x^3 + \mu x + v\tag{3.11}$$

are structurally stable and are called a cusp catastrophe.

Comparing this result with the discussion of the structural instability of the pitchfork bifurcation, one finds that the linear term of the cusp potential (or the additive constant term of the differential equation) – just the term which caused the instability – is now considered as a further control parameter. Within this extended control parameter and phase space, the pitchfork bifurcation can then be embedded in a structurally stable way (Fig. 3.11).

Figure 3.11 displays the geometry of the cusp catastrophe. The set of fixed points forms a two-dimensional manifold in a three-dimensional space. All fixed points located in the fold (dotted region) are unstable. The evolution of the bifurcation points μ_c and v_c is sketched in the μ–v control parameter plane, indicating a power law of $3/2$ characteristic of the cusp catastrophe. The bifurcation diagram of the pitchfork bifurcation can be extracted from Fig. 3.11 as a cut through the cusp catastrophe at $v = 0$. For $v \neq 0$, one obtains the "perturbed" pitchfork bifurcation (cf. Fig. 3.9f). Figure 3.11 further contains bifurcation diagrams resulting from different cuts through the μ–v control parameter plane at constant values of μ. Here v is taken as the variable control parameter. For $\mu > 0$, the stable fixed point varies continuously with changing v, i.e., it is a continuous function of v. For $\mu = 0$, there exists a vertical tangent at $x = 0$. For $\mu < 0$, the dependence of the fixed points upon v becomes ambiguous. Now one has two saddle node bifurcations located symmetrically at $v = \pm \mu^{3/2}$ such that the two stable fixed points are connected via an unstable fixed point. Variation of v from a value smaller than $-\mu^{3/2}$ up to a value larger than $\mu^{3/2}$ and back again to the starting point gives rise to two different (i.e., not coincident) unsteady jumps between the two stable branches of the bifurcation diagram. Such behavior is commonly denoted as hysteresis. One typical feature of the cusp catastrophe is represented by the fact that a hysteresis can shrink continuously as far as its final disappearance due to the variation of an independent second control parameter. A well-known example of this phenomenon from physics is given by the phase transition behavior of the van der Waals gas [3.23]. In terms of the cusp catastrophe, the van der Waals gas reflects a first-order phase transition for $\mu < 0$ that vanishes at the critical point $\mu = 0$.

We conclude with a few comments on the structurally unstable transcritical bifurcation introduced previously. The organizing center of this bifurcation can be related to a third-order potential $V(x) = x^3$. Taking into account the procedure discussed above, we immediately find a codimension of one, i.e., one needs only one control parameter to unfold this center towards structural stability. But, in the case of codimension-one bifurcations, we already know that only the saddle node bifurcation remains structurally stable. In order to see what is behind the transcritical bifurcation, we apply the linear transformation $\tilde{x} = x - \mu/2$, which leads to the differential equation $\dot{\tilde{x}} = \mu^2/4 - \tilde{x}^2$. Indeed, this equation describes the saddle node bifurcation. The change of the control parameter from μ to $\mu^2/4$ explains the linear scaling of the fixed points in the case of a transcritical bifurcation compared to their square-root scaling in the case of a saddle node bifurcation. Moreover, the symmetrical form of the transcritical behavior (with respect to $\mu > 0$ and $\mu < 0$) is also founded in the quadratic parameter term.

3.2.3   Experiments

In what follows, we want to point out that the current–voltage $(I-V)$ characteristics of the representative semiconductor system discussed in Sect. 2.4.1 can be seen in close connection with a cusp catastrophe. A voltage-controlled $I-V$ characteristic typically displays a hysteresis in the breakdown region, whereas in a current–controlled characteristic we usually have negative differential resistance. With the help of current-controlled experiments, it becomes possible to measure the unstable points on the $I-V$ characteristic that are located in the nonlinear region of negative differential resistance (i.e., negative slope dI/dV). Comparing these characteristics with a cusp catastrophe, the question arises whether they can be thought of as a cut through the extended control parameter and phase space (bifurcation diagram) of the cusp catastrophe for constant μ (cf. Fig. 3.11). As a consequence, there should exist a parameter μ, by the influence of which the $I-V$ characteristic becomes embedded in a cusp catastrophe. Experimentally, we have found different control parameters that change the physics of the semiconductor system accordingly. These are, for example, external electromagnetic irradiation, a transverse magnetic field, and the temperature (see also Sect. 2.4.1). Furthermore, the load resistor of the experimental arrangement (Fig. 2.5) can play an important role with respect to the slope of the load line in the $I-V$ characteristic and, hence, to the observability of certain (unstable) fixed points in the immediate breakdown regime.

In Fig. 3.12 three different experimental $I-V$ characteristics are shown. They were obtained at gradually increasing electromagnetic irradiation illuminating the semiconductor sample. For the weakest radiation intensity, characteristic (a) clearly displays (S-shaped) negative differential resistance in the breakdown region. It could be measured utilizing a sufficiently high load resistor. However, the characteristics (b) and (c) indicate the influence of stronger radiation, by degrees giving rise to a transition to positive differential resistance, as is known from the cusp catastrophe. Similar experimental results were also found in a CdSe system [3.24]. Analogous behavior has been predicted theoretically on the microscopic basis of generation and recombination processes that mobile charge carriers undergo with respect to the multilevel energy spectrum of extrinsic semiconductors [3.25]. Here the electromagnetic irradiation and the

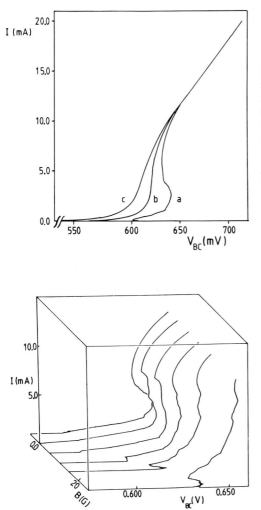

Fig. 3.12. Current–voltage character-istics obtained from the sample sketched in Fig. 2.4 at different external electro-magnetic irradiation (experimental set-up without copper shielding): *Curve a,* only due to inevitable 300 K background radiation; *curves b* and *c,* due to addi-tional illumination with increasing inten-sity by means of an electric bulb. Con-stant parameters: zero magnetic field and temperature $T_b = 4.2$ K (load resistance $R_L = 100 \, \Omega$)

Fig. 3.13. Development of a current–voltage characteristic obtained from the sample sketched in Fig. 2.4 under vari-ation of the transverse magnetic field $(1 \, \text{G} \cong 0.1 \, \text{mT})$. Constant parameters: "zero" electromagnetic irradiation (ex-perimental set-up with helium-cooled copper shielding leaves only the inevi-table 4 K background radiation) and temperature $T_b = 4.2$ K (load resistance $R_L = 100 \, \Omega$)

temperature can be regarded as appropriate control parameters. For details of semiconductor transport theory, we refer to Chap. 2.

Figure 3.13 illustrates the influence of a transverse magnetic field (applied perpendicular to the electric field and the broad sample surfaces) on the nonlinear region of negative differential resistance in the $I-V$ characteristic. Obviously, the characteristic again changes into more or less positive differential resistance behavior corresponding to that of the cusp catastrophe.

The validity of this concept for our representative semiconductor system can be demonstrated more precisely if we look at the temperature dependence of the $I-V$ characteristic [3.26]. For this purpose, instead of the integral time-averaged characteristic, the time-resolved characteristic is evaluated just at the onset point of breakdown. As already discussed in Sect. 2.4.2, the appearance of relaxation-type current and/or voltage oscillations can be attributed to stochastical firing of individual avalanche breakdown bursts. It is interesting to note that the amplitude of these current and voltage bursts directly reflects the actual elongation of the S-shaped negative differential resistance region in the time-resolved characteristic as projected onto the current and voltage axes, respectively (in contrast to that extracted from the time-averaged I–V characteristic). Hence, the expected scaling law of the cusp catastrophe (Fig. 3.11) can be confirmed if one plots, for example, the voltage amplitude ΔV (to the power of 2/3) as a function of the bath temperature T_b (Fig. 3.14).

Finally, we consider the influence of the load resistor on a given I–V characteristic with S-shaped negative differential resistance. Its form can be described approximately by

$$V = aI^3 - bI \ . \tag{3.12}$$

V represents the voltage drop at the sample and I, the current flowing through the sample. The point of inflection of the characteristic is located at the origin ($V = I = 0$) without any loss of universality. According to the basic experimental circuit used (Fig. 2.1), the bias voltage V_0 (defined in Sect. 2.3.2) embraces both the sample voltage V and the voltage drop at the load resistor, $V_L = R_L I$. Thus, we have

$$V_0 = V + V_L = aI^3 - bI + R_L I \ , \tag{3.13}$$

which corresponds to

$$aI^3 + (R_L - b)I - V_0 = 0 \ . \tag{3.14}$$

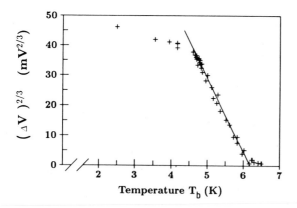

Fig. 3.14. Elongation of the S-shaped negative differential resistance region in the current–voltage characteristic (projected onto the voltage axis) as a function of temperature, obtained from a distinct sample of 3.4 mm contact distance at the following constant parameters: zero magnetic field and "zero" (4 K) electromagnetic irradiation (load resistance $R_L = 10\,\text{k}\Omega$). Note that, in the usual cases of a helium-cooled shielding arrangement, the remaining 4 K background electromagnetic irradiation is no longer stated explicitly as a control parameter

Compared to (3.11), equation (3.14) is nothing but the equation for the fixed points of the cusp catastrophe. One only has to identify the bias voltage V_0 with the control parameter v and the load resistance R_L (except for an additive constant term) with the control parameter μ.

This finding provides an understanding of the switching oscillations discussed in Sect. 2.4.2. Such a stochastical oscillatory behavior of the current between two different conducting states was found in the parameter range near the breakdown threshold, where the S-shaped nonlinearity of the I–V characteristic just vanishes, i.e. its curvature displays an infinite slope (corresponding to a just vanishing breakdown hysteresis at voltage-controlled operation). The pronounced fold topology of the cusp catastrophe immediately forces any system driven in the vicinity of its organizing center to spontaneous switching oscillations due to the influence of arbitrarily small fluctuations. We have observed that sort of current instability by changing the load resistance as long as the slope of the load line adjusts to that of the negative differential resistance part in the I–V characteristic (i.e., $R_L - b$ equals zero).

3.3 Periodic Oscillations

This section is devoted to the simplest temporal behavior, namely, the periodic oscillation. We present the related characterization methods and, subsequently, discuss how a periodic state can be created. The main emphasis is laid on three well-known bifurcations that lead from a fixed point to a periodic oscillation. These are the Hopf bifurcation, the saddle node bifurcation on a limit cycle, and the blue sky catastrophe. For a detailed discussion of the mathematical background of these bifurcations, we refer, for example, to the monographs by Guckenheimer and Holmes [3.2] and Thompson and Stewart [3.9].

3.3.1 The Periodic State

A harmonic periodic oscillation can be described by the time dependence of a variable, $x(t)$, as

$$x(t) = r \sin \omega t . \tag{3.15}$$

Here r denotes the radius and ω, the frequency. In Sect. 3.1 we have already shown that this dynamical behavior is the solution of the following linear differential equation:

$$\dot{x}_1 = \omega x_2, \qquad \dot{x}_2 = -\omega x_1 . \tag{3.16}$$

Using polar coordinates [radius $r = (x_1^2 + x_2^2)^{1/2}$, angle $\theta = \arctan(x_2/x_1)$], one obtains

$$\dot{r} = 0, \qquad \dot{\theta} = \omega . \tag{3.17}$$

From the discussion in Sect. 3.1 we know that the present description of an oscillation is structurally unstable against the smallest perturbation terms. Particularly, the radius equation $\dot{r} = 0$ shows unstable behavior. Structural stability of the differential equation can only be achieved by nonlinear terms, as is the case, for example, for

$$\dot{r} = r - r^3, \qquad \dot{\theta} = \omega . \tag{3.18}$$

So far, we have described a periodic oscillation by means of differential equations; we next turn to methods for analyzing a given time series with respect to its periodicity. The simplest way, of course, is to immediately look at the oscillatory signal, e.g., by using an oscilloscope (Fig. 3.15). Although this kind of signal presentation over a finite time interval provides the experimentalist with a subjective impression of the quality of the periodic oscillation considered, it is by no means a satisfactory characterization method. The regularity of a periodic state over an arbitrarily long time period can be documented more precisely via a phase portrait. Figure 3.16 gives such a phase portrait, which corresponds to the time trace illustrated in Fig. 3.15. It is extracted therefrom by taking into account approximately 500 oscillatory cycles. The two-dimensional phase plot directly reflects the underlying attractor, also called the limit cycle.

Another appropriate method of analyzing a periodic state is to look at its power spectrum. One defines the power spectrum of a signal $x(t)$ as the square of the Fourier amplitude of $x(t)$ per unit time,

$$S(\omega) = \lim_{T \to \infty} (1/T) \left| \int_0^T dt\, x(t) \exp(i\omega t) \right|^2 , \tag{3.19}$$

where T is the integration time. Figure 3.17 displays the power spectrum relating to the current signal of Fig. 3.15. Note that the power (ordinate)

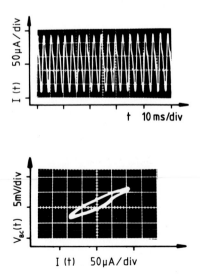

Fig. 3.15. Temporal current profile obtained from the sample sketched in Fig. 2.4 at the following constant parameters: bias voltage $V_0 = 2.616$ V, transverse magnetic field $B = 0.1$ mT, and temperature $T_b = 1.78$ K (load resistance $R_L = 50\,\Omega$)

Fig. 3.16. Phase portrait for the case of the oscillatory state of Fig. 3.15 (same sample and control parameters) obtained by plotting the partial voltage drop $V_{BC}(t)$ against the current $I(t)$

is plotted on a logarithmic scale (the value in decibels corresponds to the base 10 logarithm of the squared amplitude). From the spectral lines present, both the frequency (abscissa) and the amplitude of the fundamental oscillation (as well as that of its harmonics) in relation to the noise level can be deduced. This information, together with the width of the spectral lines, characterizes the regularity of a periodic oscillation.

Figures 3.18 and 3.19 show two additional power spectra of spontaneous current oscillations obtained at different control parameters. The first one (Fig. 3.18) represents a highly regular, periodic state. The height of the fundamental spectral line of apparently more than 60 dB means that the signal-to-noise ratio is better than 10^3. By way of contrast, the second spectrum (Fig. 3.19) gives prominence to a very noisy, chaotic oscillation. At this point, we already see that the power spectrum enables one to distinguish between a periodic (regular) and a chaotic (noisy) oscillation. But keep in mind that, by definition, the power spectrum has lost the phase information of the signal investigated.

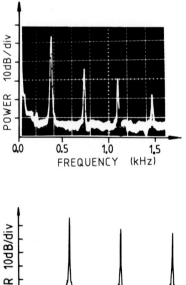

Fig. 3.17. Power spectrum for the case of the oscillatory state of Fig. 3.15 (same sample and control parameters) obtained from the current signal $I(t)$

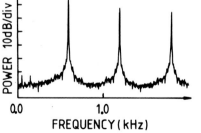

Fig. 3.18. Power spectrum of a current oscillation obtained from the sample sketched in Fig. 2.4 at the following constant parameters: bias current $I = 3.6567$ mA, transverse magnetic field $B = 1.313$ mT, and temperature $T_b = 1.97$ K. Note that, in cases of nearly perfect current control, the bias voltage V_0 and the load resistance R_L are replaced by the more convenient control parameter bias current I

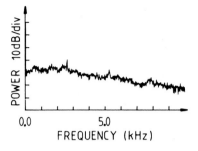

Fig. 3.19. Power spectrum of a current oscillation obtained from the sample sketched in Fig. 2.4 at the following constant parameters: bias current $I = 2.4029$ mA, transverse magnetic field $B = 2.10$ mT, and temperature $T_b = 1.95$ K

3.3.2 Bifurcations to Periodic States

In this section the problem of how a periodic state can be created is discussed. We focus on the bifurcation from a nonoscillatory state (asymptotically stable fixed point) to a periodic oscillation (limit cycle).

With the discovery made by *Hopf* in 1942 [3.27], for the first time a simple explanation was given for how a periodic oscillation can arise spontaneously. It provided an important link between statics and dynamics. The Hopf bifurcation can be described in a structurally stable way by a differential equation having only one control parameter μ. Under variation of the control parameter, an asymptotically stable fixed point continuously changes into an asymptotically stable periodic oscillation. A sequence of phase portraits obtained in this way is shown in Fig. 3.20. The use of one pair of complex conjugate eigenvalues, namely, $\lambda_1 = \bar{\lambda}_2 := \lambda$, where $\mathrm{Re}\{\lambda\} > 0$, is one way to describe the periodically oscillating state. In terms of this language, the Hopf bifurcation is characterized by a change of the real part of λ from negative to positive, while the imaginary part remains unequal to zero. A simple ansatz for such eigenvalues reads $\lambda = d\mu + \mathrm{i}(\omega + c\mu)$. Here μ and ω denote the control parameter and the frequency, respectively; c and d are constants. This ansatz yields the following differential equation:

$$\dot{x}_1 = d\mu x_1 - (\omega + c\mu)x_2 + g(x_1, x_2)\,,$$

$$\dot{x}_2 = (\omega + c\mu)x_1 + d\mu x_2 + h(x_1, x_2)\,. \tag{3.20}$$

The functions $g(x_1, x_2)$ and $h(x_1, x_2)$ contain terms higher order in x_1 and x_2. The linear part of (3.20) gives rise to a structurally unstable oscillation with frequency $\omega + c\mu$ and damping $d\mu$ [compare with the discussion of (3.4)]. The nonlinear terms g and h may cause a structural stability of the present bifurcation in that they prohibit an explosion of the undamped system (at $d\mu > 0$). The concrete form of the nonlinear terms results from the Hopf theorem (see also [3.2]).

According to the Hopf theorem, for any differential equation system $\dot{x} = f_\mu(x)$ with $x \in \mathbb{R}^n$ and $\mu \in \mathbb{R}$ that fulfills the conditions that (1) there is a fixed point x_E with a simple pair of complex conjugate eigenvalues, λ and $\bar{\lambda}$, where

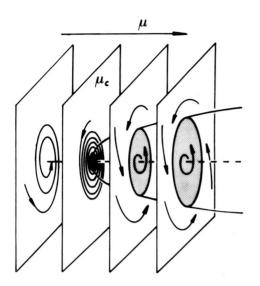

Fig. 3.20. Schematic representation of the bifurcation diagram of a Hopf bifurcation via a sequence of phase portraits obtained for different values of the control parameter μ

$\text{Re}\{\lambda\} = 0$ at $\mu = 0$ and (2) the dependence of λ upon the control parameter μ satisfies

$$(d/d\mu)\,\text{Re}\{\lambda(\mu)\}\,|_{\mu=0} = d \neq 0 , \tag{3.21}$$

there exists a uniquely determined three-dimensional center manifold through (x_E, μ) in $\mathbb{R}^n \times \mathbb{R}$ and an appropriate coordinate system such that the differential equation can be formulated as

$$\dot{x}_1 = d\mu x_1 - (\omega + c\mu)x_2 + a(x_1^2 + x_2^2)x_1 - b(x_1^2 + x_2^2)x_2 ,$$

$$\dot{x}_2 = (\omega + c\mu)x_1 + d\mu x_2 + b(x_1^2 + x_2^2)x_1 + a(x_1^2 + x_2^2)x_2 . \tag{3.22}$$

Again, μ and ω denote the control parameter and the frequency, respectively; a, b, c, and d are constants. Transforming this equation into polar coordinates yields

$$\dot{r} = d\mu r + ar^3, \qquad \dot{\theta} = \omega + c\mu + br^2 . \tag{3.23}$$

The differential equation for the radius has the well-known form of the pitchfork bifurcation or that of the subcritical pitchfork bifurcation. From the discussion of these bifurcations in Sect. 3.2, it immediately follows that for $a < 0$ the radius r bifurcates with a square-root dependence on the control parameter μ. The differential equation for the angle just describes a linear dependence of the frequency on the control parameter. These features of the Hopf bifurcation are summarized in Table 3.1.

The presentation of (3.23) in polar coordinates uncovers an essential symmetry aspect, namely, the invariance of the differential equation under the transformation $(r, \theta) \rightarrow (-r, \theta)$. This symmetry leads to the restriction that

Table 3.1. The Hopf bifurcation

Equation	$\dot{r} = d\mu r + ar^3$ $\dot{\theta} = \omega + c\mu + br^2$
Bifurcation point	$\mu_c = 0$ $r_{as} = (-d\mu/a)^{1/2}$ $f = \omega + (c - bd/a)\mu$
Characteristic features	Scaling of the radius $r \propto (\mu - \mu_c)^{1/2}$ and the frequency $f \propto \omega + A(\mu - \mu_c)$; critical slowing down of the decreasing radius for an excitation at μ close to μ_c (but $\mu < \mu_c$) due to the vanishing damping

a general ansatz like $\dot{r} = R(r)$, $\dot{\theta} = \Theta(r)$ must have a function R with only uneven terms in r as well as a function Θ with even terms in r. As a consequence of this symmetry, the structural stability of the pitchfork bifurcation is guaranteed.

As the next step, we look at the influence of terms of higher order in the above ansatz that still obey the symmetry requirement. One obtains the generalized Hopf bifurcation of the following form:

$$\dot{r} = \mu_1 r + \mu_2 r^3 + ar^5 + O_1(r^7) \,,$$

$$\dot{\theta} = \omega + b_1 r^2 + b_2 r^4 + O_2(r^6) \,. \tag{3.24}$$

Here μ_1 and μ_2 stand for two different control parameters, while a, b_1, b_2, and ω are different constants. The functions O_1 and O_2 denote all terms of higher order. Equation (3.24) gives a codimension-two bifurcation. The organizing center is located at $\mu_1 = \mu_2 = 0$. As long as the radius remains small, the constant ω dominates the angle variable. Depending upon the sign of μ_2, two different bifurcations are obtained if the other control parameter μ_1 changes from negative to positive values. In the case of $\mu_2 < 0$, we have the Hopf bifurcation described above. For $\mu_2 > 0$, the radius equation gives rise to a subcritical pitchfork bifurcation. Here the radius term of fifth order plays an important role. Taking into account the results from the discussion of this equation in Sect. 3.2.1, we expect a hysteretic bifurcation between an asymptotically stable fixed point (focus) and an oscillatory limit cycle. For this type of bifurcation, the radius and also the frequency of the just arising or disappearing oscillation have finite values. This is called a subcritical Hopf bifurcation, the characteristic features of which are summarized in Table 3.2. In this way, we have demonstrated that the generalized Hopf bifurcation is capable of explaining with the help of two control parameters how a usual Hopf bifurcation with finite frequency and zero radius at the bifurcation point continuously develops to a subcritical Hopf bifurcation.

Now we come to a second bifurcation leading, like the Hopf bifurcation, without hysteresis from an asymptotically stable fixed point to a periodic oscillation. It is named the "saddle node bifurcation on a limit cycle" [3.28], already suggesting the characteristic features. Sometimes, this bifurcation is also

Table 3.2. The subcritical Hopf bifurcation

Equation	$\dot{r} = \mu_1 r + \mu_2 r^3 - ar^5$ $\dot{\theta} = \omega + br^2$
Bifurcation point	$\mu_{1c} = \begin{cases} 0 \\ -\mu_2^2/3a \end{cases}$
Characteristic features	For $\mu_2 < 0$ Hopf bifurcation; otherwise, unsteady increase of the radius and the frequency at the bifurcation point, $r \propto \mu_1$ and $f \propto \omega + Ar^2$, respectively; hysteresis between fixed point and oscillation; fixed point lies directly on the limit cycle

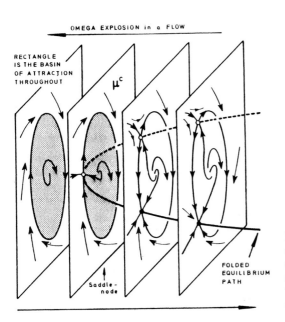

OMEGA EXPLOSION in a FLOW

RECTANGLE IS THE BASIN OF ATTRACTION THROUGHOUT

μ^c

Saddle-node

FOLDED EQUILIBRIUM PATH

Fig. 3.21. Schematic representation of the bifurcation diagram of a saddle node bifurcation on a limit cycle via a sequence of phase portraits obtained for different values of the control parameter μ (from [3.9])

called the "omega explosion" [3.29]. Its main attributes are the continuous increase of the frequency from zero, whereas the radius has a finite value at the outset that remains constant. Compared to the Hopf bifurcation, the frequency of the saddle node bifurcation on a limit cycle plays an analogous role to the radius of the Hopf bifurcation.

The bifurcation scheme of the saddle node bifurcation on a limit cycle is illustrated in Fig. 3.21 by a sequence of phase portraits obtained at different control parameter values. On the one-dimensional manifold of the limit cycle, representing the periodically oscillating state (left-hand side of Fig. 3.21), a saddle node bifurcation takes place. Hence, the oscillation vanishes. Obviously, this type of bifurcation cannot be explained by the local characteristics of phase space around the bifurcation point, but by the information of a finite phase space volume that includes the limit cycle. This is why in the present case we have a global bifurcation, as distinguished from a local bifurcation discussed previously.

The generic form of the saddle node bifurcation on a limit cycle is represented by the following differential equation for radius r and angle θ (see also [3.28]):

$$\dot{r} = r - r^3, \qquad \dot{\theta} = \mu - r\cos\theta .$$ (3.25)

The radius equation always has an asymptotically stable fixed point at $x = 1$. The whole bifurcation is expressed by the angle variable. Depending on the value of the control parameter μ, the angle equation has an asymptotically stable fixed point ($\mu < 1$) or no fixed point ($\mu > 1$). The latter corresponds to a periodic oscillation. Thus, the bifurcation point is located at $\mu_c = 1$. Due to the choice of the angle θ as a variable, the simplest nonlinear term is nothing but a trigonometrical function like $\cos\theta$. The lowest-order term in the expansion of (3.25) around the bifurcation point results in the generic quadratic form of the saddle node bifurcation, $\dot{\theta} = \mu - \theta^2$. Therefore, we expect a square-root dependence $(\mu - \mu_c)^{1/2}$ of the distance between the saddle and the node on the limit cycle. In order to obtain the dependence of the frequency upon the control parameter for $\mu > 1$, the angle equation must be integrated in the following way:

$$T = \int_{-\pi}^{\pi} d\theta/(\mu - \cos\theta) .$$ (3.26).

The lowest-order term in μ around $\mu_c = 1$ yields

$$T \propto (\mu - \mu_c)^{-1/2}$$ (3.27)

for the period T, or

$$f \propto (\mu - \mu_c)^{1/2}$$ (3.28)

for the frequency f.

It is the topology of the saddle node bifurcation which causes the same scaling laws for the radius of a Hopf bifurcation, for the laminar length of type I intermittency (Sect. 3.5), for the frequency difference at phase locking of a quasiperiodic state (Sect. 3.4), and for the frequency at the present bifurcation [3.28]. A summary of the essential characteristics of the saddle node bifurcation on a limit cycle is given in Table 3.3.

Finally, we discuss a second example of a global bifurcation, namely, the blue sky catastrophe. This results from the collision of a saddle point with a limit cycle. Its behavior in phase space is illustrated by a sequence of phase portraits in Fig. 3.22. Due to the collision of the saddle point with the limit cycle, the limit cycle attractor vanishes and, thus, the system loses its stability. That is why the metaphorical name "blue sky catastrophe" is used for the present bifurcation. The term "catastrophe" derives from the sudden disappearance of the limit cycle. It may be worth noting that this bifurcation cannot be explained by any local bifurcation, but it represents an independent, structurally stable

Table 3.3. The saddle node bifurcation on a limit cycle

Equation	$\dot{r} = r - r^3$
	$\dot{\theta} = \mu - r\cos\theta$
Bifurcation point	$\mu_c = 1$
	$r_{as} = 1$
Characteristic features	Scaling of the radius $r = $ const. and the frequency $f \propto (\mu - \mu_c)^{1/2}$; critical slowing down of the frequency for μ approaching μ_c (with $\mu > \mu_c$)

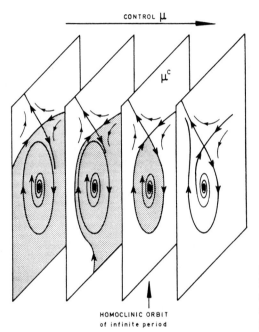

CONTROL μ

μ^c

HOMOCLINIC ORBIT
of infinite period

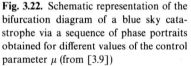

Fig. 3.22. Schematic representation of the bifurcation diagram of a blue sky catastrophe via a sequence of phase portraits obtained for different values of the control parameter μ (from [3.9])

global bifurcation in the three-dimensional extended control parameter and phase space.

As we already know from the subcritical Hopf bifurcation, the blue sky catastrophe is likewise capable of describing a hysteresis under variation of the control parameter between an asymptotically stable fixed point and a limit cycle of a stable periodic oscillation. For the hysteresis, the saddle colliding with the limit cycle must further be embedded in a saddle node bifurcation. As a result, as the control parameter μ is increased, the system ceases to oscillate at the collision between the saddle and the limit cycle (Fig. 3.22). Just at this point, the system loses its stability and jumps onto the now existing node. In the case of reversed parameter variation, the system remains in the nonoscillatory state as long as the node is still present. Only due to the saddle node bifurcation does the

Table 3.4.   The blue sky catastrophe

Phenomenon	Collision of a saddle with a limit cycle
Characteristic features	Unsteady disappearance of the oscillation; new state without any relation to the old one; possibility of hysteresis

nonoscillatory state become unstable (saddle and node annihilate one another), and the system enters the already existing limit cycle attractor. Characteristic features of this bifurcation (see also Table 3.4) are the finite radius of the oscillation at the bifurcation point and the abrupt disappearance of a formerly stable limit cycle. For a more detailed treatment, we refer to the monograph by *Thompson* and *Stewart* [3.9].

The above collision of a saddle with a limit cycle can easily be generalized to collisions with arbitrary types of attractors. The effect will always be more or less the same, namely, that the attractor becomes unstable as a consequence of the collision. Due to the saddle, there are new paths inside the attractor region of phase space by which the system may escape. It is, therefore, common to speak of a blue sky catastrophe of any attractor. Most characteristically, the new state the system jumps onto after the collision does not have any relation to the old one.

From the intrinsic features of the saddle, one should expect a distinct slowing down of the dynamics when the saddle approaches the attractor. However, the argumentation given above has used the saddle as an unstable element. Thus, if we take another unstable element, e.g., an unstable limit cycle, a slowing down of the dynamics cannot be expected any more, in general.

3.3.3   Experiments

The typical semiconductor breakdown experiment comprises a huge variety of different transitions leading from a fixed point to a spontaneous oscillation. Quite often, oscillatory behavior sets in after a discrete jump in the I–V characteristic. Here the current typically changes in the mA range, while the upcoming spontaneous oscillations have amplitudes of a few μA (Sect. 2.4). Such cases of unsteady system behavior can appropriately be described by transitions between fixed points, as was discussed in Sect. 3.2. But there are also regions in parameter space where spontaneous oscillations develop out of stable fixed points lying in the close vicinity of the oscillatory attractor. These represent good candidates to be explained by the fundamental bifurcations discussed above, especially if the emerging oscillation is periodic. A classification of the more complicated cases where the emerging oscillation already has a chaotic or noisy form requires an extended treatment in at least higher-dimensional phase space.

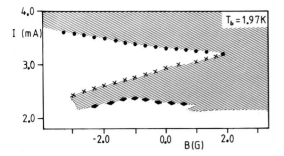

Fig. 3.23. Phase diagram of nonoscillatory and oscillatory states (*plain and hatched regions*, respectively) in the bias current vs. transverse magnetic field control parameter plane (1 G ≙ 0.1 mT) obtained from the sample sketched in Fig. 2.4 at the temperature $T_b = 1.97$ K. Three distinct transitions characterized by different types of bifurcations are marked by dots, crosses, and rhombs. For explanation, see text

Figure 3.23 displays a small part of a two-dimensional cut through the control parameter space spanned by the current and the transverse magnetic field. The plain region characterizes nonoscillatory states, whereas the hatched area stands for the observation of periodic oscillations. We emphasize that small variations of the third control parameter, namely, the bath temperature, by a few 100 mK, slightly shift the region of the periodic oscillations but keep its overall shape nearly unchanged. Similar details of the boundary line, even the same bifurcation behavior, can be observed in the whole temperature regime investigated. Representative time signals and power spectra of a nonoscillatory and an oscillatory state are shown in Figs. 3.24 and 3.25, respectively.

Upon inspecting carefully the transition from a fixed point to periodic oscillations in the phase diagram of Fig. 3.23, we find three different kinds of bifurcation behavior, as discussed in the sequel. First, at the boundary line marked by crosses, one recognizes a transition without hysteresis that can be characterized as a saddle node bifurcation on a limit cycle [3.30]. Second, at the line marked by rhombs, we have a hysteretic transition which will be interpreted in terms of a blue sky catastrophe. Here neither the amplitude nor the frequency of the oscillation display any scaling behavior announcing this transition. Third, at the line marked by dots, there is a continuous transition without hysteresis requiring a bifurcation type of a higher codimension for explanation.

To begin with the experimental evidence of the first type, the control parameter (either the bias current I or the transverse magnetic field B) is varied in such a way that the system crosses the corresponding boundary line (indicated by crosses in Fig. 3.23). The oscillatory state reflects a critical slowing down of the frequency, while at the same time the amplitude of the oscillation does not change at all. Figure 3.26 displays the scaling of both the frequency (a) and the amplitude (c) as a function of the magnetic field. As one characteristic feature of the saddle node bifurcation on a limit cycle, the square-root dependence of the frequency on the control parameter is convincingly demonstrated in

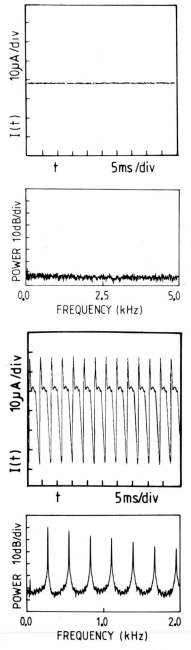

Fig. 3.24. Temporal current profile (*top*) and its power spectrum (*bottom*) for the case of a nonoscillatory state in Fig. 3.23 (same sample and temperature) obtained at the following constant parameters: bias current $I = 3.0\,\mathrm{mA}$ and transverse magnetic field $B = -0.05\,\mathrm{mT}$

Fig. 3.25. Temporal current profile (*top*) and its power spectrum (*bottom*) for the case of an oscillatory state in Fig. 3.23 (same sample and temperature) obtained at the following constant parameters: bias current $I = 2.8080\,\mathrm{mA}$ and transverse magnetic field $B = -0.033\,\mathrm{mT}$

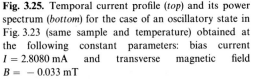

Fig. 3.26b. Except for the lowest 15 measurement points marked by dots, all other frequency data follow the square-root law. For this presentation, statistical examination yields a correlation coefficient of 0.9998. An evaluation of a double-logarithmic plot resulted in the best fit with an exponent of $a = 0.49$,

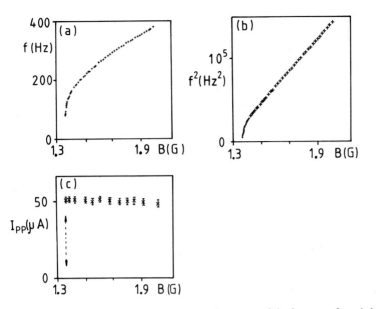

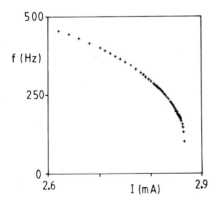

Fig. 3.26a–c. Dependence of the frequency **a**, the square of the frequency **b**, and the peak-to-peak amplitude **c** on the transverse magnetic field (1 G $\triangleq$ 0.1 mT) for the case of an oscillatory state in Fig. 3.23 (same sample and temperature) obtained at constant bias current $I = 3.1010$ mA

Fig. 3.27. Dependence of the frequency on the bias current for the case of an oscillatory state in Fig. 3.23 (same sample and temperature) obtained at constant transverse magnetic field $B = -0.033$ mT

assuming $f \propto (B - B_0)^a$. Here f denotes the frequency, B, the magnetic field, and B_0, an appropriate fit parameter. Another essential property of the saddle node bifurcation on a limit cycle is shown in Fig. 3.26c by the permanently constant peak-to-peak value of the oscillation amplitude.

Analogous behavior can be found when other paths across that boundary are selected. For example, Fig. 3.27 gives the frequency scaling under variation of the bias current. Here the regression analysis provides results similar to those for the case mentioned above. Again the amplitude of the oscillation remains constant. Moreover, this also holds for the total shape of the underlying attractor, as illustrated by means of two different oscillatory states in Fig. 3.28a and b.

Close to the bifurcation point, we observe an interesting departure from the behavior predicted. As already shown in Fig. 3.26b, the smallest values of the oscillation frequency do not obey the expected square-root law, even though the best accordance was predicted here [3.28]. Simultaneously with the deviation from square-root scaling, irregularities in the time signals are detected (Fig. 3.28c). Instead of a periodic oscillation, one obtains a stochastical spiking of single oscillatory events. The spiking rate becomes less and less as the control parameter approaches the bifurcation point. Now the frequency of this oscilla-

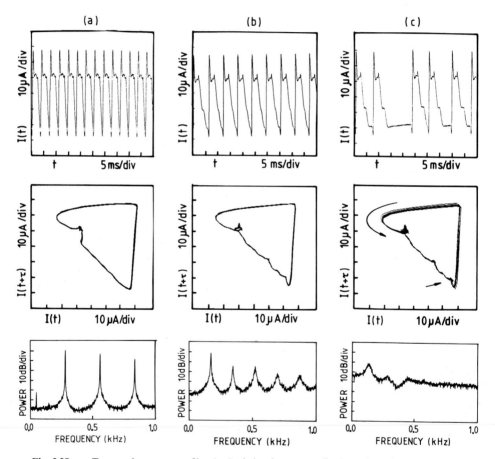

Fig. 3.28a–c. Temporal current profiles (*top*), their phase portraits (*center*), and power spectra (*bottom*) for three distinct cases of oscillatory states in Fig. 3.23 (same sample and temperature) obtained at different bias currents $I = 2.8080$ mA **a**, $I = 2.8581$ mA **b**, $I = 2.8638$ mA **c**, and constant transverse magnetic field $B = -0.033$ mT

tory behavior corresponds to the time-averaged spiking rate. Consequently, the error bars of the frequency data increase from less than 5 Hz in the nonstochastical case up to almost 50 Hz close to the bifurcation point. There are two suggestive mechanisms which might lead to the observed behavior. First, near the bifurcation point the system can be sensitively influenced by noise. For clarity, we looked at the influence of control parameter noise on the generic bifurcation equation by means of computer simulation. In this way, however, we could not model the observed departure of the frequency scaling near the bifurcation point. Second, hitherto suppressed variables may have growing influence. Thus, the observed dynamics might be the result of some underlying deterministic chaotic process. We emphasize that similar spiking behavior of single oscillatory events has been reported recently for different experimental systems. In the case of a laser experiment, these oscillation forms could be ascribed to Shilnikov chaos [3.31]. However, there definitely exists no evidence of Shilnikov-type chaotic behavior in the present semiconductor experiment. Note the unstructured form of the return map and the Poisson-like shape of the histogram displayed in Fig. 3.29a and b, respectively, both constructed from the recurrence time of the spikes of Fig. 3.28c.

The present statistical time behavior may also be discussed on the basis of semiconductor physics. During our experiments, we observed a maximum period of about 5–10 ms between the current spikes. This limit can be directly related to a maximum relaxation time to which the system is capable of slowing down. Most interestingly, this characteristic time constant agrees reasonably well with the results of pulsed current–voltage characteristics described already in Sect. 2.4.2.

Next we want to stress the fact that, besides the experimentally observed departure of the frequency scaling from the theoretically expected one, the oscillation amplitude as well as the shape of the attractor remain unchanged in accordance with the predictions (Fig. 3.28). In the reconstructed attractor of the

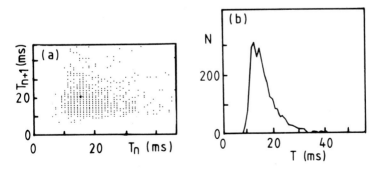

Fig. 3.29a, b. Return map **a** and histogram **b** obtained from the recurrence time of the spiking oscillatory state in Fig. 3.28c close to the bifurcation point (same sample and control parameters)

phase portrait, the location of the node where the oscillation terminates is indicated by an arrow. Bear in mind that the node lies directly on the limit cycle, indicating the correspondence of the bifurcation topology with the model. By using the topological features of the present bifurcation, we are able to give a further characterization with the help of pulse experiments. For parameter values where the saddle and the node are just existent on the limit cycle, there is still a trajectory running from the saddle to the node, displaying nearly the whole shape of the reminiscent limit cycle. Thus, we choose a nonoscillatory state close to the transition boundary to which voltage pulses are applied (see upper trace of Fig. 3.30a). The temporal current response of the system is shown in the lower trace of Fig. 3.30a. Each voltage pulse stimulates only one single oscillation cycle. Figure 3.30b presents the corresponding phase portrait. Taking into account the different scales on the axes of the phase protraits in Fig. 3.28 and Fig. 3.30b, one clearly sees the unchanged structure of the limit cycle. For a chosen control parameter, the pulse height necessary to excite such an oscillation exactly relates to the difference between the values of the actual control parameter and the critical parameter where the bifurcation occurs in the nondriven case. Obviously, these pulses act in the control parameter space, i.e., during the time the pulse is applied the system is displaced into a parameter region where no saddle and no node exist on the limit cycle. Thus, the system oscillates along the limit cycle. Switching off the pulse causes the system to fall back to parameter values where the oscillation on the limit cycle terminates through the existence of a saddle and a node. However, the oscillation has already passed the saddle during the time of the pulse applied. Consequently,

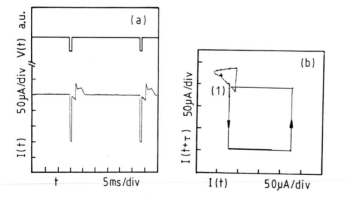

Fig. 3.30a, b. Temporal current response to periodically applied voltage pulses for the case of a formerly nonoscillatory state in Fig. 3.23 (same sample and temperature) obtained at the following constant parameters: bias current $I = 2.880$ mA and transverse magnetic field $B = -0.033$ mT: **a** time profiles of the voltage (*top*) and the current (*bottom*); **b** phase portrait of the current signal. The arrows give the direction of rotation. Symbol (1) locates the node onto which the system relaxes

only a further propagation on the reminiscent limit cycle is able to bring the system back to the node (nonoscillatory state). Our semiconductor experiment precisely reveals such behavior (cf. Fig. 3.30).

These findings are confirmed by additional pulse experiments. Voltage pulses with inverted polarity do not excite any current oscillation. During the application of longer pulses, the system elicits more oscillatory cycles accordingly. Depending on the moment when the pulses are switched off, the system falls back to different positions and again relaxes to the node along the reminiscent limit cycle. This phenomenon is demonstrated by the sequence of different pulse measurements in Fig. 3.31.

So far, we have classified the experimentally observed transition from a nonoscillatory to an oscillatory state across the boundary line marked by crosses in Fig. 3.23 as a saddle node bifurcation on a limit cycle. The theoretically predicted scaling of the oscillation frequency and amplitude has shown reasonable accordance. The invariant shape of the attractor provided information on the topology of the present bifurcation. It was even possible to demonstrate the unchanged form of the phase portrait expected for the nonoscillatory

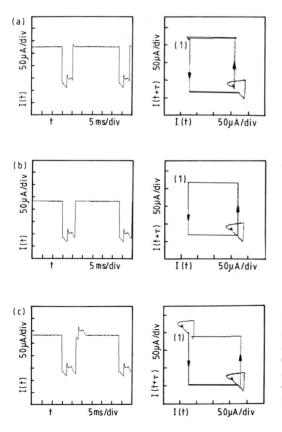

Fig. 3.31a–c. Temporal current response to periodically applied voltage pulses analogous to Fig. 3.30 (same sample, control parameters, and plots) except for different pulse durations $T = 5.5$ ms **a**, $T = 6.4$ ms **b**, $T = 6.9$ ms **c**

state. All these findings differ remarkably from the characteristic attributes of other well-known elementary bifurcations, as one can see by comparing them with the results of the previous discussions.

If once more we come back to Fig. 3.23, there are two further symbols that mark transition lines between a nonoscillatory and a periodic oscillatory state. First, we focus on the bifurcation behavior observed at the border marked by rhombs. Here the transition occurs in a hysteretic way. For a slightly different – compared to Fig. 3.23 – bath temperature, an enlarged segment of the phase diagram is presented in Fig. 3.32. Again the plain region characterizes nonoscillatory states, while the hatched area stands for the observation of periodic oscillations. The region between the two lines marked by rhombs illustrates the hysteresis range observed experimentally. The location of the symbols a to d in the phase diagram indicates different pairs of current and magnetic field control parameter values by which the time signals in Fig. 3.33 were obtained. There we present the temporal current profiles of a nonoscillatory and a periodic oscillatory state in parts (a) and (b), respectively. The hysteretic behavior is shown schematically in part (c). By decreasing the bias current, one observes a periodic oscillation up to the point where an abrupt change to a nonoscillatory state takes place. It is remarkable that neither the frequency nor the amplitude display any scaling behavior announcing this transition. Furthermore, the stable equilibrium point clearly lies outside the limit cycle attractor such that the generalized Hopf bifurcation predicting hysteretic behavior cannot be taken to explain our experiment. These findings may be described, however, by the bifurcation properties discussed in connection with the blue sky catastrophe. The fact that the frequency does not show any slowing down on approaching the bifurcation point makes us suspect that in the present case no colliding saddle, but rather a colliding unstable limit cycle destroys the stable limit cycle. Upon reversing the change of the control parameter (i.e., increasing the bias

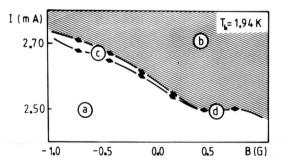

Fig. 3.32. Blow-up of part of the phase diagram in Fig. 3.23 (same sample and plot) obtained at the slightly different temperature $T_b = 1.94$ K. The boundary between nonoscillatory and oscillatory states (*plain and hatched regions,* respectively) reveals hysteretic behavior (two noncoinciding transition lines marked by rhombs). The letters a–d outline the control parameter values used in Fig. 3.33

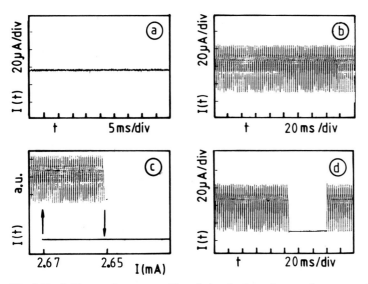

Fig. 3.33a–d. Temporal current profiles **a, b, d** and schematic control parameter dependence of the hysteresis **c** for distinct cases of nonoscillatory and oscillatory states obtained at the parameter values of the bias current and the transverse magnetic field indicated by the symbols a–d in Fig. 3.32 (same sample and temperature)

current), the asymptotically stable fixed point loses its stability, and the system jumps back to the stable limit cycle by a collision of the fixed point with an unstable element.

That there exists an unstable element that exchanges the stability of the fixed point and the limit cycle can be seen from the evolution of the hysteresis in Fig. 3.32. As the parameter regime of the phase diagram near point d is approached, the hysteretic range contracts to zero in a continuous way. Finally, a spontaneous switching between the periodic oscillation and the nonoscillatory fixed point can be observed, as illustrated in Fig. 3.33d. Such a development of the hysteresis may be set in analogy with the cusp catastrophe, but now two different states (a nonoscillatory and a periodic oscillatory one) are involved. Hence, a characteristic feature of the cusp catastrophe is violated, namely, that there always exist paths in control parameter space which continuously transform the system from one bistable state to the other in the hysteretic range. The path has just to be chosen in such a way that one goes around the organizing center. Thus, for the case of a cusp catastrophe, there is, in principle, no difference between the two states of the bistable region, in contrast to the present experimental situation.

Finally, we briefly deal with the bifurcation behavior found on crossing the transition line marked by solid dots in Fig. 3.23. This bifurcation was observed in a very small control parameter region, mainly in the corner close to the transition line marked by crosses. Figure 3.34 displays two typical current traces together with the corresponding phase portraits. Part (a) gives the periodic

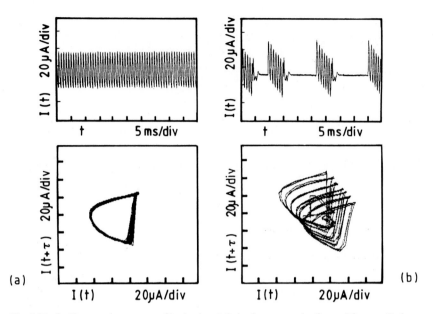

Fig. 3.34a, b. Temporal current profiles (*top*) and their phase portraits (*bottom*) for two distinct cases of oscillatory states in Fig. 3.23 (same sample and temperature) obtained at different bias currents $I = 3.23$ mA **a**, $I = 3.22$ mA **b**, and constant transverse magnetic field $B = 0.165$ mT

oscillation and part (b), the onset of this oscillation. In the latter case, we have single oscillation packets of constant shape developing statistically. The period of the oscillatory parts inside these packets approximately equals that of the periodic oscillations. Regardless of the peculiar decay in the signal form, the amplitude of the first six revolutions in each oscillation packet stays constant, i.e., there is no damping present. At the end of each packet, one recognizes a spiral-type relaxation to an asymptotically stable fixed point, i.e., there is a focus (Fig. 3.4). The corresponding phase portrait in Fig. 3.34b clearly shows that such a behavior cannot be uniquely grasped in two-dimensional phase space without any crossing trajectories. Therefore, we conclude that the present bifurcation requires at least three dimensions in phase space, i.e., a higher dimension than that occupied by the two attractors involved. As a consequence, the latter transition behavior reflects a bifurcation type having a more complex structure than those discussed so far. The following sections are, therefore, devoted to the dynamics and bifurcations in higher-dimensional phase space.

3.4   Quasiperiodic Oscillations

For the previous discussion of the characteristic properties of fixed points and limit cycles, it was adequate to describe those attractors completely in a two-dimensional phase space. The practicability of this treatment is supported by the theorem of *Peixoto* [3.32] stating that there exists no other attractor in two-dimensional phase space (for details, see also [3.2]). However, such convincing evidence of existent attractors is unknown for the case of phase spaces with a dimension larger than two. So one experiences a general interest to explore new types of attractors in higher-dimensional phase space. A well-known dynamics in three dimensions is the quasiperiodic oscillation. It belongs to an attractor having the shape of a torus. In addition to this, so-called strange attractors have been found recently in three-dimensional phase space. The dynamics on a strange attractor is characterized by a certain disorder named chaos, whereas the dynamics on a torus still behaves orderly. In what follows, we deal with different attractors in three-dimensional phase space. The present section is devoted to the ordered dynamics on a torus, while in Sect. 3.5 we turn to the chaotic dynamics on strange attractors.

3.4.1 The Quasiperiodic State

The simplest description of quasiperiodicity can be attained from the periodically driven oscillator

$$\ddot{x} - f(\dot{x}, x)\dot{x} + x = K \sin \omega t , \qquad (3.29)$$

where f is a nonlinear function and K, a constant giving the size of the driving force. This differential equation can easily be transformed into an autonomous one. We suppose that the undriven system (setting $K = 0$) displays a periodic oscillation with frequency f_1 and radius R, as illustrated schematically in Fig. 3.35a. To reveal the whole dynamics in three-dimensional phase space, the effect of the periodic driving term ($K \sin \omega t$) can be thought as a superposed oscillation with frequency $f_2 = \omega/2\pi$ and radius r. The resulting trajectory sketched in Fig. 3.35b unwinds on a torus. Its dynamics depends on the ratio of the two frequencies f_1 and f_2. If f_1/f_2 is irrational, the trajectory will never come back to the starting point, i.e., the trajectory has an infinite length covering the whole surface of the torus. This dynamics is called quasiperiodic. On the other hand, if f_1/f_2 is rational, the trajectory comes back to the starting point after a finite number of rotations. Here the path on the torus has a finite length, and the resulting dynamics is periodic. Such periodic states that represent special cases of quasiperiodic motion are dominated by phase-locking, as we will see

Fig. 3.35a, b. Schematic representation of a quasiperiodic oscillatory state: **a** composition of two different periodic oscillations; **b** resulting dynamics on a torus

later on. At any rate, the presence of nonlinearities in (3.29) can give rise to an interesting interplay between locked and quasiperiodic states under variation of the control parameters. We will come back to this point in the context of the circle map formalism.

Before nonlinear effects of quasiperiodicity are investigated, we first consider some methods to characterize a quasiperiodic state. Figure 3.36 presents some typical experimental data. The time trace of the current (a) clearly displays periodically varying amplitudes, as a consequence of two different frequencies being present at the same time. The phase portrait (b) is obtained by plotting the local voltage drop $V_{AB}(t)$ against the current $I(t)$. The shape of the phase plot shows the projection of a torus-like attractor. Here the structural difference from a limit cycle attractor of a periodic state becomes evident, although the whole structure of the torus attractor has been lost due to the projection onto the

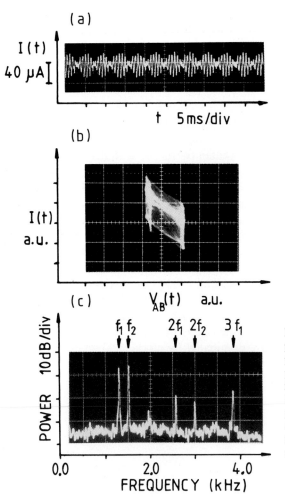

Fig. 3.36a–c. Temporal current profile **a**, its phase portrait **b**, and power spectrum **c** for a quasi-periodic oscillatory state obtained from the sample sketched in Fig. 2.4 at the following constant parameters: bias current $I = 0.811$ mA, transverse magnetic field $B = 0.722$ mT, and temperature $T_b = 1.975$ K

two-dimensional plane of presentation. The power spectrum (c) of the current signal renders prominent the two simultaneously existing frequencies f_1 and f_2 together with their harmonics.

The problem arising from the nonunique projective phase portrait presentation can be solved by means of a Poincaré cross section or a return map. The former provides a method to reduce the dimensionality of an attractor by one. Here the trajectories on the attractor are not of concern any more, but rather the set of intersection points of the trajectories with a well-chosen plane S_E, as illustrated schematically in Fig. 3.37. Note that, in general, for a d-dimensional phase space a $(d - 1)$-dimensional hyperplane S_E has to be taken. There remain only those points where the trajectory intersects the plane S_E in one defined direction. For the case of a torus, it is easily seen that the Poincaré cross section reduces the attractor to a circle-like or an elliptical set of points. Obviously, in a curtailed way the Poincaré cross section is capable of reproducing the geometrical form of the attractor in phase space.

A reduced presentation of the dynamics developing on the attractor can be achieved via the construction of the return map, sometimes also called the Poincaré map. It is given by the time sequence of the points P_n $(n = 1, 2, \ldots)$ located on the Poincaré cross section (Fig. 3.37). These points P_n are now ordered in time, i.e., each cross-section point P_n evolves immediately from its precursor P_{n-1}. The dynamics of the cross-section points can be described by the mapping equation, denoted as the return map,

$$P_{n+1} = F_\mu(P_n) , \tag{3.30}$$

where F_μ is nothing but the orbit $\Phi_\mu(P_n, T_n)$ for the recurrence time T_n necessary to come back to the plane S_E. Quite often, T_n is independent of P_n, i.e., it is constant $(T_n = T)$. This always holds, for example, for a periodically driven system, where the recurrence time equals the period of the driving frequency.

To illustrate the relationship between the Poincaré cross section and the return map, let us consider an attractor reconstructed from one variable $x_1(t)$ with the help of the Takens–Crutchfield method [3.15]. In Sect. 3.1, we have shown that a reconstructed time trace $P(t)$ in phase space has the form

$$P(t) = (x_1(t), x_1(t + \tau), x_1(t + 2\tau), \ldots) . \tag{3.31}$$

For the sake of clarity, let us restrict phase space to three dimensions. The Poincaré cross section may then be constructed on the x–y plane $(z = 0)$, for

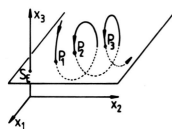

Fig. 3.37. Schematic representation of the construction principle of a Poincaré cross section

example. It, thus, suffices to describe any cross-section point by two coordinates. Note that this holds, in general, for any plane S_E provided one chooses an appropriate system of coordinates. In the case of a constant recurrence time T, the Poincaré cross section is obtained by means of a stroboscopic illumination, i.e., picking out the points

$$P_n = (x_1(nT), x_1(nT + \tau)) , \qquad (3.32)$$

where n denotes an integer. If the delay time τ of the phase space reconstruction is chosen to be equal to the recurrence time T, we have

$$P_n = (x_1(nT), x_1(nT + T)) . \qquad (3.33)$$

The Poincaré cross section one obtains in this case corresponds to the plotting of the variable $x_1((n + 1)T)$ on the y-axis against the variable $x_1(nT)$ on the x-axis. Obviously, this is nothing else but the return map of one variable $x_n = x_1(nT)$.

To summarize the discussion above, the return map has been demonstrated to provide quite often the same information as that given by the Poincaré cross section constructed via the Takens–Crutchfield technique,. We emphasize, however, that without further caution this statement cannot be extrapolated to the more complex situation of a chaotic state. It is also important to point out that the Poincaré cross section gives a topological characterization of the attractor in phase space, whereas the return map yields a dynamical characterization of the time development of a variable. In the final analysis, the thrust inherent to both methods may coincide. Anyhow, the nontrivial relation between (phase-) spatial and temporal analysis of a dynamical state comes to the rescue for the characterization of low-dimensional chaotic phenomena, as discussed later on.

A final remark concerns the fact that the return map transforms a continuous dynamics into an iterative one that is discrete in time. Nevertheless, the preceding treatment of stability and different bifurcations can be readily applied to the case of discrete dynamics (for more details, see, e.g., [3.2]).

Next we come back to the experimental situations. There it is very helpful to have access to on-line (i.e., analog) characterization methods for time signals, in order to get a first impression of the nature of the dynamics. Therefore, Poincaré cross sections and return maps represent highly significant tools, since they can easily be projected onto the screen of an oscilloscope by means of stroboscopic techniques. An example of a suitable experimental set-up is drawn schematically in Fig. 3.38. First, different time traces should be available from an experiment. In the case of the present semiconductor system, we have used the current $I(t)$ and the local voltage drops $V_i(t)$ with $i = AB, BC, CD$ according to Fig. 2.4. If, for some reason, only one signal can be extracted from an experiment, additional ones might result by applying proper phase shift procedures (similar to the Takens–Crutchfield construction) or, alternatively, by taking time derivatives of the solitary signal.

From the different signals now available, one should be filtered and amplified in such a way that it is capable of externally triggering a pulse generator. So

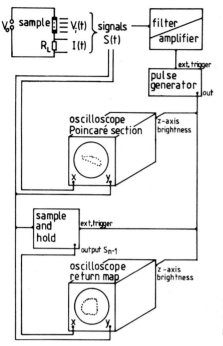

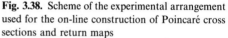

Fig. 3.38. Scheme of the experimental arrangement used for the on-line construction of Poincaré cross sections and return maps

one has gained the recurrence time T from a distinct signal present in the form of well-defined voltage pulses. For the case of nondriven (i.e., not periodically driven) experimental systems, there should also be recourse to such an intrinsic time T that is insensitive to irregularities of the system itself (e.g., fluctuations in the characteristic frequency). In fact, the recurrence time is an essential prerequisite for constructing both analog on-line mappings.

By means of the above voltage pulses, the Poincaré cross section can be displayed on the screen of an oscilloscope. First of all, one starts with a phase portrait obtained by plotting two independent signals against one another (x- and y-channel). Some examples are shown in the left column of Fig. 3.39. In the next step, the voltage pulses are used to control the brightness on the screen (z-axis). The pulse width should be chosen in such a way that hardly anything but small dots survive from the phase portrait. The resulting stroboscopic plot gives a Poincaré cross section like those displayed in the middle column of Fig. 3.39. Comparison with the phase portraits (left column) provides information on how the topology of the attractor is reflected in the Poincaré cross section. Apparently, it yields a clearer picture of the form of the attractor than that attainable by the intersecting trajectories in the two-dimensional phase portraits. All examples of Poincaré cross sections presented in Fig. 3.39 clearly demonstrate that their intersection points are located on a closed line. We already know that elliptically shaped ones are expected for quasiperiodic states. The somewhat distorted form of the experimentally obtained Poincaré cross

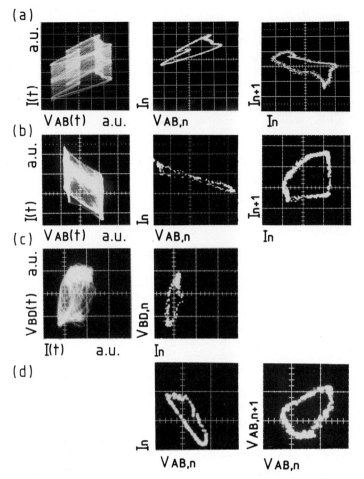

Fig. 3.39a–d. Phase portraits (*left column*), Poincaré cross sections (*middle column*), and return maps (*right column*) for a quasiperiodic oscillatory state obtained from the sample sketched in Fig. 2.4 by plotting the partial voltage drops $V_{AB}(t)$ or $V_{BD}(t)$ and the current $I(t)$, as well as their respective iterative dynamics, against one another at different sets of parameter values: **a** bias current $I = 1.224\,\text{mA}$, transverse magnetic field $B = 2.28\,\text{mT}$, and temperature $T_b = 2.005\,\text{K}$; **b** $I = 0.811\,\text{mA}$, $B = 0.722\,\text{mT}$, and $T_b = 1.975\,\text{K}$; **c** bias voltage $V_0 = 2.23\,\text{V}$, transverse magnetic field $B = 0.122\,\text{mT}$, and temperature $T_b = 4.2\,\text{K}$ (load resistance $R_L = 100\,\Omega$); **d** $V_0 = 12.0\,\text{V}$, $B = 1.656\,\text{mT}$, and $T_b = 1.975\,\text{K}$ ($R_L = 12.67\,\text{k}\Omega$)

sections may be caused by their projective mapping as well as the influence of higher harmonics present in the signals analyzed.

The relating return map results only from one signal, henceforth labelled $S(t)$, and the voltage pulses. The latter are used to externally trigger an analog sample-and-hold device (Fig 3.38). Each time a voltage pulse is applied, the instantaneous value of the signal $S(t)$ is stored. Thus, we obtain a sequence of discrete values S_n $(n = 1, 2, \ldots)$ from which the return map can be reconstructed. Simultaneously with the storage process, the trigger pulses initiate the sample-and-hold device to provide the output of the hitherto stored value S_{n-1} in the form of a voltage signal. The signal from the sample-and-hold device is sent to one channel, say, the x-axis, and the original signal $S(t)$ to the other one, say, the y-axis, of an oscilloscope. Eventually, we arrive at the return map by strobo-scopically illuminating the screen with the voltage pulses from the pulse gener-ator (Fig. 3.38). The present stroboscopic technique picks out the discrete values S_n from the analog signal trace $S(t)$ for projection in the y-direction. As a consequence, the S_{n-1} vs. S_n plot becomes visible on the screen of the oscilloscope. Some examples of return maps constructed in this way are illus-trated in the right column of Fig. 3.39.

If one compares these return maps with the related Poincaré cross sections (middle column of Fig. 3.39), there sometimes exists a striking similarity among them, e.g., in case (d). On the other hand, we can also have a clear difference in their graphs, e.g., in case (b). Here the Poincaré cross section is nearly a straight line, while the return map displays an extended loop. The interpretation of this finding may follow from an almost perpendicular orientation of the intersection plane of the Poincaré cross section compared to the plane of its presentation. In this way, the theoretically expected circle form has been projected onto a (quasi) straight line. Just at this point, we are confronted with a difficulty of the above method, namely, that there is no guarantee that the intersection plane of the Poincaré cross section (somehow defined by the recurrence time chosen) coincides with the plane of presentation (on the screen of the oscilloscope).

A common feature of all the dynamical states shown in Fig. 3.39 is that the Poincaré cross sections as well as the return maps give sets of points lying on a closed line. This result manifests an essential feature of quasiperiodic states. We already know that the Poincaré cross section causes a reduction of phase space by one dimension. From this, we conclude that the corresponding attractors are two-dimensional objects, i.e., tori, where the dynamics of quasi-periodicity develops.

Experimental experience has proved that characterization of quasiperiodic states is done more clearly and unambiguously by Poincaré cross sections or return maps than by phase portraits or power spectra. For example, structural changes, like the explosion of a torus, can hardly be identified in the phase portrait or the power spectrum, whereas the Poincaré cross section and the return map clearly reveal in such a case that their points no longer lie on a closed line. Figure 3.40 gives an impression of a structural change observed experi-mentally. Obviously, the formerly single loop in the return map (a) splits up locally into two lines (b), i.e., we have a period doubling of a torus.

At this stage, we break off the discussion of methods to characterize quasiperiodic states and turn to the investigation of the influence exerted by control parameters and nonlinearities on quasiperiodicity.

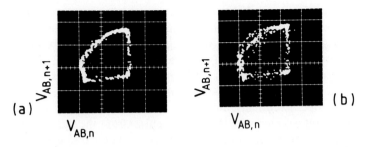

Fig. 3.40a, b. Return maps for two distinct oscillatory states obtained from the sample sketched in Fig. 2.4 by plotting the iterative dynamics of the partial voltage V_{AB} at different transverse magnetic fields $B = 0.722$ mT **a**, $B = 0.760$ mT **b**. The constant parameters are bias current $I = 0.811$ mA and temperature $T_b = 1.975$ K

3.4.2 Bifurcations to Quasiperiodic States

Quasiperiodic behavior has been characterized by the simultaneous presence of two independent frequencies. If a quasiperiodic state of a system is self-generated, i.e., created by the system itself without being influenced by any periodic driving force, then one expects a sequence of two bifurcations that lead to two independent periodic oscillations. Under variation of a control parameter, the system first undergoes a bifurcation from a fixed point to a periodic oscillation and, subsequently, another bifurcation to a quasiperiodic state. Different types of bifurcations to a limit cycle attractor have already been discussed in Sect. 3.3.2. The second bifurcation from a limit cycle to a torus can be considered in an analogous way. Now the system does not bifurcate from a fixed point, but from a limit cycle, which may be thought of as a moving point in phase space. Proceeding from this moving point, the second bifurcation gives rise to an additional limit cycle, as a whole creating the torus attractor illustrated schematically in Fig. 3.35. From the bifurcations of Sect. 3.3.2, both the Hopf bifurcation and the saddle node bifurcation on a limit cycle represent appropriate candidates to cause a quasiperiodic state. A Hopf bifurcation on a limit cycle is called a Neimark bifurcation or a secondary Hopf bifurcation. The other case may be denoted as a saddle node bifurcation on a torus. The latter designation is only correct in the case of the reduced dynamics of a Poincaré cross section. There the node reflects a periodic dynamics on a limit cycle.

If one tries to explain a quasiperiodic state by means of two successive blue sky catastrophes of limit cycles, problems arise due to the jump away from the limit cycles in phase space. Only a blue sky catastrophe of a torus attractor, by a collision with a saddle, can describe the abrupt disappearance of a torus. A last remark about the bifurcation of a torus concerns the possibility of a continuous transition leading directly from a fixed point to a torus which has nothing to do with a blue sky catastrophe. Following the discussion given above, it immediately becomes clear that such a transition form requires the two successive bifurcations to take place simultaneously. If both bifurcations possess their own independent control parameters, we have a two-dimensional parameter space. The direct bifurcation to a torus can occur only when both control parameters become critical at the same time. This behavior is typically fulfilled only for one point in control parameter space and, therefore, is very unlikely.

In what follows, we present experimental evidence of transitions from a fixed point to a quasiperiodic state for the case of our semiconductor system. Under a variation of the appropriate control parameters, we always found a first bifurcation to a limit cycle. Then, for a different parameter value, the system successively bifurcated to quasiperiodicity. Figure 3.41 shows a section of control parameter space where we have observed transitions to quasiperiodic states. The differently shaded regions correspond to the following dynamics. In the plain region (left-hand side) a stable fixed point, i.e., no oscillation, was observed. The vertically hatched region (middle part) stands for the existence of periodic oscillations. In the horizontally hatched regions (right-hand side) quasiperiodic oscillations were found, whereas in the dotted regions (upper and lower right) we have detected chaos or noisy states. At the transition lines indicated by crosses, solid triangles, and solid squares, bifurcations between the corresponding dynamical states were observed. Note that the region of quasiperiodicity never comes into contact with the region of fixed-point behavior. The reason may be found in the spatio-temporal dynamics of the present system, as discussed in Sect. 3.6.

To demonstrate the evolution of our semiconductor system from a nonoscillatory to a quasiperiodic state, we have plotted in Fig. 3.42 the dependence of the two frequencies f_1 and f_2 upon the transverse magnetic field B (in terms of the corresponding magnetic field current I_B). Obviously, two successive bifurcations can be recognized. The transition from the nonoscillatory to the periodic state

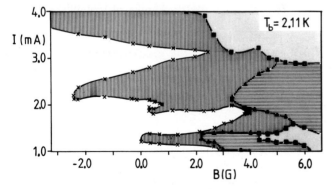

Fig. 3.41. Phase diagram of nonoscillatory, periodic oscillatory, quasiperiodic oscillatory, and chaotic oscillatory states (*plain, vertically striped, horizontally striped, and dotted regions*, respectively) in the bias current vs. transverse magnetic field control parameter plane (1 G $\cong$ 0.1 mT) obtained from the sample sketched in Fig. 2.4 at the temperature $T_b = 2.11$ K. Three distinct transitions characterized by different types of bifurcations are marked by crosses, triangles, and squares. For explanation, see text

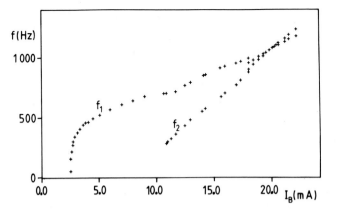

Fig. 3.42. Dependence of two intrinsic frequencies on the transverse magnetic field in terms of the corresponding magnetic field current (1 mA current relates to 0.033 mT magnetic field) for the case of the transition to a quasiperiodic state in Fig. 3.41 (same sample and temperature) obtained at constant bias current $I = 1.9836$ mA

displays a nonlinear rise of the frequency f_1 in the form of a square-root law. Here the amplitude remained constant, indicating that one has a saddle node bifurcation on a limit cycle, see Sect. 3.3. The subsequent transition to the quasiperiodic state (around $I_B = 11$ mA) is characterized by the appearance of the second frequency f_2 (in the simultaneous presence of the first frequency f_1). The smallest value of f_2 (280 Hz) markedly exceeds that of f_1. Furthermore, at first glance, the frequency f_2 seems to scale linearly with the control parameter. So far, these two features suggest a secondary Hopf bifurcation. But we also found that the amplitude behaved discontinuously at the transition point, in contradiction to the interpretation given above. Taking further into account a thorough scaling evaluation of the frequencies, described in Sect. 3.6 (yielding, among others, a nonlinear control parameter dependence of the frequency f_2), our semiconductor experiment best follows the saddle node bifurcation on a limit cycle (Sect. 3.3).

3.4.3   Influence of Nonlinearities

Thus far, we have shown how a quasiperiodic state can be characterized and how it can be created via simple bifurcations. Now we turn to the effect that nonlinearities exert on a quasiperiodic state. The appearance of additional spectral lines, the so-called mixing components, in the power spectrum will be discussed first. Then we concentrate on the evolution of the two system-immanent frequencies under a variation of the control parameters, particularly, the emergence of mode-locking phenomena. For this purpose, the sine circle map is used as an appropriate model equation.

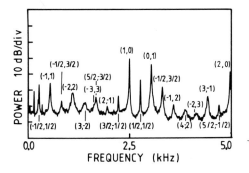

Fig. 3.43. Power spectrum of a quasi-periodic current oscillation obtained from the sample sketched in Fig. 2.4 at the following constant parameters: bias current $I = 6.1120$ mA, transverse magnetic field $B = 1.682$ mT, and temperature $T_b = 2.105$ K. The spectral lines labelled by (n, m) correspond to the frequency $f = nf_1 + mf_2$

In Fig. 3.42 we have considered the dependence of the two intrinsic frequencies on the applied magnetic field. If one also looks at the power spectra of these quasiperiodic states, there are, besides the two pronounced frequency peaks, quite often a variety of further spectral lines that have a smaller peak height. Such a power spectrum is given in Fig. 3.43. The main peaks corresponding to the frequencies f_1 and f_2 are marked by the indices $(1, 0)$ and $(0, 1)$. Following their spectral distribution, all other frequency peaks can be attributed to linear combinations of f_1 and f_2, i.e., they obey the notation $f = nf_1 + mf_2$ (n and m are integers or multiples of $1/2$) giving rise to a characterization by the tuples (n, m). These frequencies are called mixing components.

From the power spectrum in Fig. 3.43, we conclude that the underlying current signal $I(t)$ must have Fourier components like $\cos 2\pi(nf_1 + mf_2)t$. Such terms are nothing other than the result of the products of trigonometrical functions. For demonstration, let us take the following two variables:

$$y_1(t) = \sin 2\pi f_1 t, \qquad y_2(t) = \sin 2\pi f_2 t . \tag{3.34}$$

Their simplest nonlinear combination is just multiplication,

$$y_1(t) y_2(t) = (1/2)\cos 2\pi(f_1 - f_2)t - (1/2)\cos 2\pi(f_1 + f_2)t , \tag{3.35}$$

leading directly to the mixing components $f_1 - f_2$ ($n = 1, m = -1$) and $f_1 + f_2$ ($n = 1,\ m = 1$). Further mixing components are obtained analogously from higher-order products of y_1 and y_2. We, therefore, see that the corresponding signal of Fig. 3.43 can be written as

$$x(t) = \sum_n \sum_m A_{nm} y_1^n y_2^m , \tag{3.36}$$

where A_{nm} denote real numbers. Equation (3.36) represents a nonlinear expansion of the variable $x(t)$ in powers of y_1 and y_2. This result tells us the meaning of the peak height of a mixing component in the power spectrum. It reflects the magnitude of the coefficient A_{nm} and, thus, the size of the corresponding nonlinear term. So we are able to extract some information on the degree of nonlinearity involved in a quasiperiodic state by considering the height of the spectral lines resulting from mixing components in proportion to that of the fundamental frequency peaks. In what follows, we show with the help of the circle map formalism how the size of a nonlinearity present is an essential control parameter which forces the change from quasiperiodicity to mode-locked behavior. That, eventually, the strength of the nonlinearity may even cause chaos will be discussed in Sect. 3.5.

The simplest model to explain nonlinear phenomena of quasiperiodic states is given by the circle map, more precisely called the sine circle map. This model ansatz rests upon the Poincaré cross section of quasiperiodic dynamics. The cross-section points of this two-dimensional presentation can be described through polar coordinates, i.e., radius r and angle θ, in the iterative equations

$$r_{n+1} = f(r_n, \theta_n), \qquad \theta_{n+1} = g(r_n, \theta_n) . \tag{3.37}$$

Here f and g are arbitrary functions; g must be periodic in θ. To end up with the circle map, it is assumed that the radius variable can be taken as a constant. Then the system dynamics contains only the angle variable. Taking further into account that the Poincaré cross section of a quasiperiodic state has the form of a set of points located on a circle, the angle must change from iteration to iteration, in order to spread over the whole circle. The simplest iterative equation for such a periodic dynamics is

$$\theta_{n+1} = \Omega + \theta_n \,(\mathrm{mod}\,1) . \tag{3.38}$$

Here Ω stands for a constant. The modulus 1 function guarantees the required periodicity. Note that the angle variable is normalized to unity. If one chooses Ω to be rational, i.e., $\Omega = p/q$ with $p, q \in \mathbb{Z}$, we get

$$\theta_{n+q} = q\Omega + \theta_n \,(\mathrm{mod}\,1)$$
$$= p + \theta_n \,(\mathrm{mod}\,1)$$
$$= \theta_n . \tag{3.39}$$

Obviously, we no longer have quasiperiodic behavior. After q iterations, the same value of the angle is attained again. In this case, the Poincaré cross section consists of q different points such that the corresponding trajectory on a torus is recurrent after q turns. Now one has an example of a periodic dynamics. However, for the case of Ω chosen to be irrational, we obtain an infinite sequence of different angle values. The Poincaré cross section then proves to be a set of points filling up a circle. The corresponding trajectory covers a whole torus, representing the genuine case of quasiperiodicity.

Figure 3.44 illustrates four different experimentally observed dynamical states via their phase portraits. Following the discussion given above, the situations presented in parts (a), (c), and (d) relate to rational values of Ω: 2/3, 1/1, and 5/3, respectively. Unlike these locking states, the phase portrait in part (b) clearly distinguishes a quasiperiodic state. In order to demonstrate the fact that, in the case of mode-locking, the ratio of the two fundamental frequencies is nothing but the value of Ω, we have plotted the power spectrum for the state of

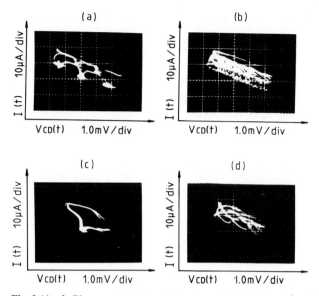

Fig. 3.44a–d. Phase portraits for distinct mode-locked (**a, c, d**) and quasiperiodic (**b**) oscillatory states obtained from the sample sketched in Fig. 2.4 by plotting the partial voltage drop $V_{CD}(t)$ against the current $I(t)$ at different transverse magnetic fields $B = 0.374$ mT **a**, $B = 0.413$ mT **b**, $B = 0.570$ mT **c**, $B = 0.753$ mT **d**, and the following constant parameters: bias voltage $V_0 = 2.3280$ V and temperature $T_b = 2.10$ K (load resistance $R_L = 100 \, \Omega$). Note that the partially visible bright spots on the trajectories correspond to stroboscopic Poincaré cross sections

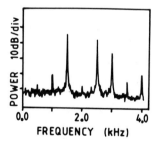

Fig. 3.45. Power spectrum for the case of the mode-locked oscillatory state in Fig. 3.44d (same sample and control parameters) obtained from the current signal $I(t)$

part (d) in Fig. 3.45. Indeed, the two highest spectral lines are located at 2.5 kHz and 1.5 kHz, yielding a frequency ratio of 5/3.

So far, we have tried to describe quasiperiodicity by means of the simple linear ansatz of (3.38). There exists an easy argument that this approach is not a good one. The value of Ω being rational or irrational determines whether we have a periodic or a quasiperiodic dynamics, respectively. But it is well known that the subset of all rational (irrational) numbers has the measure zero (one) in the set of real numbers. This measure can be taken as the probability of finding these numbers by accident. As a consequence, one should expect to detect only quasiperiodic states in the case of an experimental system that obeys (3.38). Here we assume the experimental control parameter to affect Ω in a defined way. On the other hand, periodic states should be found only if there are finite parameter regions where they exist. It is common to say that the phase (or the mode) of the system must lock to a periodic state over an extended control parameter range. Next we point out the way the occurrence of mode-locked states can be explained by means of a nonlinear function $g(\theta)$ in the angle equation.

A nonlinear iterative angle equation is easily constructed by adding a periodic nonlinear term to the right-hand side of (3.38). In the simplest case, one might take a sine function, leading to

$$\theta_{n+1} = \Omega + \theta_n + (K/2\pi)\sin 2\pi\theta_n \;(\mathrm{mod}\; 1) \; . \tag{3.40}$$

Here K denotes an arbitrary constant. The factor 2π guarantees that the angle variable is normalized to the interval $(0, 1)$. Equation (3.40) gives the sine circle map. In what follows, we often just say circle map. The size of the constant K provides the strength of the nonlinearity. It is known from experiments and computer simulations that the coupling strength of a periodically driven oscillator [3.33] can be described analogously by a constant K which exerts the same influence on the related equation (3.29) as the present constant K on (3.40). This is why the constant in (3.29) has already been designated as K. In that sense, K can also be called the coupling constant. We, therefore, have two essential control parameters in (3.40), namely, the ratio Ω of the fundamental frequencies defined for the linear model and the coupling constant K characterizing the degree of the prevailing nonlinearity.

The dynamics of the circle map (just as for all one-dimensional iterative equations) is usually visualized by a graphical representation like the one shown in Fig. 3.46. The solid curve indicates the graph of the return map $\theta_{n+1} = f(\theta_n)$. Furthermore, the construction principle to reach the angle value θ_n due to the nth iteration of an arbitrarily chosen initial condition θ_0 has been elucidated. The solid step-like line gives the iterative dynamics of the angle variable (with the direction marked by arrows). Under variation of the control parameters, the graph changes its shape, affecting the resulting dynamics accordingly. Figure 3.47 illustrates the angle dynamics for the case of two different coupling constants. Upon increasing the value of K, one obtains a transition from

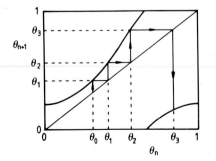

Fig. 3.46. Schematic representation of the return map obtained by plotting the iterative dynamics of the angle variable θ in (3.40) at subcritical coupling constant $K < 1$

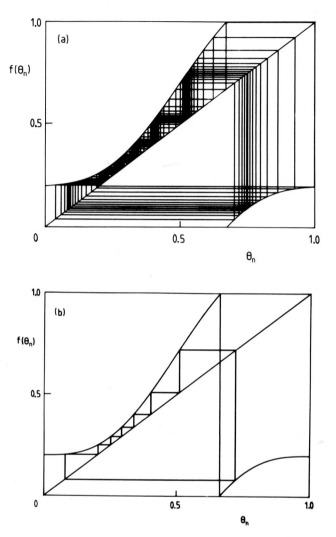

Fig. 3.47a, b. Schematic representation of return maps obtained by plotting the iterative dynamics of the angle variable θ in (3.40) at different coupling constants $K = 0.9$ **a**, $K = 1.0$ **b**, and constant frequency ratio $\Omega = 0.2$ (from [3.34])

a quasiperiodic (a) into a mode-locked state (b). Obviously, the latter has a period of eight, which can be recognized from the fact that the initial condition is recurrent after eight iterations. For a detailed treatment of the circle map formalism, we refer to the literature [3.33–36].

An overview of the dynamics of the circle map can be obtained from the phase diagram in Fig. 3.48. In the (Ω, K) control parameter plane the different regions of mode-locking are marked by their frequency ratio. Note that mode-locked states are also possible for irrational values of Ω, usually, if K is

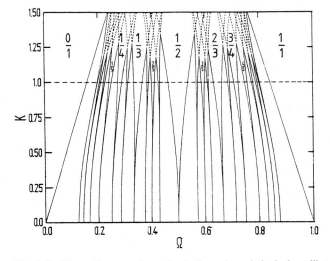

Fig. 3.48. Phase diagram of quasiperiodic and mode-locked oscillatory states in the coupling constant vs. frequency ratio control parameter plane obtained from the sine circle map in (3.40). The different locking states are labelled by their actual frequency ratio (from [3.34])

sufficiently large. At this point, we again stress the fact that Ω represents the frequency ratio of only the linear model (i.e., in the case of $K = 0$). With increasing K, the adjusting frequency ratio differs from Ω. Then the actual frequency ratio is given by the winding number

$$W = \lim_{n \to \infty} (\theta_n - \theta_0)/n \ . \tag{3.41}$$

Here θ_n stands for the nth iteration of the initial condition θ_0 by the function g (e.g., the sine circle map), in the sense that $\theta_1 = g(\theta_0)$, $\theta_2 = g(g(\theta_0))$, etc. The winding number W is, thus, defined as the average rotation per iteration. It does not depend on the choice of the initial condition θ_0. If W is rational, $W = p/q$ with $p, q \in \mathbb{Z}$, there exists a periodic cycle on the circle such that $\theta_{n+q} = \theta_n + p$. Consequently, a mode-locked state is characterized by a rational value of W and, otherwise, a quasiperiodic state by an irrational value of W.

Coming back to Fig. 3.48, one recognizes the pronounced tongue-like shape of the locking regions, the so-called Arnold tongues [3.37]. Let us consider, for example, the Arnold tongue that corresponds to the mode-locked state with the rational winding number $W = 1/2$. It comes out of the Ω-axis with zero width, i.e., for small values of K, there are quasiperiodic states even at values of Ω close to 1/2. Upon increasing K, the tongue gets wider such that an extended interval of different parameter values develops around $\Omega = 1/2$ which all belong to the $W = 1/2$ locking state. A characteristic feature of the circle map is that the dominance of the quasiperiodic behavior in the linear case ($K = 0$) vanishes at increased K, while mode-locking just takes place the other way round. In other words, the measure of the locked states continuously increases from zero at

$K = 0$ to unity at $K = 1$. Thus, for $K = 1$, the quasiperiodic states have zero measure. Along the so-called "critical line" of $K = 1$, any two neighboring quasiperiodic points are separated by distinct mode-locking intervals with finite length. The structure of locking states forming along this line displays similarities to the Cantor set and can, therefore, be characterized as a fractal [3.34].

Criticality in the circle map model derives from the fact that beyond the threshold $K = 1$ the Arnold tongues start overlapping which gives rise to the quasiperiodic approach to chaos discussed in Sect. 3.5. The pattern of overlaps is quite complicated. Only the tiniest tongues overlap near the critical line; the bigger ones overlap at larger values of K. Also, any given tongue is overlapped for the first time by the thinnest neighboring tongues some finite distance above criticality. As soon as two tongues overlap, there is an infinity of tongues in between which are overlapped too, since between each pair of rational numbers there are, of course, an infinity of other rationals. The winding number then becomes nonunique, replaced in general by a winding interval [3.33]: different initial conditions can lead to all different winding numbers within an interval.

Apart from the growing tendency to find locking states with increasing coupling constant K, there exists a hierarchical order with respect to their size by variation of the frequency ratio Ω via a horizontal cut through the control parameter plane in Fig. 3.48 (i.e., with K kept constant). Obviously, the widths of locking states characterized by winding numbers with rather "simple" fractions (e.g., 0/1, 1/2, 1/1, etc.) always exceed those having more "complicated" fractions (e.g., 1/4, 2/5, 3/5, 3/4, etc.). This size effect is closely related to the Farey tree ordering of rational numbers. Figure 3.49 indicates the underlying construction principle. Rational numbers in a certain horizontal line correspond to different locking states of comparable size. To obtain additional numbers from those of the line above, the following general rule has to be applied: the "old" rational numbers p/q and r/s give the "new" rational number $(p + r)/(q + s)$.

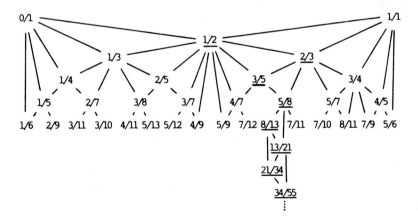

Fig. 3.49. Schematic representation of the Farey tree yielding a hierarchical order of the rational numbers (or the locking states). The Fibonacci numbers that converge toward the golden mean are underlined

Finally, we have underlined the sequence of Fibonacci numbers in the Farey tree of Fig. 3.49. These numbers approximate the value of the "golden mean", $(\sqrt{5} - 1)/2$. The golden mean is commonly known as the "most" irrational number in the sense that it has the largest possible distance from a simple rational number with reference to the Farey tree configuration. As an outstanding peculiarity, we emphasize that the quasiperiodic behavior just developing at the coupling constant equal to unity and the winding number equal to the golden mean elicits a universal dynamics [3.33–35]. So far, this singular situation could be realized with an astonishingly high precision by a variety of different experimental systems (mainly from hydrodynamics, chemistry, and solid state physics).

3.4.4 Experiments

In what follows, we report some experimental observations obtained from our semiconductor system with respect to the circle map formalism introduced above. The appropriate continuous model of the circle map is a periodically driven, damped oscillator. Thus, the best conformity with numerical results of the circle map can be attained in experiments where external periodic forcing is applied to a single internal oscillatory mode of the system. As an advantage, the amplitude and frequency of the driving force represent adequate control parameters which are simply related to the standard parameters of the circle map. However, external periodic forcing does not allow certain types of quasiperiodicity. Due to the simultaneous presence of two or more competing natural oscillatory modes intrinsic to our semiconductor system as a consequence of the spatio-temporal nonlinear dynamics in the breakdown regime (Sect. 2.4), self-generated quasiperiodic and mode-locked states can also emerge without being forced by any periodically changing external parameter. We, therefore, pay only little attention to the former case of an externally driven semiconductor experiment, while the main focus is on the question of whether self-generated quasiperiodicity obeys the circle map behavior as well. The latter case has also been investigated experimentally in a Rayleigh–Benard convection system [3.38].

Let us first consider briefly the periodically forced semiconductor breakdown. For this purpose, the constant (d.c.) voltage source V_0 has superimposed on it a temporally alternating (a.c.) voltage signal $V_{ext} \sin 2\pi f_{ext} t$, the amplitude and frequency of which can be varied appropriately. Bear in mind that the driving frequency f_{ext} plays the role of an external parameter, by no means being susceptible to any nonlinear reaction of the semiconductor system. Only the intrinsic system frequency f_{int} can be affected accordingly. Figures 3.50 and 3.51 illustrate the influence of the driving force on a periodic state (adjusted in the undriven case) by changing its amplitude V_{ext} and frequency f_{ext}, respectively.

Starting from the initial state at $V_{ext} = 0$, in Fig. 3.50 the amplitude of the a.c. voltage component is increased gradually under constant driving frequency. The

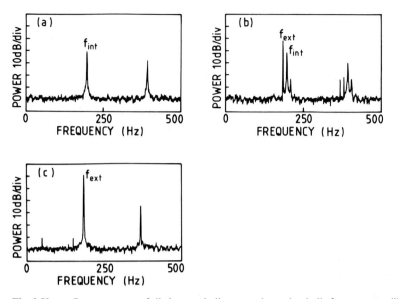

Fig. 3.50a–c. Power spectra of distinct periodic **a, c** and quasiperiodic **b** current oscillations obtained from the sample sketched in Fig. 2.4 when driven by an alternating voltage bias at different amplitudes $V_{ext} = 0$ **a**, $V_{ext} = 40$ arbitrary units **b**, $V_{ext} = 200$ arbitrary units **c**, and the following constant parameters: external frequency $f_{ext} = 180$ Hz, bias current $I = 2.5171$ mA, transverse magnetic field $B = 0.164$ mT, and temperature $T_b = 1.95$ K

power spectrum in part (a) displays a spontaneous periodic current oscillation of the semiconductor with the system-immanent frequency f_{int}. Now pursue the development of such an undriven situation when applying a nonzero driving force ($V_{ext} \neq 0$). The resulting power spectra in parts (b) and (c) clearly reveal a quasiperiodic state (due to the presence of two spectral lines located at the frequencies f_{int} and f_{ext}) and a periodic state (due to the presence of only one spectral line located at the frequency f_{ext}), respectively. The latter corresponds to a 1/1 locking state, just as is expected from the circle map theory. Of course, increasing the driving amplitude V_{ext} means nothing other than an increased coupling (characterized by the constant K) between the driving force and the prevailing dynamical state of the system. Thus far, locking states are favored. In the present case, we have chosen a ratio of the external frequency f_{ext} to the intrinsic frequency f_{int} (related to $V_{ext} = 0$) close to unity. In this sense, our experiment complies with a distinct path in the (Ω, K) control parameter plane, extending parallel to the K-axis for a value of Ω near to one until finally reaching the 1/1 Arnold tongue (Fig. 3.48).

The influence of a varying external frequency f_{ext} exerted on the intrinsic frequency f_{int} is demonstrated in Fig. 3.51. Here the amplitude V_{ext} of the driving force was kept constant. Part (a) directly shows f_{int} vs. f_{ext}. The horizontal dashed line at $f_{int} = 440$ Hz indicates the actual frequency value of the spontaneous current oscillation present in the undriven case. The solid line plots the diagonal,

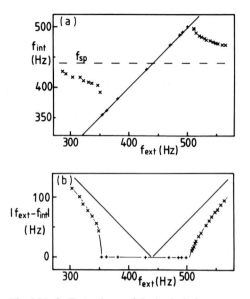

Fig. 3.51a, b. Dependence of the intrinsic frequency **a** and the difference between the external and the intrinsic frequency **b** on the external frequency for the case of a quasiperiodic current oscillation in the neighborhood of a 1/1 mode-locked state obtained from the sample sketched in Fig. 2.4 when driven by an alternating voltage bias at the following constant parameters: voltage amplitude $V_{ext} = 100$ arbitrary units, bias current $I = 4.9230$ mA, transverse magnetic field $B = 1.05$ mT, and temperature $T_b = 1.79$ K. Experimental data are marked by crosses. The 1/1 mode-locked state is indicated by the diagonal solid line in **a** and the horizontal solid line in **b**. The undriven periodic case follows the horizontal dashed line in **a** and the piecewise linear solid curve in **b**. Here the intrinsic frequency remains unchanged at the value f_{sp} of the undriven spontaneous oscillation

thus, reflecting the case $f_{int} = f_{ext}$ (i.e., the 1/1 locking state). The diagram gives prominence to how the system-inherent frequency f_{int} is attracted onto the diagonal by the driving frequency f_{ext}. In terms of the circle map behavior, the variation of f_{ext} corresponds to a change of Ω, while K remains constant. Note that Ω is defined as the ratio of the driving frequency to the intrinsic frequency of the undriven oscillation, hence, f_{ext}/f_{sp}. Part (b) displays the difference between f_{ext} and f_{int} as a function of f_{ext} (or Ω). The solid curve describes the linear case where both frequencies are uncoupled ($K = 0$) and, thus, the intrinsic frequency stays constant ($f_{int} = f_{sp}$). It is evident, however, that the experimental data clearly deviate from such linear behavior. Theory predicts the frequency difference to follow a square-root law. The reason derives from the fact that these locking phenomena are nothing else but a saddle node bifurcation on a torus, which reduces the torus to a limit cycle. As a consequence, one obtains the same topology as that of the saddle node bifurcation on a limit cycle (Sect. 3.3).

Henceforth, we concentrate on the self-organized (i.e., system-immanent) emergence of quasiperiodic and mode-locked oscillatory behavior where no external periodic driving force is applied (i.e., all control parameters are kept constant in time). In the authors' opinion, this fascinating dynamics represents the more "natural" approach to quasiperiodicity and low-dimensional chaos. As we have already seen, our semiconductor experiments deal with a synergetic system consisting of spatially separated oscillation centers arising spontaneously and interacting in a nonlinear way. Depending sensitively upon such control parameters as the applied d.c. electric and magnetic field, the resulting spatio-temporal transport behavior can give rise to the simultaneous (per se) existence of different fundamental oscillatory modes with originally incommensurate frequencies. At this point, the question – to what extent the circle map model is capable of describing, for the case of an undriven system, the mutual competition between at least two of these oscillators under different strengths of their coupling (i.e., the possible phase transition between two-frequency quasiperiodic and the corresponding mode-locked states) – becomes interesting.

Recall the phase portraits of different experimentally observed dynamical states already shown in Fig. 3.44. The ordering of the diverse locking intervals within the region of quasiperiodicity is schematized in the synopsis of Fig. 3.52. Here we have examined a transition from an initial nonoscillatory state (fixed point) to a final noisy (chaotic) state as a function of the transverse magnetic field applied. Obviously, the present control parameter influences the system frequencies in such a way that not only the sequence but also the width of the mode-locked states (given by the width of the arrows) are in accordance with the circle map predictions. Two examples of magnetic field induced variation of the intrinsic frequencies f_1 and f_2 in the neighborhood of the 1/1 locking state are presented in Figs. 3.53 and 3.54. Both diagrams clearly reveal a mutual attraction of the two frequencies when approaching the mode-locked parameter interval. We emphasize that just the active participation of both fundamental oscillatory modes (i.e., the parameter-induced change of both frequencies) constitutes a remarkable difference to the former case of periodically driven experiments (cf. Fig. 3.51). A more detailed treatment of the quasiperiodic phase transition outlined above can be found elsewhere [3.39].

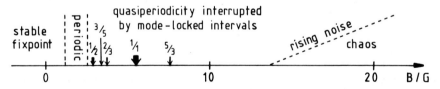

Fig. 3.52. Schematic representation of the quasiperiodic transition to chaos via a nonoscillatory stable fixed point, a periodic oscillatory state with a single intrinsic frequency, and a quasiperiodic oscillatory state with two incommensurate intrinsic frequencies interrupted by different mode-locked oscillatory states obtained from the sample sketched in Fig. 2.4 under variation of the transverse magnetic field (1 G ≙ 0.1 mT) at the following constant parameters: bias voltage $V_0 = 2.3280$ V and temperature $T_b = 2.10$ K (load resistance $R_L = 100 \, \Omega$). The major stable resonances locked at the commensurate frequency ratios 1/2, 3/5, 2/3, 1/1, and 5/3 over finite regions of the applied magnetic field (i.e., 4 µT, 0.5 µT, 1 µT, 8 µT, and 1 µT, respectively) are indicated by arrows

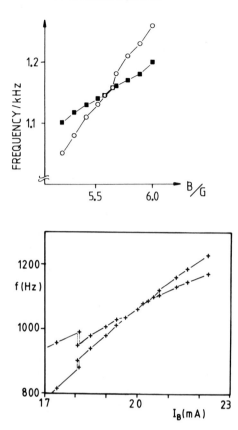

Fig. 3.53. Dependence of two intrinsic frequencies on the transverse magnetic field (1 G $\simeq$ 0.1 mT) for the case of a quasiperiodic current oscillation in the neighborhood of a 1/1 mode-locked state in Fig. 3.52 (same sample and control parameters)

Fig. 3.54. Blow-up of the frequency vs. transverse magnetic field (in terms of the magnetic field current) diagram in Fig. 3.42 (same sample and control parameters) in the neighborhood of a 1/1 mode-locked state

Next we turn to another exemplary scenario of self-generated quasiperiodic and mode-locked behavior associated with a characteristic self-similar emergence of high-order mixing frequencies which takes place when applying a magnetic field in the longitudinal direction (i.e., parallel to the electric field). For a typical scan through control parameter space, the longitudinal magnetic field was slightly increased by degrees, while the bias voltage (or the electric field) always remained at a constant value. The resulting overall system flow again consists of two competing intrinsic oscillatory modes, where the originally incommensurate frequencies (say, f_0 and f_1) change from $f_0 = 1.15$ kHz and $f_1 = 0.30$ kHz (at $B = 7.2$ mT) to $f_0 = 1.05$ kHz and $f_1 = 0.90$ kHz (at $B = 8.2$ mT), respectively. The frequency bifurcation diagram in Fig. 3.55 gives a schematic synopsis of the examined transition as a function of the applied magnetic field. Quasiperiodicity was found to be interrupted by distinct stable resonances locked at the rational frequency ratios $f_1/f_0 = 1/2, 2/3, 3/4, 4/5, 5/6$, and $6/7$ over the finite parameter intervals of $\Delta B = 0.090$ mT, 0.045 mT, 0.015 mT, 0.020 mT, about 0.005 mT, and less than 0.005 mT, respectively. Locking states $f_1/f_0 < 1/2$ could not be resolved. At longitudinal magnetic fields

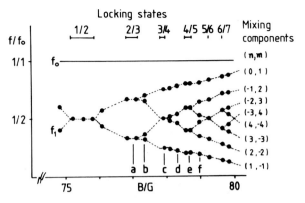

Fig. 3.55. Dependence of two intrinsic frequencies and their mixing components on the longitudinal magnetic field (1 G ≅ 0.1 mT) for the case of a quasiperiodic current oscillation interrupted by different mode-locked states obtained from a distinct sample of 2.8 mm contact distance (with capacitor-like geometry) at the following constant parameters: bias voltage $V_0 = 1.788$ V and temperature $T_b = 4.2$ K (load resistance $R_L = 100\,\Omega$). All frequencies of the spectral bifurcation diagram are normalized to f_0. The mixing components labelled by (n,m) are subject to the frequency $f = nf_0 + mf_1$. The letters a–f outline the control parameter values used in Fig. 3.56

below $B = 7.2$ mT, quasiperiodicity becomes unstable (rising noise). Most strikingly, an increasing number of high-order mixing components (n, m), defined as the linear combinations $nf_0 + mf_1$ (n, m integers) of the two fundamental frequencies f_0 and f_1, gradually arises with increasing magnetic field at particular lockings of odd-numbered denominators (i.e., 2/3, 4/5, 6/7). We point out that, due to the continuous presence of the low-order mixing frequency $f_0 - f_1$, the second fundamental frequency f_1 is not uniquely defined, since f_1 and $f_0 - f_1$ may be interchanged.

The apparent self-similar formation of stable high-order mixing frequencies beyond the 2/3 locking state becomes clearly manifest in the different power spectra of Fig. 3.56 obtained for distinct control parameter values (indicated by the letters a–f in Fig. 3.55). Part (a) gives the power spectrum of the 2/3 locking state. Increasing the magnetic field from $B = 7.70$ mT to $B = 7.73$ mT leads to a bifurcation of the two spectral lines $(0, 1)$: $f_1 = f_0/3$ and $(1, -1)$: $f_0 - f_1 = 2f_0/3$ into the four components $(0, 1)$, $(2, -2)$ and $(1, -1)$, $(-1, 2)$, respectively [part (b)]. Note the emergence of the second-order mixing frequencies $2f_0 - 2f_1$ and $-f_0 + 2f_1$ that are preserved without further bifurcation up to the 4/5 locking state at $B = 7.87$ mT [part (e)]. The intervening 3/4 locking state at $B = 7.78$ mT [part (c)] is a result of their crossover and is not followed by a bifurcation of new mixing components. For clarity, see part (d) obtained at slightly increased magnetic field $B = 7.83$ mT. However, the system flow again bifurcates beyond the 4/5 locking state via the introduction of other two third-order mixing frequencies $3f_0 - 3f_1$ and $-2f_0 + 3f_1$ in an analogous self-similar way, as indicated in the power spectrum of part (f) obtained at $B = 7.90$ mT. The

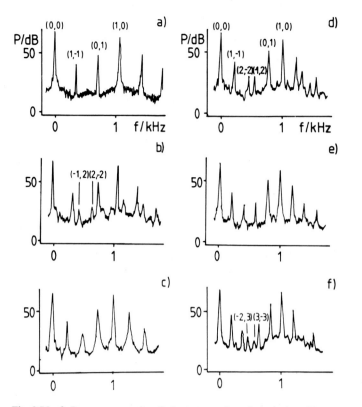

Fig. 3.56a–f. Power spectra for distinct cases of mode-locked oscillatory states with high-order mixing frequencies obtained from the current signal $I(t)$ at different longitudinal magnetic fields $B = 7.70$ mT **a**, $B = 7.73$ mT **b**, $B = 7.78$ mT **c**, $B = 7.83$ mT **d**, $B = 7.87$ mT **e**, $B = 7.90$ mT **f**, as indicated by the corresponding symbols a–f in Fig. 3.55 (same sample and control parameters)

same bifurcation procedure recurs at the higher 5/6 and 6/7 locking states, see Fig. 3.55.

The pattern of the frequency bifurcation diagram together with the power spectra clearly shows an increasing number and strength of high-order mixing frequencies with increasing longitudinal magnetic field. A simple interpretation is possible assuming that the degree of nonlinear coupling between the two fundamental oscillatory modes of the quasiperiodic state increases with increasing magnetic field. Comparing the experimental locking sequence with the resonance structure of the (Ω, K) parameter space of the circle map (cf. Fig. 3.48), one might presume that our experimental system, upon variation of the single control parameter B, follows a distinct path through the two-parameter mapping space, as indicated schematically in Fig. 3.57a. Thus, variation of the longitudinal magnetic field B simultaneously changes both the frequency ratio Ω and the coupling strength K. The circles in Fig. 3.57a point out the intersection intervals of the system path $K = \Omega$ with the Arnold tongues. An

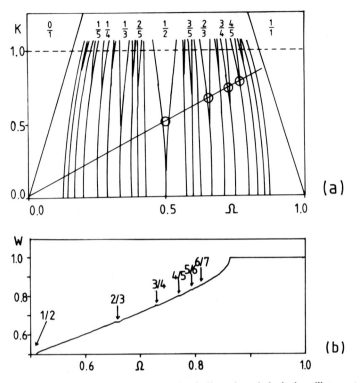

Fig. 3.57a, b. Phase diagrams of quasiperiodic and mode-locked oscillatory states in the coupling constant vs. frequency ratio **a** and the winding number vs. frequency ratio **b** control parameter plane obtained from the sine circle map in (3.40). The different locking states are accentuated by their actual frequency ratio. The diagonal solid line $K = \Omega$ in **a** indicates an exemplary path followed by the experimental system. The intersection intervals with the Arnold tongues marked by circles correspond to the experimentally observed locking sequence. The winding number W in **b** was calculated from (3.41) for $n = 500$ iterations assuming $K = \Omega$

approximate reproduction of the experimentally observed locking sequence is also demonstrated in Fig. 3.57b. Here the "effective" frequency ratio W [termed winding number above and defined by (3.41)] determining the average twist of the phase space trajectory as a consequence of the nonlinear coupling is plotted for the circle map (3.40) as a function of the "bare" frequency ratio Ω under the assumption $K = \Omega$. It appears worth noting that analogous behavior can be derived from a simple mechanical oscillatory system of two weakly coupled, undamped pendula by increasing, for example, the mass of one pendulum as a relevant physical control parameter. For a more comprehensive study of the latter quasiperiodic phase transition, we refer to [3.40].

The last example that the experimentally observed self-generated quasiperiodicity can be modelled by the circle map is presented in Fig. 3.58. Here we have calculated an iterative angle dynamics from the return map shown

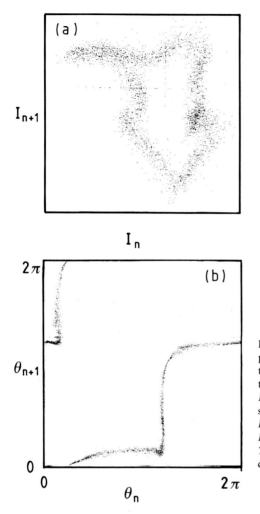

Fig. 3.58a, b. Return maps for a quasi-periodic oscillatory state obtained from the sample sketched in Fig. 2.4 by plotting the iterative dynamics of **a** the current I and **b** the angle θ at the following constant parameters: bias current $I = 1.2094$ mA, transverse magnetic field $B = 2.28$ mT, and temperature $T_b = 2.005$ K. The angle variable in **b** is derived from the current in **a** via (3.42)

in part (a). For comparison, see Fig. 3.39a. With the assumption that the return map of the current I_n corresponds to a Poincaré cross section, the angle variable θ_n can be derived via the relation

$$\theta_n = \arctan((I_{n+1} - I_n)/I_n) .\tag{3.42}$$

The function $\theta_{n+1} = f(\theta_n)$ displayed in Fig. 3.58b undoubtedly demonstrates that the present quasiperiodic behavior is founded in a simple angle dynamics. Moreover, this graph unveils another interesting feature, namely, that the return map almost reflects a local maximum. Thus, the underlying dynamics would not be invertible in time any more. Consequences of this finding will be the subject of the following section.

3.5 Chaotic Oscillations and Hierarchy of Dynamical States

The present section is devoted to a totally new type of dynamics that emerges in phase space with dimensions larger than three. For systems whose description requires a three-dimensional phase space, i.e., three independent variables x_1, x_2, and x_3, we have already introduced the quasiperiodic oscillation as a new quality of dynamical behavior. Thus, up to now, the following hierarchical order of dynamical states with different attractors can be set up. In one-dimensional phase space, there exist only fixed points, i.e., temporally constant states. In two-dimensional phase space, infinitely long periodically changing dynamics may further occur, i.e., limit cycles are possible in addition to point attractors. In three-dimensional phase space, one can also have an infinitely long temporally varying state with a well organized quasiperiodic dynamics. In all these cases, the dimensions of the underlying attractors are integer values and, in general, one dimension lower than the corresponding phase space.

Up to recent times, it was the common opinion that three-variable dissipative systems should not give rise to any further complicated type of dynamics. In the 1970s, the pioneer work done by *Lorenz* in 1963 [3.41] attracted a lot of attention. There for the first time a system of three autonomous deterministic nonlinear ordinary differential equations was demonstrated to exhibit the peculiarity that "all of the solutions (trajectories in phase space) are found to be unstable, and almost all of them are nonperiodic" [3.41]. In contrast to quasiperiodicity, these solutions reflected some kind of disorder "so that slightly different initial states evolve into considerably different states" [3.41]. It was *Li* and *Yorke* [3.42] who introduced the name "chaos" for such a form of dynamics. Hence, in three-dimensional phase space, two qualitatively new types of attractors appear, namely, toroidal (quasiperiodic) and chaotic. In four variables, the system dynamics is expected analogously to consist either of a quasiperiodic oscillation with three incommensurate frequencies on a hypertorus or ordinary chaos on a torus – or hyperchaos [3.43]. Moreover, we propose the possible existence of a further dynamical state distinguished by qualitatively new features which cannot be derived from the attractor analogs in lower-dimensional phase space. The main difference between the ordinary chaotic and the hyperchaotic state is that the former can be reduced to a three-variable dynamics, while the description of the latter necessarily requires the total number of variables available to be four. Again, a single quadratic nonlinearity suffices [3.43]. Finally, let us remark that, of course, all dynamical states possible in lower-dimensional phase space as well as their combinations, the so-called mixed cases, may also emerge in higher-dimensional phase space.

Before chaotic phenomena are treated in more detail, we call the reader's attention to some references in the literature. For example, there are comprehensive monographs by *Guckenheimer* and *Holmes* [3.2], *Schuster* [3.4], *Devaney* [3.8], *Thompson* and *Stewart* [3.9], and *Berge* et al. [3.10]. A collection of important papers on this topic can be found in [3.5–7].

3.5.1   The Chaotic State

The two fundamental elements of a chaotic dynamics are the sensitive dependence on initial conditions and the boundedness of expanding nearby trajectories on an attractor via a backfolding process. One can easily recognize how such a sequence of stretching and folding in turn is capable of creating a certain degree of disorder. As an example, we regard the shuffling of playing cards. To begin with, let there be two piles of playing cards that have a distinct order. To shuffle, we take one pile and separate its playing cards (stretching process) to gain spaces that allow one to put cards of the other pile in between. In this way, one obtains a single pile containing all the playing cards which, in the next step, is divided up into two new piles (folding process). If we repeat this shuffling procedure again and again, the formerly well-ordered playing cards finally end up in complete disorder.

Considering the key features of stretching and folding, *Rössler* [3.44] constructed the simplest possible differential equation capable of modelling chaos. To approach the chaotic dynamics of the Rössler system, we start with an oscillation of the form

$$\dot{x}_1 = -x_2,$$
$$\dot{x}_2 = x_1 . \tag{3.43}$$

The stretching mechanism can be achieved by adding an explosion (i.e., negative damping) term,

$$\dot{x}_1 = -x_2,$$
$$\dot{x}_2 = x_1 + ax_2 . \tag{3.44}$$

Figure 3.59 gives the corresponding phase portrait. Obviously, from an unstable focus all trajectories run away on spiral lines. Note that the distance between nearby initial conditions grows exponentially with time. Thus, the present dynamics displays a sensitive dependence on initial conditions – a characteristic feature of chaos. In order to arrive at a chaotic attractor, the trajectories must be prevented from running to infinity. They rather have to stay bounded in a finite region of phase space. The only way this can be realized is if the trajectories are folded back to the vicinity of the origin. Due to the

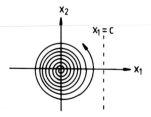

Fig. 3.59. Schematic representation of the phase portrait obtained from the exploding oscillator in (3.44)

uniqueness of the solution of differential equations, any crossing of (expanding and backfolding) trajectories is generally forbidden. The latter condition can only be met in three-dimensional phase space where the trajectories have the possibility to turn away from the two-dimensional plane of the spiral by taking advantage of the third dimension in the vertical direction. We introduce such a backfolding process into the dynamics of (3.44) by defining a boundary $x_1 = c$. If the spiral-type evolution of a dynamical state crosses this boundary, then a third variable x_3 immediately starts growing such that the trajectory leaves the (x_1, x_2) plane as a prerequisite of the subsequent backfolding process. The latter can be accomplished by a negative feedback of the variable x_3 onto x_1. In this way, we end up with the following differential equation system, called the Rössler model:

$$\dot{x}_1 = - x_2 - x_3 \,,$$

$$\dot{x}_2 = x_1 + ax_2 \,, \tag{3.45}$$

$$\dot{x}_3 = b + x_3(x_1 - c) \,.$$

This prototype equation has only one nonlinear term, namely, $x_3(x_1 - c)$. If x_1 is larger than c, this nonlinear term becomes positive and, thus, x_3 grows exponentially. On the other hand, the increase of x_3 gives rise to a negative value of $\dot{x}_1$, i.e., a decrease of x_1. As a consequence, the nonlinear term changes its sign, leading the trajectory back to small values of x_1. The resulting attractor dynamics is sketched in Fig. 3.60. The basic elements that generate the present (and, of course, any) chaotic attractor are separately illustrated in Fig. 3.61. These are the attraction of initial points in phase space towards the attractor [part (a)] and the stretching and folding of nearby trajectories on the attractor [parts (b) and (c), respectively].

So far, one might get the impression that the local structure of the Rössler attractor is two-dimensional and, hence, there exists no difference in topology compared to the quasiperiodic torus. However, upon a more careful inspection, the Rössler attractor reveals a finite thickness. This finding is necessary to guarantee that no crossing of trajectories takes place. While, on the one hand, two dimensions are by no means sufficient to embed the trajectories of the chaotic attractor, no three-dimensional volume element in phase space, no matter how small it is, can, on the other hand, be occupied completely by them. This follows from the dissipative nature of the system under consideration. All volume elements must shrink to zero under the flow generated by (3.45). So we deal with the strange case of an attractor being too thick for a (local) two-dimensional phase space (manifold) but at the same time too thin for a (local) three-dimensional phase space. In this sense, chaotic attractors are often called strange attractors. With the help of an abstract definition of the term "dimension" first introduced by *Hausdorff* [3.46], it can be demonstrated that a noninteger dimensionality is capable of providing a quantitative measure for characterizing the strange attractor and, hence, the corresponding system-immanent degree of chaos.

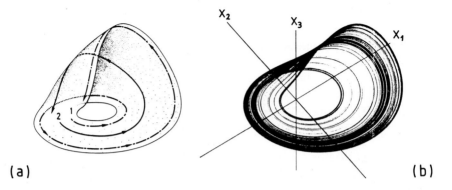

Fig. 3.60a, b. The Rössler system. **a** Schematic representation of the prototype chaotic attractor in phase space accentuating the divergence of two neighboring trajectories (denoted by 1 and 2). **b** Dynamics of the prototype chaotic attractor in phase space obtained from (3.45) at the constant parameters $a = 0.398$, $b = 2$, and $c = 4$ (from [3.1, 9])

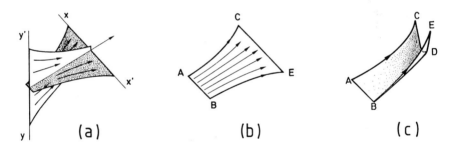

Fig. 3.61a–c. Schematic representation of the fundamental elements of a chaotic attractor in phase space: **a** attraction; **b** divergence; **c** folding (from [3.45])

In the next step, we discuss the temporal evolution of neighboring trajectories on an attractor, making use of the return map. This technique favorably applies to the almost two-dimensional structure of the Rössler attractor. As a suitable plane of intersection, one might choose a half-plane containing the x_3-axis. Figure 3.62 gives the return map extracted from the x_1-components of the intersection points (Poincaré points). Now the time behavior of nearby trajectories can be illustrated as follows. Let us consider an arrow which, starting from the x_3-axis, traverses the attractor. By further means of the return map, the iteration of all points on this arrow developing under the flow of the attractor is bound to be reconstructed. That the resulting dynamics extends over

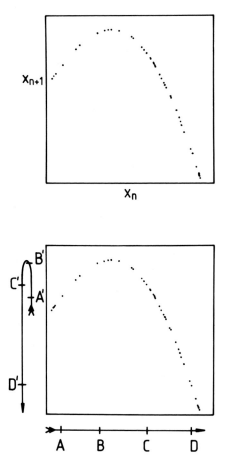

Fig. 3.62. Return map for the case of the Rössler attractor in Fig. 3.60b obtained by plotting the iterative dynamics of the variable x_1 in (3.45) (from [3.9])

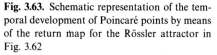

Fig. 3.63. Schematic representation of the temporal development of Poincaré points by means of the return map for the Rössler attractor in Fig. 3.62

a total phase angle of 360° is directly revealed in Fig. 3.63. Obviously, the distance between these points changes considerably during the mapping procedure (e.g., the region between C and D or C′ and D′ becomes larger due to stretching, while the region between B and C or B′ and C′ becomes smaller due to folding). The crucial point is whether convergence or divergence represents the dominating mechanism. In the former case, all trajectories come closer and closer as time goes by, i.e., they are attracted to a limit cycle. We then have a periodic oscillation. In the latter case, the prevailing divergence of nearby trajectories will lead to a chaotic dynamics. The average amount of convergence or divergence gives a quantitative measure to be expressed in the form of an exponential factor $\exp(\lambda t)$, where t denotes time and λ, the Lyapunov characteristic exponent. Thus, a positive exponent can be attributed to the stretching mechanism of a chaotic motion, as discussed later on in more detail. It turns out that, besides the probabilistic concept of the generalized Hausdorff dimension, the present dynamical approach of Lyapunov exponents represents a second fundamental characterization method for a chaotic state.

So far, we have demonstrated that the return map is capable of grasping the complex dynamics on the Rössler attractor. But keep in mind that some information gets lost as a consequence of the reduction to the iterative mapping. While the dynamics on the attractor is invertible in time, this no longer holds for the case of the return map due to its characteristic maximum (Fig. 3.62). Just in the region of the maximum corresponding to the fold of the attractor, there always exist two different original Poincaré points x_n that after one iteration are mapped onto one single point x_{n+1}. By way of contrast, it is generally possible to follow any trajectory on the attractor backwards in time, i.e., one still has an overall uniqueness in the case of a continuous dynamics. This invertibility can be guaranteed by the finite thickness of the attractor. Therefore, the only point where we have lost information is in the reduction of the Poincaré cross section to one line (determined by the x_1-components), the noninvertible dynamics of which is described by the return map. Note that such a problem of nonunique reduction, of course, did not arise in the case of regular quasiperiodicity (Sect. 3.4). Nevertheless, the reduced dynamical possibilities of a discrete return map suffice to model the chaotic motion on an at least low-dimensional strange attractor. The particular shape of the return map still contains the essential features of the system dynamics investigated.

In the work of *Feigenbaum* [3.47] it has been shown explicitly how the dynamics of an iterative system can be classified by the order of the maximum. Thus, for a quadratic maximum it is sufficiently described by the following equation:

$$x_{n+1} = \mu x_n(1 - x_n) . \tag{3.46}$$

This equation, called the logistic map, has acquired a considerable reputation due to the underlying universality. We should point out that the universality of (3.46) results from a generic function $g(x)$ determined by a sequence of infinite iterations of a return map with a simple quadratic maximum. Starting from the logistic map, several universal bifurcations from periodic states to chaos were discovered. However, before the emphasis is finally laid on these different chaotic scenarios, we first present the most important methods for characterizing chaotic dynamics in a more quantitative way.

3.5.2   Characterization Methods

Already during the discussion on the different mechanisms of chaos given above, we have mentioned some quantitative characteristic features, the background of which will henceforth be in the center of interest. For a more rigorous and extensive mathematical treatment, we refer to Chap. 4. The intention of the present section is rather to provide the reader with a vivid idea of these methods and, most particularly, to demonstrate their applicability to experimental data.

For this purpose, we have picked out in the following two representative experimental situations of our semiconductor system, obviously exhibiting a different quality of chaotic dynamics (Fig. 3.64). For each of the two states, the analog signals immediately obtained from the semiconductor were digitized with a sampling frequency of 100 kHz into 80 000 data points. The efficiency of all numerical analysis techniques to be discussed in this section will be demonstrated just on the basis of the above experimental data set. Importance is attached not only to the characterization of a particular dynamical state with respect to the possible existence of a chaotic motion involved, but also to provide a conclusive criterion that allows one to distinguish between different chaotic states with respect to their degree of inherent complexity. Even though remarkable progress has been made lately in the analysis of chaotic time signals, to date there does not exist any universal investigation method capable of revealing all fundamental aspects of a chaotic dynamics. On the contrary, each characterization technique gives prominence to one particular aspect. A first theoretical approach towards more generality will be presented in Chap. 4.

In a rough scheme, all methods can be divided into two subgroups. On the one hand, we have the analysis of chaos on the basis of the topology (i.e., the geometrical structure) of an attractor and, on the other hand, the analysis of the dynamics (i.e., the temporal behavior) of a chaotic state. In this sense, the construction of a Poincaré cross section yields a topological analysis, while the return map enables a dynamical analysis. In Sect. 3.4, we already have shown that both approaches may be equivalent, but this does not at all mean that they must be equivalent in every case.

Figure 3.64 displays the time signals of different partial voltage drops across the semiconductor sample, $V_{AB}(t)$ and $V_{BC}(t)$, observed for the two representative experimental situations, henceforth denoted as state A and state B. The corresponding power spectra are shown in Fig. 3.65. Due to the high noise level present in both cases, one certainly has neither a periodic nor a quasiperiodic dynamics. Upon closer inspection of the power spectra as well as of the time traces, it becomes evident that state A is less noisy than state B. The pronounced wideband spectral distribution indicates that there exist infinitely many different frequency components or oscillatory modes. Under the assumption of a linear model conception (known as the Landau–Hopf model of noise), common until recently, one might get the misleading idea of demanding for each frequency at least one further degree of freedom. The resulting dynamics should, thus, derive from an overall system possessing an infinite number of internal degrees of freedom. Imagine the case of a semiconductor where the multitude of individual charge carriers would have the potential to give rise to such a number of degrees of freedom. In what follows, we point out that this ansatz leads to a wrong understanding of the dynamics present in our system. If one rather takes into account the general existence of nonlinear phenomena, it immediately becomes obvious to what extent even a small number of degrees of freedom is capable of

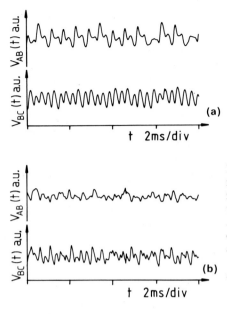

Fig. 3.64a, b. Temporal partial voltage profiles for two distinct chaotic oscillatory states obtained from the sample sketched in Fig. 2.4 at different transverse magnetic fields $B = 3.15$ mT **a** (state A), $B = 4.65$ mT **b** (state B), and the following constant parameters: bias voltage $V_0 = 2.145$ V and temperature $T_b = 4.2$ K (load resistance $R_L = 100 \, \Omega$). The partial voltages are plotted in arbitrary units

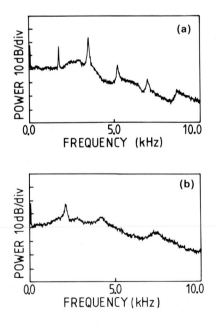

Fig. 3.65a, b. Power spectra for the two distinct cases of chaotic oscillatory states in Fig. 3.64 (same sample and control parameters) obtained from the partial voltage signal $V_{BC}(t)$

producing such complex-looking forms of noisy states. The necessity of applying novel analysis methods to chaos as a distinct peculiarity of nonlinear dynamics can, therefore, be elucidated.

First, we focus on topological characterization methods which quantitatively look at the geometrical structure of an attractor in phase space. So far, we have seen that the Rössler attractor reveals some strange kind of geometry. This usually holds true for any chaotic attractor. The Rössler attractor was too thick for a two-dimensional plane and too thin to fill up a three-dimensional volume element in phase space. Such a behavior can be grasped via attributing a fractal dimension of a noninteger value that ranges between two and three. The term "fractal" was introduced by *Mandelbrot* [3.48] for objects with noninteger dimensions.

In order to discuss the idea behind such an abstract definition of a dimension, let us start with a given set M. To obtain the true shape of this set, its presentation must be exempted from any projection (if one part of the set is put onto another one), i.e., the set M should be embedded in a space of at least the same dimension as M already has or of a higher dimension. Note that the dimension of the embedding space (sometimes called the topological dimension) is always an integer value. In such an embedding space one chooses a set of volume elements $V_i(l)$ with the characteristic size l (e.g., bowls of diameter l). This set of volume elements should cover the whole set M. Now the definition of a dimension is based on the following idea. Starting from a suitably defined property of M which somehow depends on $V_i(l)$, we look at the scaling of this property with the size of the volume elements. The resulting scaling law provides an abstract definition of a dimension.

As the first example, let us take a one-dimensional curve as set M. This curve may be embedded in a two-dimensional space. We then select squares of side l as volume elements $V_i(l)$. A property of M is given by the length of the above curve inside such a square. Here one must be cautious, because the length of a curve already implies that the curve has a one-dimensional structure. Therefore, we intend to construct a property that reflects our idea of the length, but is independent of the actual shape of the set M. One convenient way proceeds by reducing the curve (set M) to a set of equidistant points. The distance between these points should be measured on the set M itself. The above decomposition procedure is easily done if the set can be parametrized. Now we take as the property of M the number of points, $N_i(l)$, in each volume element $V_i(l)$. Obviously, this property scales linearly with the length l,

$$N_i(l) \propto l .   \tag{3.47}$$

We point out that this scaling does not depend on the embedding space chosen, as long as M is presented without any projection of one part of the set onto another. Moreover, the scaling law (3.47) does not depend upon the shape of the volume elements.

As a next example, let us briefly regard a plane as set M, the dimensionality of which should be determined. For equidistant points, we end up with the following scaling law:

$$N_i(l) \propto l^2 . \tag{3.48}$$

Well, now we can generalize this concept and define for any set M the dimension by the expected scaling law:

$$N_i(l) \propto l^d . \tag{3.49}$$

Here the dimension d of the set M does not necessarily have to be an integer value. In our scaling hypothesis, we assume (3.49) to hold for any set M when calculating its dimensionality in this way.

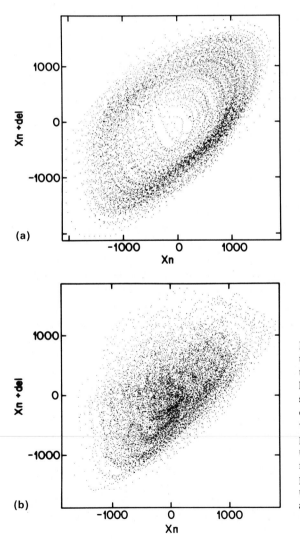

(a)

(b)

Fig. 3.66a, b. Phase portraits of the stroboscopic points for the two distinct cases of chaotic oscillatory states in Fig. 3.64 (same sample and control parameters) obtained by digitizing the partial voltage signal $V_{BC}(t)$ with a sampling frequency of 100 kHz and using a delay time of 50 μs (corresponding to a shift of 5 data points). Note that 1000 arbitrary units relate to about 5 mV signal amplitude

The current definitions of the dimension differ in the chosen properties whose scaling behavior is investigated. It has proven to be conducive to understanding these properties if one decomposes the set into points. In the case of an attractor, the points are usually constructed by taking stroboscopic snapshots of the trajectory. The numerical investigation per se provides stroboscopic points due to the integration over discrete stepwidths. From experimentally obtained signals, they are extracted via the finite sampling frequency of the digitization process applied. In contrast to the Poincaré cross section, these time intervals are usually independent of the system dynamics. The stroboscopic points are, therefore, distributed over the whole attractor. For instance, Fig. 3.66 gives the stroboscopic phase portraits of the two experimental states A and B defined above. Here we have plotted the digitized signal x_n against its delayed analog $x_{n+\text{del}}$. This presentation clearly features the different shape of the two chaotic attractors.

On the basis of such a pointwise method of data preparation, it is possible to define a "natural measure" on the attractor by looking at the relative density of the points. Note that this natural measure follows only from the attractor. It does not depend upon the choice of the embedding space. The natural measure $\mu_i(l)$ of a place i with a volume element $V_i(l)$ containing $N_i(l)$ points is

$$\mu_i(l) = N_i(l)/N , \tag{3.50}$$

where N denotes the total number of points. For a more accurate treatment, one would have to consider the limiting value for N approaching infinity. The measure $\mu_i(l)$ can also be regarded as the probability $p_i(l)$ of finding an arbitrarily chosen point within the volume element $V_i(l)$. By means of that measure or the probability, the current definitions of dimensionality are deducible.

To get the capacity dimension (usually, identical with the Hausdorff dimension), one has to find the minimum number $A(l)$ of volume elements $V_i(l)$, $i = 1, \ldots, A(l)$, that are necessary to cover the whole attractor [3.49]. Since $A(l)$ increases with decreasing l, we expect the following scaling law:

$$A(l) \propto l^{-d} . \tag{3.51}$$

Therefrom, the capacity dimension can be defined as

$$d_c = \lim_{l \to 0} \frac{\log A(l)}{\log(1/l)} . \tag{3.52}$$

To obtain the correlation dimension, one starts with the spatial correlation $C(l)$ as the characteristic property [3.50], which is given by

$$C(l) = \lim_{N \to \infty} \frac{1}{N^2} \{\text{Number of pairs of points } (i,j) \text{ with separation smaller than } l\}$$

$$= \lim_{N \to \infty} \frac{1}{N^2} \sum_{i,j=1}^{N} H(1 - |x_i - x_j|) , \tag{3.53}$$

where H stands for the Heaviside function, x_i and x_j are the coordinates of two points on the attractor. With the scaling hypothesis $C(l) \propto l^{d_{cor}}$, we end up with the correlation dimension

$$d_{cor} = \lim_{l \to 0} \frac{\log C(l)}{\log l} .$$ (3.54)

To obtain the information dimension, one has to ask for the information to be gained if the system is known with the accuracy l. Provided that we have covered the attractor with $A(l)$ volume elements $V_i(l)$ and that we know the resulting probabilities $p_i(l) = N_i(l)/N$, the information can be defined as

$$I(l) = - \sum_{i=1}^{A(l)} p_i(l) \log p_i(l) .$$ (3.55)

This quantity, sometimes called Shannon information, also refers to an entropy. Therefrom, the information dimension can be derived as

$$d_I = \lim_{l \to 0} \frac{I(l)}{\log(1/l)} .$$ (3.56)

To finally come to the generalized dimension, we start from the generalized information introduced by *Renyi* [3.51] as

$$I^q(l) = \frac{1}{(1-q)} \log \sum_{i=1}^{A(l)} p_i^q(l) .$$ (3.57)

The generalized (Renyi) information directly yields the generalized (Renyi) dimension via

$$D(q) = \lim_{l \to 0} \frac{I^q(l)}{\log(1/l)} .$$ (3.58)

This concept was first applied to chaotic attractors by *Grassberger* and *Procaccia* [3.52]. It can easily be shown that $D(0) = d_c$, $D(1) = d_I$, and $D(2) = d_{cor}$ [3.53]. For a lucid explanation, let us put forward the following arguments. In the case of $q = 0$, all terms p_i^q of the Renyi information are equal to unity if $p_i \neq 0$, otherwise equal to zero. Thus, $I^0(l)$ is nothing other than the number of volume elements where $p_i \neq 0$, i.e., the number of volume elements necessary to cover an attractor. In the case of $q = 1$, the Renyi information I^q corresponds to the Shannon information, see [3.52]. Finally, in the case of $q = 2$, the terms p_i^2 of the Renyi information are in keeping with the probability of finding two points within one volume element $V_i(l)$ at once. Then (3.53) and (3.57) coincide except for a minus sign.

When applying these topological concepts of determining a fractal dimension to the two experimental situations A and B defined above, we obtain the following results: $D(0) = 2.6 \pm 0.1$ and $D(1) = 2.5 \pm 0.1$ for state A, $D(0) = 3.6 \pm 0.1$ and $D(1) = 3.5 \pm 0.1$ for state B. For numerical calculation, we have used the procedure established by *Badii* and *Politi* [3.54]. Instead of investigating the scaling of the probability $p(l)$ as a function of l, for a given constant probability p they investigated the minimum size l of a volume element necessary to attain p. Following (3.50), the constant probability $p\ (=N_i/N)$ is obtained by finding k nearest neighbors (where $k = N_i$). Variation of the probability can be accomplished by changing the total number of points, N. To do this, one chooses N arbitrary points out of the whole set of data points available from the digitization process (i.e., $N < 80\,000$). As N is varied at fixed k, the probability $p = k/N$ changes accordingly. With the proposed scaling $p \propto l^d$, we then have

$$\log l \propto (1/d)\log p = -(1/d)\log(N/k) \ . \tag{3.59}$$

The dimension d obtained by this method corresponds to the information dimension $D(1)$. The results of the numerical calculations are presented in Fig. 3.67. The distinct curves refer to different embedding dimensions used for the Takens–Crutchfield reconstruction technique. As long as the embedding procedure has not unfolded the actual shape of the attractor, i.e., some parts of the attractor are still projected onto one another, one finds a changing slope with increasing embedding dimension. Only when the whole structure of the attractor is reconstructed by a sufficiently high embedding dimension, do the slope and, thus, the scaling cease to change anymore under a further increase of the embedding dimension. For the present evaluation, we have exploited embeddings of dimension up to 30. With the help of a linear regression analysis, the slopes of the double-logarithmic plots and, hence, the dimensions of the two experimental data sets have been determined (for details, see [3.55]). A more

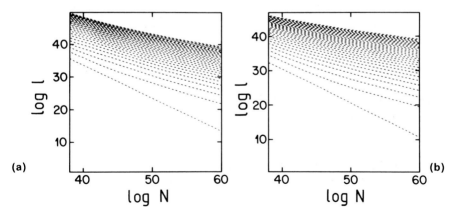

Fig. 3.67a, b. Scaling behavior of the next neighbor distance vs. the number of trial points according to (3.59) for the two distinct cases of chaotic oscillatory states in Fig. 3.64 (same sample and control parameters) obtained by digitizing the partial voltage signal $V_{BC}(t)$ and reconstructing the attractor dynamics with increasing dimension of the embedding phase space. Following the increase of embedding dimensionality, the ensuing constant slopes of the upper curves yield the fractal dimensions of the two chaotic states

comprehensive treatment of the diverse dimensionality approaches and, particularly, of the problems that arise when applying numerical algorithms to
chaotic experimental systems can be found elsewhere [3.56].

That the dimensions calculated for the two representative states A and B do
not have integer values reflects the fractal nature of the underlying attractors.
Such a geometrical property, in general, distinguishes a chaotic dynamics, even
though there are exceptions [3.57]. At this point, we again come across the
interplay between the behavior in phase space and that in time. From the
dimensions evaluated, one recognizes that state A is lower dimensional than
state B. Their noninteger values also provide information on the minimal
number of degrees of freedom inherent to each dynamics. Thus, state A requires
at least three and state B, four independent variables, i.e., state B has more
degrees of freedom than state A. The physical origin of this quantitative
difference between the two experimental situations will be discussed later on.

Let us briefly compare the results of the nonlinear analysis of chaos presented above with that based on a "linear point of view". Recall the power spectra
shown in Fig. 3.65. There the noise level was explained as resulting from an
infinite number of different oscillatory modes, i.e., an infinite number of different
degrees of freedom (Landau–Hopf model of noise). In discrepancy with this
linear interpretation, the dimension analysis gives a totally different picture of
the underlying dynamics, namely, that we have a low-dimensional nonlinear
dynamics of only a few degrees of freedom.

Next we concentrate on the fact that one obtains different values of the
dimensions $D(0)$ and $D(1)$, even though they are calculated from the same
experimental data. Let us consider the case of a uniform set where the probabilities $p_i(l)$ are everywhere equal to

$$p(l) = p_i(l) = N_i(l)/N .$$ (3.60)

Due to

$$\sum_{i=1}^{A(l)} p_i(l) = 1 ,$$ (3.61)

we end up with $p(l) = 1/A(l)$. Following (3.57) and (3.58), the generalized
information and dimension then have the form

$$I^q(l) = [1/(1 - q)] \log[A(l) A(l)^{-q}]$$
$$= \log A(l)$$ (3.62)

and

$$D(q) = \lim_{l \to 0} \frac{\log A(l)}{\log(1/l)} ,$$ (3.63)

respectively. Apparently, all $D(q)$ coincide to one distinct value, the capacity
dimension, see (3.52). Conversely, we can analogously conclude that different
values of $D(q)$ indicate the data set considered not to be uniform in the sense

above, i.e., there exists an inhomogeneous distribution of locally different probabilities $p_i(l)$ on the attractor. If the $p_i(l)$ have different values, one expects the sum over all $p_i^q(l)$ in (3.57) to depend on q. Due to $0 < p_i(l) < 1$, we know that for large (small) values of q the sum of all p_i^q is dominated by values of p_i close to unity (zero). Note that q, thought of as a real number, can also become negative. According to the statement above, for large values of q mainly volume elements containing a lot of points (i.e., densely occupied regions of the attractor) contribute to the evaluation, while for small values of q rather sparsely occupied volume elements are taken into account. Such a vivid interpretation of the generalized dimension $D(q)$ leads to the next method of characterization.

In contrast to the techniques presented hitherto which proceed on the assumption that the whole attractor yields a uniform scaling law, we now focus on the more realistic case of a local scaling behavior [3.58]. We again start with the probability

$$p_i(l) = N_i(l)/N \propto l^{\alpha(i)} \tag{3.64}$$

regarded as a local quantity displaying a local scaling law with the scaling index $\alpha(i)$. It is possible to look at the attractor with respect to its local scaling behavior and, thus, to collect regions which are characterized by the same scaling index. We assume α to vary continuously between two values. Regions on the attractor with the minimal scaling index α_{min} have the maximal probability, whereas, analogously, regions with the maximal scaling index α_{max} are of minimal probability, see (3.64).

As a further step, let us consider the number $n(\alpha, l)$ of regions with the same scaling index. We suppose that, as l gets smaller, more and more regions having the same scaling index α can be found. It is a fundamental feature of fractals that they reveal similar structures when parts of these objects are enlarged again and again [3.48]. Hence, one starts with the following ansatz:

$$n(\alpha, l) = \rho(\alpha) l^{-f(\alpha)}, \tag{3.65}$$

where $\rho(\alpha)$ gives a density and $f(\alpha)$ a scaling index. Equation (3.65) supposes that the number of regions characterized by a distinct scaling index α follows a scaling law with the exponent $f(\alpha)$. In this sense, $f(\alpha)$ is nothing other than a dimension of these regions. For the case of a uniform set, we obtain the trivial result that all regions obey the same scaling and, therefore, $f(\alpha)$ provides the scaling index of all regions, i.e., $f(\alpha) = \alpha$.

We have already mentioned above that different values of α characterize different regions on an attractor. Consequently, the sum over the general momenta of the probabilities p_i^q in (3.57) can be written as

$$\sum_{i=1}^{A(l)} p_i^q(l) = \int d\alpha' \, \rho(\alpha') l^{-f(\alpha')} p^q(\alpha'). \tag{3.66}$$

Instead of covering the attractor with $A(l)$ volume elements, it is now partitioned into regions with the same scaling index α which leads to the new interpretation $\int d\alpha' \, \rho(\alpha')$. With (3.64) one obtains

$$\sum_{i=1}^{A(l)} p_i^q(l) = \int d\alpha' \, \rho(\alpha') l^{-f(\alpha')} l^{q\alpha'} \ . \tag{3.67}$$

When using (3.67) to calculate the generalized dimension $D(q)$ defined by (3.58), one must consider the limiting case for $l \to 0$. Here the integral is dominated by the minimal exponent of l provided that ρ differs from zero. It reads

$$\sum_{i=1}^{A(l)} p_i^q(l) = K l^{q\alpha - f(\alpha)} \ , \tag{3.68}$$

where K means a real constant if α obeys the conditions

$$(d/d\alpha') \{q\alpha' - f(\alpha')\}_{\alpha' = \alpha} = 0 \ , \tag{3.69}$$

$$(d^2/d\alpha'^2) \{q\alpha' - f(\alpha')\}_{\alpha' = \alpha} > 0 \ . \tag{3.70}$$

Equations (3.69) and (3.70) state that the exponent of l is minimal just for the distinct scaling index α. It follows that

$$(d/d\alpha) f(\alpha) = q \ , \tag{3.71}$$

$$(d^2/d\alpha^2) f(\alpha) < 0 \ . \tag{3.72}$$

The scaling function $f(\alpha)$ must, therefore, be a concave curve.

To obtain the generalized dimension $D(q)$, we insert (3.68) into (3.58) and end up with

$$D(q) = \lim_{l \to 0} \frac{1}{(1-q)} \frac{\log(K l^{q\alpha - f(\alpha)})}{\log(1/l)}$$

$$= \frac{1}{(q-1)} [q\alpha - f(\alpha)] \tag{3.73}$$

or

$$q\alpha - f(\alpha) = (q-1) D(q) \ . \tag{3.74}$$

From the definition

$$\tau(q) = (q-1) D(q) \ , \tag{3.75}$$

equation (3.74) yields

$$(d/dq) \tau(q) = \alpha \tag{3.76}$$

and

$$f(\alpha) = q\alpha - \tau(q)$$
$$= q(d/dq) \tau(q) - \tau(q) \ , \tag{3.77}$$

providing the relationship between $D(q)$ and $f(\alpha)$. Equations (3.71), (3.76), and (3.77) clearly demonstrate that the functions $f(\alpha)$ and $\tau(q)$ represent nothing other than a Legendre transformation of each other.

There is another derivation of these functions via a thermodynamical approach (Chap. 4) where one starts by defining a general partition function

$$\Gamma(q, \tau) = \sum_{i=1}^{A(l)} p_i^q(l)/l_i^\tau .$$ (3.78)

Here q and τ denote any real exponents. The size l_i of the volume element V_i can be chosen arbitrarily, but it should be smaller than an upper limit l. Now it depends upon the values of q and τ whether the partition function diverges or converges. There are distinct values where the behavior of Γ just changes from divergence to convergence. Such exponents can be used to characterize a partition function. Hence, one defines a set $\tau(q)$ with

$$\Gamma \to \infty \quad \text{for} \quad \tau > \tau(q) ,$$
$$\Gamma \to 0 \quad \text{for} \quad \tau < \tau(q) .$$ (3.79)

For the special case that $l_i = l$ is constant for all volume elements, the connection to $D(q)$ becomes obvious as

$$\Gamma(q, \tau) = l^{-\tau} \sum_{i=1}^{A(l)} p_i^q(l)$$
$$= l^{-\tau} l^{(q-1)D(q)} .$$ (3.80)

Here we have inserted (3.57) and (3.58). For decreasing l, the partition function fulfills condition (3.79) only if the exponent becomes zero, i.e.,

$$\tau = (q - 1)D(q) .$$ (3.81)

By substituting the sum over the probabilities p_i^q by the integral over the scaling index α as stated in (3.67), the relation

$$\tau(q) = q\alpha - f(\alpha)$$ (3.82)

is obtained again. For a more detailed discussion of this approach and its connection to thermodynamics, we refer to Chap. 4.

Figure 3.68 displays the scaling functions $f(\alpha)$ calculated for the two experimental states A and B. It becomes evident that both attractors possess multifractal structures. Furthermore, all components of the generalized dimension $D(q)$ can be extracted from these curves by simply making use of (3.71) and (3.74). For the cases $q = 0$ and $q = 1$, the corresponding dimensions are easily found. The maximum of the $f(\alpha)$ curve projected onto the ordinate gives the value of $D(0)$, the tangential point with the diagonal, $f(\alpha) = \alpha$, projected onto the abscissa the value of $D(1)$. Since the scaling function $f(\alpha)$ represents nothing other than the spectrum of scaling indices α, the form of the $f(\alpha)$ curve provides information on the extent to which the values of α are scattered. In this sense, the difference $\alpha_{max} - \alpha_{min}$ is a measure of nonuniformity inherent to an attractor. If one compares the spectra of the two experimental situations, the higher-dimensional state B apparently displays a more pronounced spreading of scaling

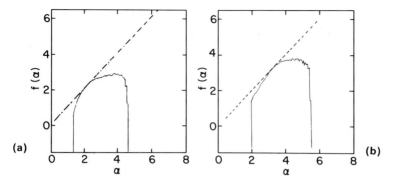

Fig. 3.68a, b. Scaling functions of generalized dimensions according to (3.64–82) for the two distinct cases of chaotic oscillatory states in Fig. 3.64 (same sample and control parameters) obtained by digitizing the partial voltage signal $V_{BC}(t)$ and reconstructing the attractor dynamics with an appropriate embedding technique

indices, i.e., it has a larger fractal inhomogeneity with respect to the lower-dimensional state A. Finally, we briefly direct the reader's attention to another interesting phenomenon to be recognized on the $f(\alpha)$ curve of state B. The graph looks piecewise linear on the left-hand side, it seems to touch the diagonal in more than one point. Therefore, we conclude that for one distinct value of q the attractor possesses two scaling indices simultaneously. Using the language of the thermodynamical formalism for multifractals, the linear part of the scaling function in Fig. 3.68 can be taken as suggesting a phase-transition-like behavior (for a thorough treatment, see Chap. 4).

So far, we have dealt with topological characterization methods featuring the geometrical structure of an attractor in phase space. In what follows, we concentrate on dynamical analysis techniques, in order to quantitatively look at the temporal development of a chaotic state. Following *Eckmann* and *Ruelle* [3.57], one should distinguish between a fractal and a strange attractor. Fractality is determined by a noninteger dimension. They call an attractor strange if its dynamics reflects a sensitive dependence on the initial conditions characterized by the exponential divergence of nearby trajectories under time development. The latter property represents the more fundamental one for describing chaos. Keep in mind that there are exceptional cases where a chaotic attractor is strange but not fractal, i.e., its dimension has an integer value. However, we should remark that such phenomenon can be expected to take place very rarely in dissipative systems, especially if experimental situations are looked at. Nevertheless, other cases may occur where it is important to have access to an independent characteristic of a chaotic system. For example, imagine the possibility that the fractal dimension of an attractor is close to an integer value. With the uncertainty in the calculation of the fractal dimension, it might be impossible to say whether an attractor has an integer or a noninteger dimensionality. Trouble of this kind is rather likely to arise during the analysis of higher-dimensional systems where the applicability of characterization methods becomes noticeably more difficult and, thus, their error limits turn out to be of great weight.

To consider the dynamical properties of chaos, we begin with the phase portraits of the trajectories demonstrated in Fig. 3.69 on the basis of the two experimental states A and B. Now we keep track of the temporal evolution of an attractor point x_n to the succeeding ones $x_{n+1}, x_{n+2}, \ldots$. The time development can be described by the mapping law

$$x_{n+1} = f(x_n) , \tag{3.83}$$

where the function f conceals the system dynamics and the index n indicates the discrete time steps. In order to comprehend the overall dynamical behavior of an attractor, one also compares how any neighboring point y_n proceeds relatively to x_n as time goes by (Fig. 3.70). If y_n stems from a sufficiently small neighborhood, then (3.83) may be solved by a linear ansatz. In analogy with the linear stability analysis and the center manifold theorem discussed in Sect. 3.1, the dynamics around the attractor point x_n reduces to certain characteristic

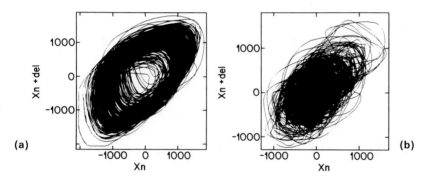

(a)   (b)

Fig. 3.69a, b. Phase portraits of the trajectories for the two distinct cases of chaotic oscillatory states in Fig. 3.64 (same sample and control parameters) obtained according to the stroboscopic plots in Fig. 3.66 except that any two succeeding points of the attractor are connected by a solid line

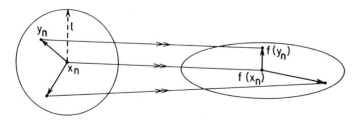

Fig. 3.70. Schematic representation of the temporal development in the neighborhood of an attractor point described by the mapping in (3.83)

directions. Accordingly, small distances l_i in these directions develop for some time as

$$l_i(t) = l_i \exp(\lambda_i t) \ . \tag{3.84}$$

By means of the so-called Lyapunov exponent λ_i it is possible to distinguish among stable ($\lambda_i < 0$), unstable ($\lambda_i > 0$), and central ($\lambda_i = 0$) directions (see also Fig. 3.61). Therefore, a negative Lyapunov exponent describes the attraction of trajectories, and a positive exponent, the exponential divergence of neighboring trajectories. In the case of a zero Lyapunov exponent, the distance between two attractor points remains constant in time. That holds for any two points on the same trajectory. A mathematically more rigorous derivation of Lyapunov exponents can be found in Chap. 4.

Just to briefly give the idea behind the numerical evaluation of the Lyapunov exponents that characterize the two different chaotic states A and B, one starts by calculating a Jacobian matrix of the mapping function f illustrated in Fig. 3.70. The logarithm of its eigenvalues $j_k^{(n)}$, with $k = 1, 2, \ldots$ denoting the dimensionality and n the time step of the attractor point x_n, yields the temporal scaling behavior around x_n. The Lyapunov exponents represent the time average of these exponents,

$$\lambda_k = \lim_{n \to \infty} (1/n) \sum_{s=1}^{n} \log j_k^{(s)} \ . \tag{3.85}$$

It is essential to take the Lyapunov exponents as time-averaged values for $n \to \infty$. A positive exponent $\log j_k^{(s)}$ in the short time interval between n and $n + 1$ does not say anything about chaos. In the concrete case of the two experimental situations considered, we have obtained the following Lyapunov exponents: $\lambda_1 = 0.095 \pm 0.005$, $\lambda_2 = 0.003 \pm 0.005$, and $\lambda_3 = -0.72 \pm 0.02$ for state A, and $\lambda_1 = 0.159 \pm 0.005$, $\lambda_2 = 0.076 \pm 0.005$, $\lambda_3 = -0.021 \pm 0.005$, and $\lambda_4 = -0.77 \pm 0.03$ for state B. For the present evaluation of the relevant exponents, embeddings of dimension up to 11 were used. The results above clearly confirm that both states possess at least one positive Lyapunov exponent, the presence of which suffices to characterize a dynamics as chaotic. In contrast to the analysis of the fractal dimension, it does not play any role whether the exponents are of integer or noninteger values. Practical implications immediately become obvious.

In accordance with the gradually increasing dimensionality, the two chaotic states are further discriminated by a different number of positive Lyapunov exponents ($\lambda_1 > 0$ for state A, $\lambda_1 > 0$ and $\lambda_2 > 0$ for state B), determining mutually independent directions of stretching and folding-over of nearby trajectories in phase space and, thus, reflecting the order of chaos. Adopting the terminology introduced by *Rössler* [3.43], the first state is called "ordinary chaotic" (i.e., three-variable chaos defined by one positive exponent), and the higher-order analog of the second state is called "hyperchaotic" (i.e., four-variable chaos defined by two positive exponents). Indeed, the Lyapunov exponent analysis has the potential to reveal the minimal number of system-immanent degrees of freedom giving rise to the low-dimensional chaotic dynamics. It is, therefore, capable of distinguishing the hierarchical order inherent to chaos, as proposed already at the beginning of this section.

One conjecture that unifies topological and dynamical properties of an attracting set follows from the *Kaplan–Yorke* relationship [3.59]. Accordingly proceeding from the spectrum of Lyapunov exponents, we have calculated the corresponding dimension

$$D = j + \sum_{i=1}^{j} \lambda_i / |\lambda_{j+1}|, \tag{3.86}$$

where j is defined by the condition that

$$\sum_{i=1}^{j} \lambda_i > 0 \tag{3.87}$$

and

$$\sum_{i=1}^{j+1} \lambda_i < 0. \tag{3.88}$$

The results obtained for the two chaotic attractors are the following: $D = 2.1 \pm 0.2$ for state A, $D = 3.3 \pm 0.1$ for state B. Comparison between the Lyapunov dimension D and the information dimension $D(1)$ evaluated independently shows satisfactory agreement within an experimental accuracy of one standard deviation. Obviously, the present accordance between the results obtained from different numerical concepts, i.e., the verification of the Kaplan–Yorke conjecture, yields a self-consistent picture of the complex dynamics underlying the physical situation investigated.

The analysis of the topology of an attractor was based on the definition of a probability corresponding to the natural measure in (3.50). It simply took advantage of the geometrical shape of the attractor. One can, analogously, introduce a probability for the dynamical attractor characterization, too [3.52]. Here we no longer ask for the probability of finding a trial point within a volume element V_i (for simplicity, denoted by i in the following). Instead, we look for the probability that an initial point after n time steps will successively have passed through the volume elements $i_1, \ldots, i_n$. According to (3.57), this probability, written as $p_{i_1 \ldots i_n}$, can also lead to a generalized information of the form

$$\tilde{I}_n^q(l) = [1/(1-q)] \log \sum_{i_1 \ldots i_n} p_{i_1 \ldots i_n}^q(l), \tag{3.89}$$

where l denotes the size of the volume elements and q is an arbitrary exponent. The summation must be performed over all possible sequences of volume elements that can be covered by a trajectory during n time steps. Following *Grassberger* and *Procaccia* [3.52], a generalized entropy can be derived from (3.89):

$$K(q) = \lim_{l \to 0} \lim_{n \to \infty} \tilde{I}_n^q(l)/n. \tag{3.90}$$

The particular cases of q equal to zero and unity, $K(0)$ and $K(1)$, are known as topological entropy and Kolmogorov–Sinai entropy, respectively.

To give a vivid idea of what is meant by the entropy $K(1)$, we ask for the information $I_e(t_1, t_2)$ necessary to determine the course of a trajectory in the time interval ranging from t_1 to t_2 with an accuracy of l. This information divides up into, first, the information necessary to determine the initial point of the trajectory and, second, the information necessary to keep track of the uncertainty resulting from the time development of an initial condition known with accuracy l. It gives rise to the ansatz [3.60]

$$I_e(t_1, t_2) = I_1 + (t_2 - t_1) K^1 . \tag{3.91}$$

Recall that, with the introduction of the Lyapunov exponents, we already have access to a quantity reflecting the degree of uncertainty which is generated by the temporal development. The sum of all positive Lyapunov exponents (indicated symbolically by $\sum^{(+)} \lambda_i$) can be regarded as the growth rate of a volume element spanned by the corresponding eigenvalues. This growth rate causes a drastic increase of uncertainty with time and, therefore, relates to the quantity K^1 in (3.91). It is immediately evident that K^1, in principle, represents nothing other than the Kolmogorov–Sinai entropy $K(1)$ if one looks at the influence exerted by the sum of all positive Lyapunov exponents $\sum^{(+)} \lambda_i$ on the dynamical probability $p_{i_1 \ldots i_n}$ introduced above. The growth rate of the volume element, expressed by $\sum^{(+)} \lambda_i$, leads to an exponential decay of the probability,

$$p_{i_1 \ldots i_n} \propto \exp(-n \sum^{(+)} \lambda_i) . \tag{3.92}$$

Note that the number n of time steps compares with the time interval $t_2 - t_1$. As time progresses, the probability of a trajectory running through a particular sequence of volume elements will decrease under the assumption that positive Lyapunov exponents are present. Upon inserting (3.92) into (3.89) and (3.90), we obtain

$$K(1) \le \sum^{(+)} \lambda_i , \tag{3.93}$$

where usually the equality sign holds [3.57].

The above conjecture combining the Kolmogorov–Sinai entropy with the lower bound of the sum of all positive Lyapunov exponents will be checked when applied to the two limiting cases. First, for a regular (i.e., nonchaotic) dynamics not having any positive exponent, it is trivial that the entropy $K(1)$ must be zero. In terms of the dynamical probability, we find that there remains only one finite sequence of volume elements to be passed again and again by the trajectory. Such a behavior becomes obvious, for example, in the case of a limit cycle attractor. Here $p_{i_1 \ldots i_n}$ does not depend on time (i.e., the number n of time steps). From (3.89) and (3.90) we then obtain $K(1) = 0$. Second, for a totally stochastical dynamics (e.g., in the case of Brownian motion), the entropy $K(1)$ becomes infinite [3.52, 60]. This also fits in with the idea that a stochastical dynamics derives from a continuum of degrees of freedom giving rise to a divergence of $\sum^{(+)} \lambda_i$. Therefore, a chaotic dynamics can be attributed to a finite positive entropy $K(1)$.

Next we deal with a further aspect of the entanglement between the topological and the dynamical analysis that leads to a practical numerical evaluation of the entropies. The Takens–Crutchfield construction of phase space is again taken into account. Starting from the time series of a (one-dimensional) variable $x(t)$, one obtains points x_i in d-dimensional phase space via (Sect. 3.1)

$$x_i = (x(t_i), x(t_i + \tau), \ldots, x(t_i + (d-1)\tau)),\qquad (3.94)$$

where t_i denotes a distinct time event fixing x_i and τ represents an arbitrarily chosen delay time. For the topological analysis, like the calculation of the fractal dimension, we consider distances of points in phase space, in order to decide, e.g., whether or not these points are located inside the same volume element. Conditions fulfilling a defined norm

$$\|x_i - x_j\| < l \qquad (3.95)$$

have to be evaluated. Note that the phase space construction per se embraces distances of the components like $|x(t_j) - x(t_i)|$, $|x(t_j + \tau) - x(t_i + \tau)|, \ldots,$ $|x(t_j + (d-1)\tau) - x(t_i + (d-1)\tau)|$, too. This means that the topological analysis also answers the question of whether two signals $x(t_i)$ and $x(t_j)$ will diverge by no more than the distance l during $(d-1)$ time steps. One can show that the double-logarithmic plot drawn for the determination of the fractal dimension obeys the linear equation $y(l) = D(q) \log l - d\tau K(q)$ [3.60, 61]. Here d gives the embedding dimension. The above relation corresponds to a scaling like $l^{D(q)} e^{-d\tau K(q)}$. The slope of the limiting curve (in a double-logarithmic plot) provides the generalized dimension $D(q)$, while the distance between two curves obtained for different embedding dimensions is proportional to the generalized entropy $K(q)$; for a graphical representation, see Fig. 3.67. With the help of the characterization method outlined above, we have obtained the following results for the two experimental situations considered: $K(0) = 0.09 \pm 0.01$ and $K(1) = 0.09 \pm 0.01$ for state A, $K(0) = 0.15 \pm 0.01$ and $K(1) = 0.15 \pm 0.01$ for state B. According to the evaluation of the generalized dimension, the generalized entropy is computed employing embeddings of dimension up to 30. Just as in the case of the Lyapunov exponents, the values of the entropy are in units of the sampling rate. As expected, both chaotic states possess positive entropies. In particular, the higher-dimensional and more complex hyperchaotic state B displays a larger generalized entropy. Even though it is obvious that the determination of the generalized entropy has been performed totally independent of the determination of the Lyapunov exponents, one finds reasonable agreement between the results obtained from the two methods in terms of the conjecture (3.93), thus providing again a consistent picture in the numerical analysis of the two experimental situations investigated.

In the case of the generalized entropy $K(q)$, we proceed in close analogy with the way the generalized dimension $D(q)$ led to the corresponding spectrum of invariant static scaling indices, $f(\alpha)$. As demonstrated more comprehensively in Chap. 4, it is also possible to derive from the generalized Lyapunov exponents λ_i the corresponding spectrum of invariant dynamical scaling indices, $\Phi(\Lambda)$ [3.62, 63]. Recall that, in the topological analysis, the scaling function $f(\alpha)$ was obtained by replacing the global dimensionality as an average fractal measure of the whole attractor by the distribution of local dimensions or scalings α on the multifractal attractor, see (3.64). In the dynamical analysis, the Lyapunov exponents were accordingly defined in (3.85) as time-averaged attractor quantities. Moreover, in (3.92) we have demonstrated how the Lyapunov exponents can yield a dynamical probability. Following the derivation of $f(\alpha)$, we introduce a local dynamical scaling index Λ via the probability ansatz

$$p_{i_1 \ldots i_n} \propto \exp[-n\Lambda(i_1 \ldots i_n)] , \tag{3.96}$$

where n denotes the discrete time steps. If one defines the time as

$$T = \exp(-n) , \tag{3.97}$$

the dynamical probability reads

$$p_{i_1 \ldots i_n} \propto T^{\Lambda(i_1 \ldots i_n)} \tag{3.98}$$

analogous to the scaling ansatz of the topological probability in (3.64). In keeping with (3.65), we define a corresponding density of regions with the same scaling index as

$$n(\Lambda, T) \propto T^{-\Phi(\Lambda)} . \tag{3.99}$$

On the basis of the arguments used for the static scaling function $f(\alpha)$, the following relations can be deduced for the dynamical scaling function $\Phi(\Lambda)$:

$$K(q) = [1/(q-1)] [\Lambda q - \Phi(\Lambda)] , \tag{3.100}$$

$$\tau'(q) = (q-1) K(q) , \tag{3.101}$$

$$(d/d\Lambda) \Phi(\Lambda) = q , \tag{3.102}$$

$$(d/dq) \tau'(q) = \Lambda , \tag{3.103}$$

$$\Phi(\Lambda) = \Lambda q - \tau'(q) . \tag{3.104}$$

Figure 3.71 displays the scaling functions $\Phi(\Lambda)$ calculated for the two experimental states A and B. Their graphs clearly demonstrate the different dynamical complexity of the attractors. The spreading and the shifted position of the $\Phi(\Lambda)$ spectra are qualitatively seen in $f(\alpha)$. Again, the hyperchaotic state B reflects a more extended distribution of scaling indices compared to the ordinary chaotic state A. Finally, let us focus on a particularity of the dynamical scaling function of state B that has already been indicated in the corresponding $f(\alpha)$ spectrum. As can be clearly recognized from a larger-scaled presentation of the $\Phi(\Lambda)$ curve of state B in Fig. 3.72, there also exists an as well piecewise linear part on the left-hand side of the graph. The concrete interpretation of this

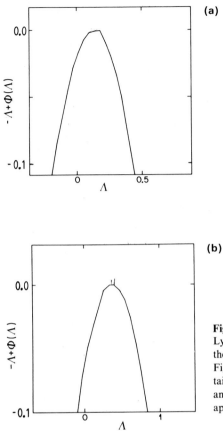

(a)

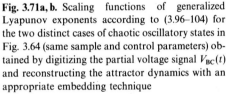

(b)

Fig. 3.71a, b. Scaling functions of generalized Lyapunov exponents according to (3.96–104) for the two distinct cases of chaotic oscillatory states in Fig. 3.64 (same sample and control parameters) obtained by digitizing the partial voltage signal $V_{BC}(t)$ and reconstructing the attractor dynamics with an appropriate embedding technique

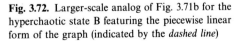

Fig. 3.72. Larger-scale analog of Fig. 3.71b for the hyperchaotic state B featuring the piecewise linear form of the graph (indicated by the *dashed line*)

phenomenon as dynamical phase-transition-like behavior will be given in Chap. 4 on the basis of a thermodynamical ansatz.

3.5.3 Bifurcations to Chaotic States

While, so far, characteristic features of chaos together with their quantitative evaluation on the basis of a representative experimental situation have been in the center of interest, we now emphasize the dynamical possibilities of a nonlinear system to bifurcate from a regular (i.e., ordered) to a chaotic oscillatory state. One of the most exciting results of nonlinear dynamics was the discovery of universal routes leading to chaos, also called scenarios. Such system-independent bifurcations could be verified in a huge variety of computer model simulations and real-world experiments. Today, even though a rigorous proof is still lacking, it is commonly accepted that there exists only a limited number of universal scenarios. The currently best-known ones are the period-doubling cascade to chaos, the intermittent switching between different oscillation modes, and the quasiperiodic routes to chaos. After their inventors, the first type of transition is named the Feigenbaum–Grossmann scenario [3.47, 64] and the second type, the Pomeau–Manneville scenario [3.65]. The third type, the more original Ruelle–Takens–Newhouse scenario [3.66, 67] proclaiming the instability of a quasiperiodic state with more than three incommensurate frequencies, has lately aroused controversial discussion. Providing counterexamples, computer experiments have convincingly demonstrated that such states may be astonishingly stable, even in the presence of nonlinear coupling [3.36, 68]. Therefore, the scenario via suppression of quasiperiodicity through frequency-locking giving prominence to the circle map formalism outlined in Sect. 3.4 is nowadays believed to represent the most important quasiperiodic route to chaos, not least supported by an overall experimental verification. These typical universal scenarios will be discussed more or less comprehensively in the sequel.

We start with a brief introduction to the Feigenbaum–Grossmann scenario. The reader interested in more details is referred to the original literature [3.5, 6, 47, 64] or to the monographs by *Schuster* [3.4], *Devaney* [3.8], and *Leven* et al. [3.11]. In order to explain the period-doubling route to chaos, one again considers the logistic map defined by (3.46). Although up to now we did not deal with features of iterative mappings, it is helpful to briefly look at the topology of chaos in a return map. Recall the discussion of the circle map in Sect. 3.4 and, particularly, the iterative dynamics of the Rössler attractor in Figs. 3.62 and 3.63. For the return map $x_{n+1} = f_\mu(x_n)$, a fixed point obeys the condition $x_{n+1} = x_n$, a q-periodic point $x_{n+q} = x_n$. A chaotic state never comes back to any previous point in phase space. If this were not fulfilled, one would have a q-periodic motion. Thus, the return map of a chaotic state must contain nothing but infinitely periodic points. However, such behavior does not already imply chaos, as we have seen in the case of quasiperiodicity (Sect. 3.4). Chaos in iterative mappings is further distinguished by the exponential divergence of two neighboring attractor points under iteration. Therefore, the existence of at least one unstable (repelling) point located in between two such neighboring points can be concluded. So any bifurcation towards chaos must give rise to a transition from an originally finite number of attractor and repeller points to a set of an infinity of these points. If a single bifurcation increases the attractor points by only a finite number, then there is, of course, a need for an infinite sequence of such kind of bifurcations.

The Feigenbaum–Grossmann scenario just consists of an infinite sequence of pitchfork bifurcations realized by the successive, ever closer-spaced appearance of higher and higher periodic solutions ending in chaos. The bifurcation diagram in Fig. 3.73 displays the iterative dynamics of the attractor points x_n as a function of the control parameter μ obtained from the logistic map (3.46). Obviously, with each bifurcation event doubling the number of attractor points, the sequence scales with the control parameter μ in such a way that period doubling eventually terminates at a finite critical value $\mu_c = 3.5699\ldots.$ The scaling of bifurcation points that becomes evident if one considers the development of forkings when approaching μ_c provides a quantitative measure to characterize the Feigenbaum–Grossmann scenario. Furthermore, there also exists a scaling law predicted with respect to the amplitudes of the attractor points (i.e., the sizes of the forks in Fig. 3.73). A vivid idea for the appearance of chaos above a critical parameter threshold does not follow only from the increase of attractor points towards infinity. We can also take advantage of the following explanation. From (3.46), it is well known that the maximum of the parabolic curve has the value $\mu/4$. Allowing for the discussion of Fig. 3.63, we see that an increase of the control parameter μ places a stronger emphasis on the folding region in the return map. That is, small values of μ favor a larger proportion of convergence in the total dynamics, thus, preventing chaos. On the other hand, for $\mu \geq \mu_c$, divergence becomes the dominating mechanism providing the basic prerequisite of a chaotic dynamics.

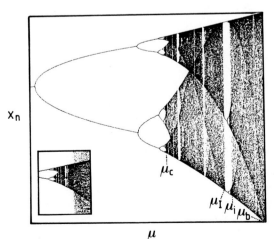

x_n

μ_c

$\mu_I \; \mu_i \mu_b$

μ

Fig. 3.73. Bifurcation diagram of the logistic map (3.46) in the extended control parameter and phase space. The critical points $\mu_c, \mu_I, \mu_i,$ and μ_b indicating different bifurcations to chaos are explained in the text. The inset gives a blow-up of the bifurcation diagram in the vicinity of μ_i

A central point of the Feigenbaum–Grossmann scenario is its universality. We emphasize that the present transition to chaos does not at all require the presence of an iterative dynamics which exactly follows the logistic map (3.46). It is already sufficient to demand the existence of a quadratic maximum. Herewith, all scalings proposed can be evaluated. Concerning an experimental verification of the Feigenbaum–Grossmann scenario, we point out that a tremendous diversity of physical systems (not least, of course, impact ionization semiconductor breakdown [3.69, 70]) has proven the predictions outlined above.

Another very common way a periodic state bifurcates to chaos is via intermittency. This kind of scenario reflects several times an abrupt switching from periodic oscillation to irregular bursts. Figure 3.74 gives the temporal profile of an intermittent signal. Under variation of an appropriate control parameter, the repeated emergence of burst interrupts takes place more and more often such that the time span of the periodic flow (also called the laminar phase) decreases accordingly. Therefore, the chaotic dynamics present during the burst events more and more dominates the whole system behavior. On the other hand, if one looks at the dynamics of a system that intermittently returns from a chaotic to a periodic state, then the participating share of chaos declines gradually. Here we have, so to speak, a slowing down of chaotic bursts. Such a simple description of intermittency lets us think of a saddle node bifurcation on a chaotic attractor characterized by a critical slowing down behavior. For the sake of clarity, recall the corresponding bifurcations from quasiperiodicity to a periodic state and from periodicity to a node, discussed previously. As we will see in the following, there is a close relationship among all these bifurcations. To end up with a more thorough comprehension of intermittency, the three essential model assumptions leading to somewhat different transitions to chaos are discussed successively. They were all introduced by *Pomeau* and *Manneville* [3.65]. An overview can also be found in the monographs by *Schuster* [3.4], *Berge* et al. [3.10], and *Leven* et al. [3.11].

Type I intermittency follows from a close analogy with the saddle node bifurcation on a limit cycle and that on a torus mentioned above. In the present

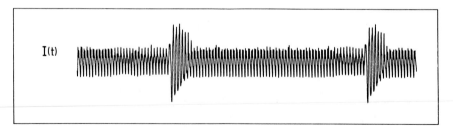

Fig. 3.74. Temporal profile of an intermittent current oscillation obtained from the sample sketched in Fig. 2.4 at the following constant parameters: bias current $I = 4.53$ mA, transverse magnetic field $B = 0.644$ mT, and temperature $T_b = 1.91$ K. The amplitude is in the range of a few μA. The total time span amounts to 240 ms

case, a saddle node bifurcation takes place on a chaotic attractor. We, thus, observe a critical slowing down of chaos expressed via an increasing length of laminar phases. This type of intermittency can easily be explained on the basis of the return map. Here a node stands for a periodic dynamics. Note that, without loss of generality, the node can be interpreted as a q-periodic point if the map $x_{n+q} = f_\mu(x_n)$ is considered. In the return map, we find a node as a crossing of the diagonal where the absolute value of the slope of the mapping function f_μ has to be smaller than unity. As illustrated in Fig. 3.75, the neighboring points are then attracted towards the node under iteration. A saddle is, consequently, characterized by the absolute value of the slope of the mapping function being larger than unity when crossing the diagonal. The return map displays a saddle node bifurcation if the graph of the mapping function changes upon variation of the control parameter as follows. First, for $\mu < \mu_c$, two crossings give rise to a saddle and a node. Second, for $\mu > \mu_c$, no crossing remains (Fig. 3.76). The simplest possible equation describing such a behavior would have the form

$$x_{n+1} = x_n + \mu + x_n^2 , \qquad (3.105)$$

where the saddle node bifurcation occurs for $\mu = 0$ at $x_n = 0$. It provides the lowest-order nonlinear description of the local system dynamics restricted to a small part of the return map (recall the discussion in Sect. 3.2).

Next we turn to the dynamics following the annihilation of the saddle and the node. As can be clearly seen in Fig. 3.77, the concrete form of the return map relative to the nearly tangential diagonal for several time steps causes the

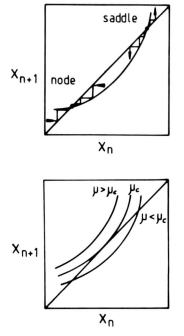

Fig. 3.75. Schematic representation of the development of neighboring points around a saddle and a node via the return map

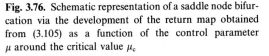

Fig. 3.76. Schematic representation of a saddle node bifurcation via the development of the return map obtained from (3.105) as a function of the control parameter μ around the critical value μ_c

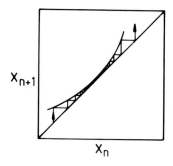

X_{n+1}

X_n

Fig. 3.77. Schematic representation of the iterative dynamics in (3.105) slightly above the critical point of a saddle node bifurcation ($\mu > \mu_c$) via the return map

amplitude of x_n to change only by a small amount, i.e., to stay almost constant. This laminar phase does not terminate before the iterative dynamics has left the narrow channel between the mapping graph and the diagonal. Unless the system is capable of performing any continuous large-amplitude oscillation, it is immediately reinjected into the lower entry of the channel, announcing the beginning of a new laminar phase. Depending upon the prevailing dynamics, one obtains different kinds of critical processes. If, for example, the node describes a fixed point, and the reinjecting dynamics, a periodic oscillation, we have a saddle node bifurcation on a limit cycle. Note that this unusual case cannot, of course, be concluded for a data set obtained from a Poincaré cross section of a continuous dynamics, but for any stroboscopic view of a signal still containing the information on the whole system dynamics. If the node stands for a periodic state and the reinjection is done by a quasiperiodic iterative mapping, we have a saddle node bifurcation on a torus. In the last case, the reinjection results from a chaotic attractor; we then have type I intermittency in its original meaning as introduced by *Pomeau* and *Manneville* [3.65]. For example, Fig. 3.78 illustrates an experimentally obtained time series (a) together with the corresponding iterative dynamics of consecutive signal maxima (b). The laminar phases are clearly visible.

In what follows, we focus on the length of the laminar phases, in short, also referred to as the laminar length. It can be determined by the distance of the graph of the mapping function from the diagonal, called the width of the channel. The smaller this width, the longer the time the system stays in the channel. Since the width of the channel depends on the actual value of the control parameter relative to the critical one (i.e., $\mu - \mu_c$), the time average of the laminar lengths must scale with $\mu - \mu_c$. More precisely, the scaling has again the form $(\mu - \mu_c)^{-1/2}$, which was already found in (3.27) and confirmed experimentally in Fig. 3.78c. For a thorough mathematical derivation of this scaling behavior, we refer to the literature [3.4, 10, 65]. As another characteristic feature, the laminar phases of type I intermittency are expected to exhibit an upper bound in their statistical distribution of laminar lengths [3.10]. The mean peak of the distribution function is located close to this upper bound.

Apart from our experimental semiconductor system (for details, see [3.71]), such a kind of intermittent behavior can also be found in the logistic map. In

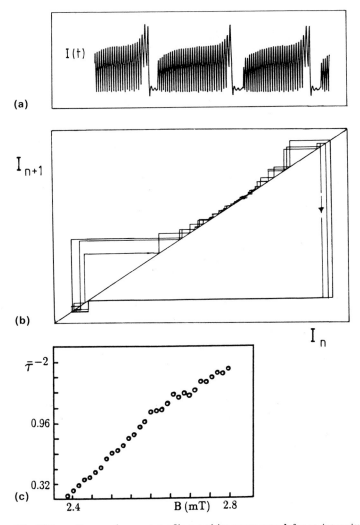

Fig. 3.78a–c. Temporal current profile **a** and its return map **b** for an intermittent oscillatory state obtained from the sample sketched in Fig. 2.4 at the following constant parameters: bias current $I = 0.826$ mA, transverse magnetic field $B = 2.67$ mT, and temperature $T_b = 2.12$ K. The amplitude is in the range of a few μA. The total time span amounts to 30 ms. The scaling of the mean laminar phase length becomes obvious on plotting its inverse square versus the transverse magnetic field **c**

Fig. 3.73, the emergence of intermittency at the control parameter $\mu = \mu_i$ is labelled accordingly. There the system dynamics undergoes a transition from chaos to period 3 (left- and right-hand sides of the bifurcation diagram, respectively). If one considers the mapping of the third iteration, the corresponding return map just displays the behavior stated above.

Type II intermittency will be treated only very briefly, because it has been found to represent a rather exceptional route to chaos. Up to now, the existence of such bifurcation seems to result mainly from theoretical arguments [3.65]. There are not too many hints that it may really be observable in an experiment [3.72, 73]. The dynamics of type II intermittency may be assumed to take place on a torus, giving rise to the description by an angle variable

$$\theta_{n+1} = \theta_n + \Omega \tag{3.106}$$

and by a radius variable

$$r_{n+1} = \mu r_n + \gamma r_n^3 , \tag{3.107}$$

with the parameters Ω, μ, and γ. Apparently, the angle equation does not contain any nonlinearity, in contrast to the radius dynamics. Depending upon whether the control parameter μ is larger or smaller than unity, one observes or does not observe, respectively, a laminar phase (Fig. 3.79). This laminar phase results from a continuously growing radius variable. Together with the angle dynamics, some kind of symmetrically winding up oscillation is expected. The mean length of the laminar phases should scale with $|\mu - \mu_c|^{-1}$.

Type III intermittency, on the contrary, has been observed in a variety of different experimental systems [3.74–76]. The present bifurcation is again characterized by a stable fixed point in the iterative dynamics which becomes unstable. This time, we start from a functional form of the return map displaying a negative slope with the absolute value being smaller than unity (like the one shown in Fig. 3.80a). In the sense of Sect. 3.1 (particularly, the synopsis of different fixed points given in Fig. 3.4), here we have a focus since all neighboring points approach the fixed point via a spiral line. However, if the absolute value of the negative slope of the graph exceeds unity at the crossing with the diagonal (Fig. 3.80b), the focus becomes unstable. Thus, neighboring points slowly spiral away from the formerly stable focus. Such a behavior can easily be described by the equation

$$x_{n+1} = -\mu x_n - \mu x_n^3 . \tag{3.108}$$

The bifurcation occurs when the control parameter μ gets larger than unity. The dynamics is then characterized by an alternating change of the values of x_n with

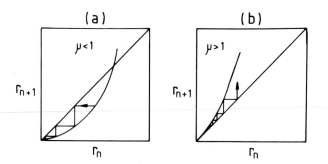

Fig. 3.79a, b. Schematic representation of the iterative dynamics of the radius variable in (3.107) below **a** and above **b** the critical control parameter $\mu_c = 1$ via the return map

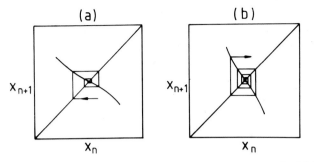

Fig. 3.80a, b. Schematic representation of the development of neighboring points around a focus **a** and an unstable focus **b** via the return map

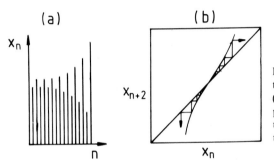

Fig. 3.81a, b. Schematic representation of the iterative dynamics in (3.108) above the critical control parameter $\mu_c = 1$ via the time trace of the variable **a** and the return map of the second iterate **b**

time, as shown schematically in Fig. 3.81a. For simplification, one commonly looks at the dynamics of the second iterate. From Fig. 3.81b, we see that there are two branches where the maxima develop to either larger or smaller values than that of the original fixed point. Obviously, this picture has a shape similar to the one of type II intermittency (Fig. 3.79b), except that now the system evolves in two directions away from the intersection with the diagonal. For the case of a continuous dynamics, the iterative development of the variable x_n may correspond to the sequence of successive maxima. Then one has to expect the maxima of a continuous flow to exhibit alternating behavior.

Figure 3.82 displays three different time traces obtained from our semi-conductor experiment that could be classified as type III intermittency [3.76]. In what follows, we concentrate on this particular experimental situation more comprehensively. The current oscillation shown in Fig. 3.82a is even periodic. With increasing magnetic field, the oscillatory form displays more and more intermittent bursts, as seen in Fig. 3.82b and c. Looking more carefully at the signals, in particular, at the segments that are located immediately prior to each intermittent event (see the inset of Fig. 3.82b), one recognizes the typical time development of the maxima discussed above. Clearly, we have a different evolution of two successive maxima. While the lower maxima marked by circles increase, the upper maxima marked by crosses decrease. From the previous

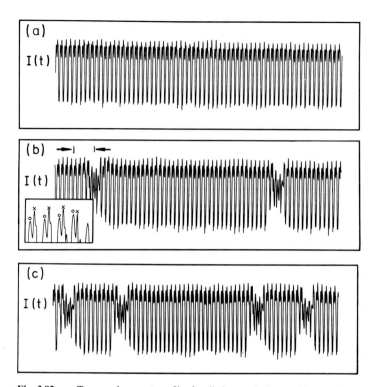

Fig. 3.82a–c. Temporal current profiles for distinct periodic **a** and intermittent **b, c** oscillatory states obtained from the sample sketched in Fig. 2.4 at different transverse magnetic fields $B = 1.922$ mT **a**, $B = 1.950$ mT **b**, $B = 1.958$ mT **c**, and the following constant parameters: bias voltage $V_0 = 12.0$ V and temperature $T_b = 1.88$ K (load resistance $R_L = 2.0$ kΩ). The amplitude is in the range of a few μA. The total time span amounts to 15 ms. The inset in **b** gives a blow-up of the segment of the time trace indicated by the arrows. The development of the maxima $I_n, I_{n+2}, \ldots$ and $I_{n+1}, I_{n+3}, \ldots$ is marked by circles and crosses, respectively

discussion of the iterative dynamics, it would be expected that the maxima start on a distinct level, in order to split up afterwards (Fig. 3.81a). But the time traces presented in Fig. 3.82 result from a continuous system flow. As we already know, this system is governed by at least three different independent variables. Upon measuring one signal, for example, the current, one can anticipate having a projection of the whole dynamics onto one (measurable) observable that may lead to the observed behavior of the maxima. This conjecture was verified by constructing another time signal as follows. We added to the original one a new variable that was obtained by a time shift. Then the two successive maxima of the resulting signal changed their position relative to each other.

Next we discuss the scaling of the mean laminar lengths $\bar{\tau}$ (measured as the number of revolutions [3.71]) with the experimental control parameter. For type III intermittency, it is expected from theory that $\bar{\tau}$ scales proportional to $(\mu - \mu_c)^{-1}$, where $\mu - \mu_c$ stands for the distance of the control parameter μ from the bifurcation point μ_c, the latter being defined by the first occurrence of an intermittent burst. In contrast to this scaling, for type I intermittency a scaling of $\bar{\tau}$ with $(\mu - \mu_c)^{-1/2}$ was predicted. To check the scaling with the experimental control parameter, data were detected for 20 different values of the experimental control parameter (here, the transverse magnetic field). Each time a set of 200 000 points corresponding to about 4000 revolutions was recorded. In Fig. 3.83, the inverse of $\bar{\tau}$ is plotted as a function of the magnetic field. The experimental data are marked by crosses. We find a clear linear scaling of $\bar{\tau}^{-1}$, just as is characteristic for type III intermittency. Only very close to the bifurcation point can a departure be seen. On the one hand, the laminar phases in this region increased drastically (up to 130 revolutions for the mean length), and, thus, the error bars become larger due to the poorer statistics. On the other hand, we expect that close to the bifurcation point noise has an increasing influence, as was found for type I intermittency [3.77]. This problem has also been addressed in Sect. 3.3 in connection with the experimental results of Fig. 3.26.

Let us now consider the distribution $P(\tau)$ of the laminar lengths τ for a constant control parameter. One typical histogram is shown in Fig. 3.84. In contrast to type I intermittency, both type II and type III intermittency still contain an unstable fixed point inside the intermittent region. Assuming a random and uniform reinjection of the system (close to the unstable fixed point) into the laminar phase, one expects more or less elongated laminar lengths that just depend on the distance of the reinjection point from the unstable fixed point. Here we do not have any limitation on the maximum length of the laminar phases, as was the case for type I intermittency. In the rather exceptional case where the system is reinjected directly onto the unstable fixed point, it would take infinite time to escape. On the basis of these arguments, the distribution of

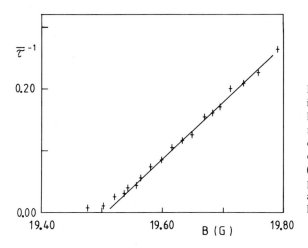

Fig. 3.83. Dependence of the inverse of the mean laminar length on the transverse magnetic field $(1\ \mathrm{G} \cong 0.1\ \mathrm{mT})$ for distinct cases of intermittent oscillatory states as in Fig. 3.82 (same sample and constant parameters). Experimental data are marked by crosses, their linear scaling by a solid line

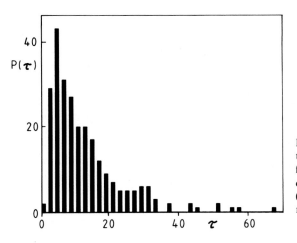

Fig. 3.84. Histogram of the distribution of the laminar lengths for the case of the intermittent oscillatory state in Fig. 3.82c (same sample and control parameters)

the laminar lengths should, therefore, decay as

$$P(\tau) \propto \exp(-m\tau)/[1 - \exp(-2m\tau)]^{3/2}, \tag{3.109}$$

reflecting an exponential decay for $\tau \gg m^{-1}$ and a power-law decay $P(\tau) \propto (4\pi\tau)^{-3/2}$ for $1 \ll \tau \ll m^{-1}$ [3.10]. Note that it only makes sense to evaluate the large laminar lengths, because only these are determined by the local structure of the universal map. The short laminar lengths correspond to reinjections far away from the unstable fixed point and mainly reflect the chaotic reinjecting dynamics for which intermittency does not predict anything. The coefficient m of the exponent in (3.109) should scale linearly with $\mu - \mu_c$, more precisely, it should be equal to $-2(\mu - \mu_c)/\mu_c$.

To gain an appreciation of the exponential decay, it seems to be more suitable to look at the integral distribution of the laminar lengths, namely,

$$N(\tau_0) = \int_{\tau_0}^{\infty} d\tau \, P(\tau)$$

$$\propto \exp(-m\tau_0)/[1 - \exp(-2m\tau_0)]^{1/2}, \tag{3.110}$$

as was proposed by *Manneville* [3.78]. Besides the experimental data obtained for different values of the control parameter (marked by crosses), Fig. 3.85 shows the best fits (solid curves) to these results, calculated on the basis of (3.110). Curve (c) corresponds to the histogram of Fig. 3.84. In Fig. 3.85, the best fit has been achieved with the coefficient $m = 0.090 \pm 0.002$. One can see that curve (a) satisfies only a part of the experimental data (about one third). On the other hand, the quality of the theoretical fit can be estimated only by taking into account the whole integral distribution $N(\tau_0)$; in the inset, therefore, we have shown the corresponding larger-scaled analog, with τ_0 extending up to 350. We clearly see that in the decay region the experimental data follow quite well the predicted scaling of (3.110). Note that, for the sake of adaptation to the

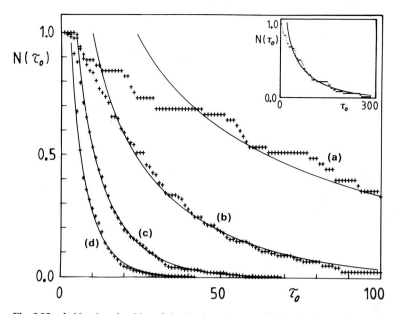

Fig. 3.85a–d. Number densities of the laminar phases with length larger than τ_0 (normalized to unity) for distinct cases of intermittent oscillatory states as in Fig. 3.82 (same sample and constant parameters) obtained at different transverse magnetic fields $B = 1.950$ mT **a**, $B = 1.956$ mT **b**, $B = 1.958$ mT **c**, $B = 1.965$ mT **d**. Experimental data are marked by crosses, and their best fits according to (3.110), by solid curves. The inset gives a larger-scale analog of **a**

experimental situation, always different starting values of τ_0 were chosen (e.g., between 5 and 15 in the case of Fig. 3.85c), in order to ensure evaluation of the proper decay region.

In Fig. 3.86, the coefficient m, which represents the main fitting parameter (in addition to a proportionality factor), is plotted as a function of the experimental control parameter. As expected, the values obtained from the experimental data (marked by crosses) show a definite linear scaling behavior. Again, close to the bifurcation point, a departure from linearity can be seen, as was the case for the scaling of the mean laminar length (Fig. 3.83). These scalings are anticipated to hold only in the vicinity of the bifurcation point. Thus, the departure from the linear scaling is unimportant for large values of m. Keep in mind that here the mean laminar length decreases to less than five revolutions. As predicted above, the decay exponent should be $m = -2(\mu - \mu_c)/\mu_c$. This relation could not be verified. For the present experimental situation, we believe that it is not possible to equate the experimental control parameter directly with the theoretical bifurcation control parameter μ. It appears that the experimental control parameter (the magnetic field) somehow changes the bifurcation control parameter. Due to the smallness of the parameter range, a linear relation between the magnetic field B and the effective parameter $\mu - \mu_c$ is reasonable on the basis of a first-order approximation.

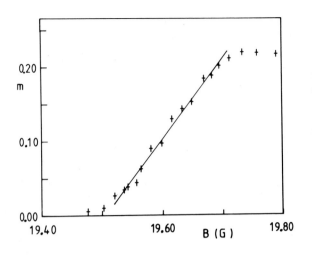

Fig. 3.86. Dependence of the decay exponent according to (3.109) and (3.110) on the transverse magnetic field (1 G $\cong$ 0.1 mT) for distinct cases of intermittent oscillatory states as in Fig. 3.82 (same sample and constant parameters). Experimental data are marked by crosses, and their linear scaling, by a solid line

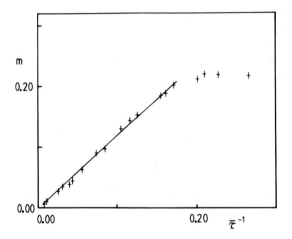

Fig. 3.87. Dependence of the decay exponent according to (3.109) and (3.110) on the inverse of the mean laminar length for distinct cases of intermittent oscillatory states as in Fig. 3.82 (same sample and constant parameters) obtained at the different transverse magnetic fields of Figs. 3.83 and 3.86. Experimental data are marked by crosses, and their linear scaling, by a solid line

If the two scalings of Figs. 3.83 and 3.86 are compared, one finds that the distributions of the experimental data points along the linear fit are highly correlated. Plotting m as a function of $\bar{\tau}^{-1}$ (Fig. 3.87) yields a linear scaling with a slope of 1.2. An interesting point is that this scaling also holds in the region very close to the bifurcation point.

All three types of intermittency discussed so far possess the common characteristic feature that a stable fixed point corresponding to a stable periodic orbit becomes unstable through collision with a saddle. This phenomenon seems to be one fundamental mechanism giving rise to bifurcations and, hence, to instabilities, as we have seen already in the previous sections. Recall, for example, the saddle node bifurcation on a limit cycle or the blue sky catastrophe. In analogy with these bifurcations, the conceptual meaning of a crisis can easily be understood. Here a chaotic attractor becomes unstable due to the collision with a saddle point. Following *Grebogi* et al. [3.79], two different cases are distinguished. First, one has a boundary crisis if a saddle point approaches the attractor from outside such that the stability of the attractor abruptly vanishes, as if it took place at the blue sky catastrophe of a limit cycle. This type of a crisis is found in the logistic map at the control parameter value $\mu = \mu_b \, (= 4)$. For a graphical representation, see Fig. 3.73. Just when μ slightly exceeds μ_b, the states diverge to infinity. Due to the boundary crisis, the system apparently leaves the phase space region of the formerly stable attractor. Second, one has an inner crisis if the collision of a saddle point with a chaotic attractor immediately changes the internal structure of the attractor. In the case of the logistic map, this leads to a sudden blow-up of the chaotic bands. As can be clearly recognized from Fig. 3.73, at the control parameter $\mu = \mu_i$ the threefold band structure extends to a single large interval comprising a continuum of states in phase space (thought of as a vertical cut through the bifurcation diagram). The inset gives prominence to the blow-up of the middle band. Obviously, the inner crisis somehow represents a bifurcation between different chaotic states. A more detailed treatment can be found elsewhere [3.79].

The last universal scenario is provided by the quasiperiodic route to chaos. Thereby, we concentrate on the – from the experimental point of view – more relevant transition via suppression of quasiperiodicity through frequency-locking. From Sect. 3.4, we already know that the circle map formalism has the potential to appropriately model the underlying nonlinear quasiperiodic and mode-locked behavior of the present semiconductor experiment. So far, the dynamics of the circle map has been restricted to parameter values where chaos definitely does not occur.

In analogy with Figs. 3.46 and 3.47, Fig. 3.88 illustrates the graph of the return map $\theta_{n+1} = f(\theta_n)$ obtained from (3.40) in the case when the coupling constant K exceeds unity. However, the supercritical angle dynamics is distinguished by the existence of a local maximum (as well as a local minimum), which is responsible for the approach to chaotic behavior. Beyond the critical line $K = 1$ (Fig. 3.48), only mode-locking or chaos can take place; quasiperiodicity is no longer possible. It follows from the fact that the maximum of the return map gives rise to a folding process, the consequences of which were discussed at the beginning of the present section. Accordingly, we can have either a periodic orbit or a chaotic motion. Moreover, also the requisite stretching mechanism is manifest in some parts of the mapping graph where the slope has a value larger than unity. Evidence of the chaotic dynamics may also be found when considering two neighboring points θ_m and $\theta_{m'}$ in Fig. 3.88. For the subcritical case $K < 1$, the order $\theta_m < \theta_{m'}$ is not affected by the iteration process (Figs. 3.46

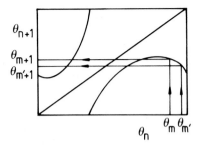

Fig. 3.88. Schematic representation of the return map obtained by plotting the iterative dynamics of the angle variable θ in (3.40) at supercritical coupling constant $K > 1$. Due to the local maximum of the graph, two neighboring points $\theta_m < \theta_{m'}$ develop to a reversed order $\theta_{m+1} > \theta_{m'+1}$ under iteration

and 3.47). But for the supercritical case $K > 1$, such an ordering can be reversed due to the local maximum of the graph and the existence of two different slopes on the left- and right-hand sides of the maximum (Fig. 3.88).

So far, we have demonstrated the possible appearance of chaos in a particular control parameter range of the circle map. The next question to be posed is: Which of the universal routes to chaos are expected to take place? If one starts from a locking state, only the common scenarios, like the period-doubling cascade and intermittency, can be found [3.36]. Of course, at the critical line $K = 1$ the system will nearly always proceed to a locking state again, since there – mathematically speaking – the measure of the locking states is unity (recall the discussion of the Arnold tongues in Sect. 3.4). Nevertheless, there remain infinitely many distinct points on the critical line that favor a direct transition from quasiperiodicity to chaos, see Fig. 3.48. Much effort has been made in systematic investigations of the transition point at the critical coupling constant $K = 1$ and the winding number W [defined by (3.41)] equal to the golden mean ($\sqrt{5} - 1)/2$ [3.80, 81]. As already outlined in Sect. 3.4, the golden mean represents the most irrational number, in the sense that it is farthest from any simple rational number following the hierarchical Farey tree ordering (Fig. 3.49). This immediately corresponds to locations in the phase diagram of the circle map (Fig. 3.48) that are farthest from any larger locking state.

Just to analyze the transition point at the golden mean, it was customary to apply an externally driven nonlinear dynamical system. Here the amplitude and the frequency of the forcing provide two important experimental control parameters that are closely related to the model parameters K and Ω. Due to the possibility of a direct parameter adjustment from outside, the quasiperiodic approach to chaos can be verified experimentally with extremely high accuracy [3.82–86]. Some remarkable features have been used to characterize the particular transition point. For example, the power spectrum displays a self-similar pattern of the mixing components [3.82]. The Poincaré cross section becomes wrinkled and highly nonuniform due to the onset of chaos. The corresponding multifractal structure of the underlying attractor has been evaluated convincingly through the scaling function of generalized dimensions [3.83–86].

In what follows, we report on the experimental observation of self-generated (i.e., without any external periodic forcing) quasiperiodic behavior of our semiconductor system when arriving at criticality. Therefore, the stroboscopic data of the current signal already shown in the return map of Fig. 3.58a have been examined. The present state was selected taking into account the just beginning wrinkling in the structural form of the return map and the appearance of certain subharmonics in the power spectrum, unless these made the dynamics too noisy. We postulate that the circle map formalism is adequate to model the results of our experiment. Indeed, it will be demonstrated that for a distinct set of parameter values reasonable agreement between experimental data and theoretical calculations comes to light. A more detailed treatment of this study can be found elsewhere [3.87].

First, we start with the numerical evaluation of the experimental data, the results of which are given in the left columns of Figs. 3.89 and 3.90 and the solid curves in Fig. 3.91 (all labelled by the symbol E). The time series of the current signal I_n is shown in Fig. 3.89a, the phase portrait I_n vs. $I_{n+\tau}$ corresponding to the stroboscopic points in Fig. 3.89c. Note that the portrait for $\tau = 1$ (i.e., the return map) looks almost identical to that for $\tau = 2$, with mirror symmetry about the diagonal. If one repeats the procedure for $\tau = 4, 5$, and 6, the three new portraits are almost identical to those for $\tau = 1, 2$, and 3, respectively. This period 3 cycle will be discussed later on in the theoretical part. Also, the histogram of the experimentally observed current signal I_n is plotted in Fig. 3.89e.

The time series consists of two seemingly different partial sequences (Fig. 3.89a), namely, a poorly periodic part at the beginning (t and n from 1 to 3000) followed by a more definite periodic part at the end (t and n from 3001 to 4033). The whole time series was Fourier analyzed. The resulting power spectrum is presented in Fig. 3.90a. Upon further Fourier analyzing both partial sequences separately, one obtains surprisingly similar results. The power spectrum has a peak at the frequency $\omega = 693$ with a broad distribution, superposed on which are several sharp peaks of equal spacing of $\Delta\omega = 21$ (for conversion to actual frequency values, see the figure legend). The correlation function displayed in Fig. 3.90c and e is calculated using two methods, namely, the Fourier transform of the power spectrum,

$$C(n) = \text{FT}(P^I(\omega)) \,, \tag{3.111}$$

and the standard expression for the correlation function,

$$C(n) = \sum_{i=1}^{N} (I_i - \bar{I})(I_{i+n} - \bar{I}) \bigg/ \sum_{i=1}^{N} (I_i - \bar{I})^2 \,, \tag{3.112}$$

where $\bar{I} = \sum_{i=1}^{N} I_i/N$ with the total number N of stroboscopic current data points I_i. Both methods yield the same result. The experimental data analyzed

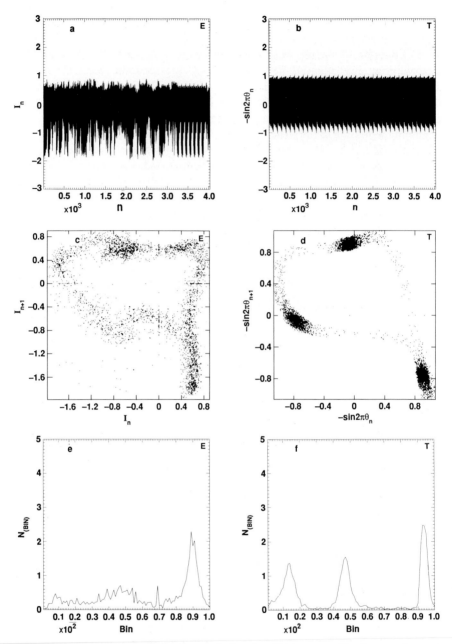

Fig. 3.89a–f. Temporal profiles **a, b**, phase portraits **c, d**, and histograms **e, f** for the case of a quasi-periodic oscillatory state obtained from the iterative dynamics of the current signal I_n in Fig. 3.58 (same sample and control parameters) and the current variable $I_n = -\sin 2\pi\theta_n$ calculated from (3.40) at the constant model parameters $K = 1$ and $\Omega = 0.66355$. The experimental curves **a, c**, and **e** in the left column are indicated by the label E, and the theoretical curves **b, d**, and **f** in the right column, by the label T. The current values are plotted in arbitrary units, and the time values, relative to the number of sampling intervals performed

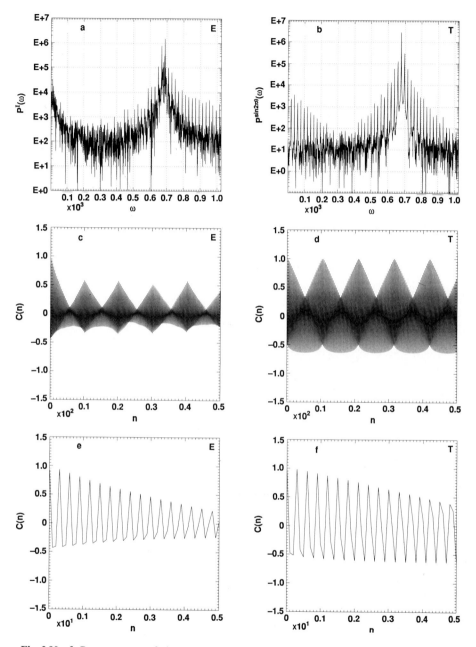

Fig. 3.90a–f. Power spectra **a, b**, long-range correlation functions **c, d**, and short-range correlation functions **e, f** for the case of a quasiperiodic oscillatory state analogous to Fig. 3.89 (same sample, model equation, and control parameters) obtained as outlined in the text. The experimental curves **a, c**, and **e** in the left column are indicated by the label E, and the theoretical curves **b, d**, and **f** in the right column, by the label T. The spectra are plotted in arbitrary units. Note that the frequency $\omega = 1024$ relates to 1.05 kHz, representing half of the sampling frequency

exhibit two types of correlations, a long-range correlation characterized by the slow modulations (Fig. 3.90c) and a short-range correlation characterized by fast oscillations (Fig. 3.90e). In both cases, positive and negative correlations are observed.

Finally, the multifractal structure of the data is shown in Fig. 3.91. We, therefore, started again from the generalized partition function $\Gamma(q, \tau)$ defined by (3.78) and solved for the functional relationship between the exponents q and τ. Taking further into account (3.75–77), the generalized dimension $D(q) = \tau(q)/(q - 1)$, the scaling exponents $\alpha = (d/dq)\, \tau(q)$, and their spectrum $f(\alpha) = q\alpha - \tau(q)$ are subsequently calculated. The experimental results of these quantities are shown in Fig. 3.91a, b, c, and d, respectively (solid curves with label E).

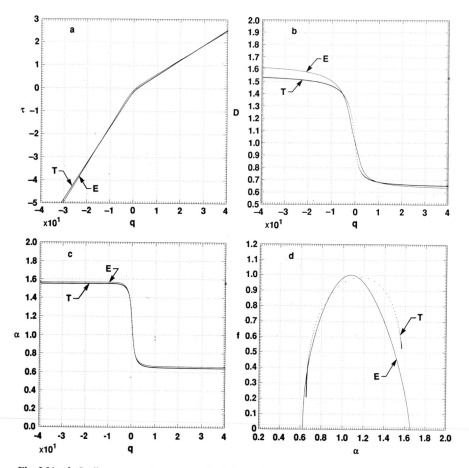

Fig. 3.91a–d. Scaling exponents **a, c**, generalized dimensions **b**, and their scaling functions **d** according to (3.64–82) for the case of a quasiperiodic oscillatory state analogous to Fig. 3.89 (same sample, model equation, and control parameters) obtained as outlined in the text. The experimental (*solid*) curves are labelled E, and the theoretical (*dotted*) curves, T

Next we turn to the interpretation of the experimental data using the physics of the circle map theory. In periodically driven nonlinear dynamical systems, there are two distinct frequencies, the external driving and the intrinsic system frequency (Sect. 3.4). It then immediately becomes obvious that the circle map is an appropriate model to describe the data. For the case of our undriven semiconductor experiment, we developed a method of internal triggering where a self-generated periodic signal plays the role of the clock (Fig. 3.38). Thus, an attempt to make use of the circle map formalism for the present experimental situation may be justified. We explicitly found that (3.40) with the parameters $K = 1$ and $\Omega = 0.66355$ yields results that agree well with the experimental data. The theoretical calculations are shown in the right columns of Figs. 3.89 and 3.90 and the dotted curves in Fig. 3.91 (all labelled by the symbol T).

When applying the circle map (3.40) to analyze our data, we must first decide whether θ or $\sin 2\pi\theta$ gives the proper variable to be identified with the current I. Here it should be pointed out that this choice is not arbitrary. In particular, power spectra are extremely sensitive to this choice. We have obtained the self-similar pattern of the mixing components expected when passing criticality at the golden mean ($K = 1$ and $\Omega = 0.60608$) in a double-logarithmic plot of $P^\theta(\omega)/\omega^2$ (i.e., for the case of the variable θ) and $P^{\sin 2\pi\theta}(\omega)/\omega^4$ (i.e., for the case of the variable $\sin 2\pi\theta$). In the first instance, the current I is chosen to be like $\sin 2\pi\theta$. Starting from this identification, we set $K = 1$ and search for an appropriate value of the parameter Ω, in order to provide a reasonable fit to the power spectrum of our data (Fig. 3.90a) by adapting the theoretical winding number to the experimental one. For $\Omega = 0.66355$, one gets the spectrum in Fig. 3.90b. This value of Ω is reasonable, because the phase portraits exhibit a period of three, as mentioned earlier. The value of $K = 1$ deserves comment. A continuous scan of the parameter K with $\Omega = 0.66355$ produces interesting results. For example, the spacing between the sharp peaks in the power spectrum at $K = 0.95$ is twice as large as the value $\Delta\omega = 21$ obtained from the experimental data (Fig. 3.90a). When K exceeds unity, these peaks disappear. Only at $K = 1$ do we recover the exact value $\Delta\omega = 21$ (Fig. 3.90b). The time series of the variable $-\sin 2\pi\theta$ is given in Fig. 3.89b. Obviously, one obtains a striking similarity of the experimental (Fig. 3.89c) and theoretical (Fig. 3.89d) phase portraits. Previous workers in this field have observed that the $f(\alpha)$ curve found experimentally can be well described by the results of the circle map, while the experimental phase portrait is twisted and contorted in a complicated way and looks nothing like the orbits of the circle map. In our analysis, because we use $-\sin 2\pi\theta$ as the proper variable to represent the current, the phase portrait of the circle map displays a form remarkably similar to the experimental one. The broadening of the attractor becoming visible in Fig. 3.89c also deserves some comment. Keep in mind that our experiment does not utilize any external frequency to drive the system investigated (commonly performed with a typical precision of 10^{-5} [3.85], for example). We have simulated this limited precision imperfection by adding a random noise of about 2% to the generated $-\sin 2\pi\theta_n$ time series. Therewith, one obtains a broadening of our generated attractor as shown in Fig. 3.89d. Moreover, the low-amplitude noise can be found superimposed on the power spectrum (Fig. 3.90b). The minus sign included in our variable identification $-\sin 2\pi\theta$ simply refers to an arbitrary phase assignment, in order to keep the orientation of the phase portrait consistent with the experimental data. The histogram of the theoretically generated current variable is plotted in Fig. 3.89f. Apparently, the main peaks coincide with those of the experimental histogram (Fig. 3.89e). The correlation function of the generated time series, calculated using the two methods described above, is juxtaposed to the relating experimental one (Fig. 3.90c and e) in Fig. 3.90d and f, respectively. Both long-range and short-range correlations are well represented theoretically.

Finally, the graphs of the functions $\tau(q)$, $D(q)$, $\alpha(q)$, and $f(\alpha)$ obtained from the circle map are given in Fig. 3.91 (dotted curves with label T). There is excellent agreement with the experimental results (solid curves with label E) for large values of q (right-hand side of Fig. 3.91a–c). This range corresponds to the low α (high density) part of the attractor (left-hand side of Fig. 3.91d). However, for small and negative values of q and the high α (low density) portion, respectively, the disagreement becomes appreciable, as expected. It is a common fact that the error bars increase in the scaling region in question.

Previous experiments testing the applicability of the circle map have adjusted their control parameters in such a way as to produce the critical value of the coupling constant ($K = 1$) and the distinct winding number of the golden mean. In our experiment, on the other hand, both K and Ω are deduced by fitting one set of experimental data (the power spectrum). It is surprising that all aspects of the experimental quantities studied are well modelled by the circle map. Furthermore, we find that, with the assignment of $-\sin 2\pi\theta$ as the current, not only the experimental and theoretical sets are the same, from the metric point of view, but also the configuration of the set is similar. Such a result provides a firmer foundation for the use of the circle map formalism to describe the type of nonlinear dynamical system considered here.

3.6 Spatio-Temporal Dynamics

In previous sections, we have shown that the present semiconductor system is capable of eliciting a wide variety of different dynamical states upon slightly changing the relevant experimental control parameters (see, for example, the phase diagram in Fig. 3.41). One important result of our nonlinear analysis via distinct characterization methods was that the minimum number of independent system variables determining the degree of complexity of a dynamical state and, thus, providing the basis of its mathematical description could be deduced systematically. In the concrete case of our experimental system, we have disclosed low-dimensional dynamical behavior embracing up to four degrees of freedom. Therefore, any attempt at theoretical modelling requires at least the corresponding number of system variables (e.g., four variables to describe the most complex dynamics, i.e., hyperchaos).

The crucial question concerning where these independent variables come from has not yet been addressed. In principle, there are two possibilities. Either they result from different particles (e.g., mobile charge carriers, phonons) which obey a homogeneous distribution across the sample medium and interact with one another in a nonlinear way. Or they result from different spatially localized and mutually interacting subsystems, each of which possesses its own set of variables. The latter case is called a synergetic system, following the terminology introduced by *Haken* [3.3].

While the treatment of a spatially nonlocalized system always necessitates a new theoretical model if one finds a dynamical state with higher dimensionality, the simple dynamics of any subsystem suffices to explain a whole "zoo" of dynamical possibilities that can arise in a coupled multicomponent synergetic system. Here the gradual increment of actively participating degrees of freedom leading to a growing diversity of low-dimensional dynamical modes is founded in the enslaving of subsystems to more or less extended cluster configurations or their changing mutual relationship. Such an approach reflects a close connection to the well-known models of spin systems. Imagine that any nonlinear element (for example, the logistic map) is placed on each lattice point and, further, that neighbors are interacting with one another. Recent computer simulations have demonstrated convincingly how whole regions of subsystems undergo an enslaving process and give rise to a coherent spatio-temporal dynamics of low dimensionality, the detailed structure of which sensitively depends upon the control parameters applied [3.88–91].

For the case of our semiconductor experiment, Sect. 2.4 has already shed some light on the interwoven relationship between the temporal and spatial phenomena of the nonlinear dynamics. With the help of low-temperature scanning electron microscopy, we have proved that the self-generated oscillatory behavior results from breathing current filament structures, the boundary regions of which are constituting localized oscillation centers. The long-range coupling among these spatially separated subsystems takes place via lattice heat diffusion. Beyond these – from the viewpoint of the underlying system-immanent semiconductor physics – protruding scientific findings, the present section is devoted to the more general properties of the system-independent nonlinear dynamics, particularly, its spatio-temporal aspects. Therefore, we investigate the spatial components of the dynamics by means of local contact probes attached to the broad surface of the germanium crystal (see the sample geometry in Fig. 2.4). The experimental set-up is sketched in Fig. 3.92. Analogous to the arrangement discussed in Sect. 2.3, a d.c. bias voltage V_0 supplies the series combination of the sample and the load resistor R_L. In addition to the total voltage V, the partial voltage drops V_i ($i = 1, 2, 3$) can be measured independently along the sample. They provide spatially localized system response of the corresponding sample parts. On the other hand, the resulting

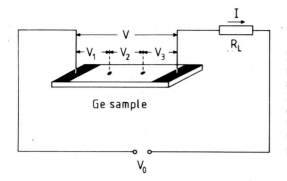

Fig. 3.92. Scheme of the experimental arrangement applied to the semiconductor sample sketched in Fig. 2.4. Note that the partial voltage drops V_1, V_2, and V_3 as well as the total sample voltage V correspond to the voltages V_{AB}, V_{BC}, V_{CD}, and V_{AD}, respectively

current I recorded via the voltage drop at the load resistor gives the whole integral response of the semiconductor system. We point out that phase portraits constructed from different partial voltage drops not only reflect a two-dimensional projection of the attractor, but can also be interpreted as direct measurement of the temporal coherence between distinct localized regions of the sample investigated. If there were no spatially inhomogeneous dynamics present, the time signals of these partial voltage drops would have to be proportional to each other. Then the resulting phase portraits would represent no more than a straight line. From this, we conclude that phase portraits of two distinct partial voltage signals displaying any structural difference from a straight line indicate the existence of different dynamical behaviors in different spatial regions of the system. The simplest case would be a pure phase shift.

Figure 3.93 gives a sequence of phase portraits that are constructed from different pairs of partial voltage drops V_i ($i = 1, 2, 3$). The corresponding dynamics ranges from a stable fixed point to hyperchaos. It immediately becomes obvious how the oscillatory behavior in distinct sample parts gradually decorrelates with increasing degrees of dynamical complexity (i.e., with increasing dimensionality of the overall system). For example, the phase portrait of the periodic state (Fig. 3.93b) has nearly the form of a straight line due to the strong coherence between both partial voltage signals. Note that the sign of its slope refers only to the signal polarity used for presentation. In the case of hyperchaos, no order can be recognized any more. The phase portrait (Fig. 3.93e) displays trajectories running in all directions. Hence, one partial voltage may increase, decrease, or stay constant, while the other changes in one particular way. That is, both signals develop totally independently of one another in time. Compared to the hyperchaotic behavior, the phase portrait of the chaotic state (Fig. 3.93d) still indicates some measure of correlation between the corresponding two sample parts. It may be worth noting that the numerical analysis of those partial voltage signals with respect to their fractal dimension and their Lyapunov exponents yielded no significant deviations from the present results. So far, we conclude that there exists a close connection between structural changes in phase space and in the number of independent variables. The apparent loss of spatial coherence among different sample parts indicates the break-up of our

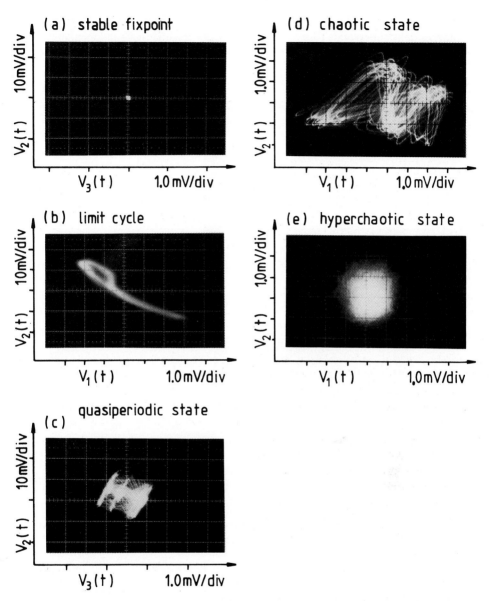

Fig. 3.93a–e. Phase portraits for distinct nonoscillatory **a**, periodic **b**, quasiperiodic **c**, chaotic **d**, and hyperchaotic **e** oscillatory states obtained from the experimental set-up sketched in Fig. 3.92 by plotting pairs of partial voltage drops V_i $(i = 1, 2, 3)$ against one another at different sets of parameter values: **a** bias voltage $V_0 = 2.220$ V, transverse magnetic field $B = -0.4$ mT, and temperature $T_b = 4.2$ K (load resistance $R_L = 100\,\Omega$); **b** $V_0 = 2.184$ V, $B = 0$ mT, and $T_b = 4.2$ K $(R_L = 100\,\Omega)$; **c** $V_0 = 2.220$ V, $B = 0.21$ mT, and $T_b = 4.2$ K $(R_L = 100\,\Omega)$; **d** $V_0 = 2.145$ V, $B = 3.15$ mT, and $T_b = 4.2$ K $(R_L = 100\,\Omega)$; **e** $V_0 = 2.145$ V, $B = 4.65$ mT, and $T_b = 4.2$ K $(R_L = 100\,\Omega)$

multicomponent semiconductor system from coherently coupled into more independent subsystems (the partial voltages V_i become more independent). In this way, new actively participating degrees of freedom are gained gradually.

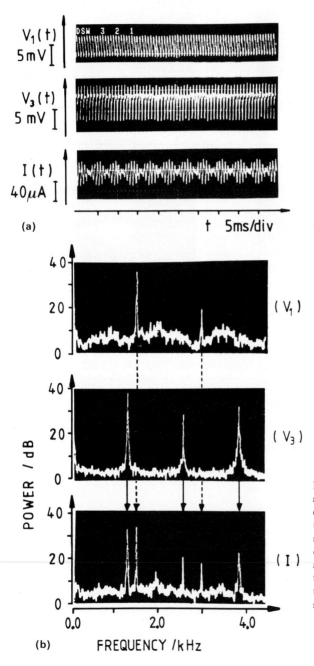

Fig. 3.94a, b. Temporal profiles **a** and power spectra **b** for the case of the quasiperiodic oscillatory state in Fig. 3.36 (same sample and control parameters) obtained from the partial voltage drops $V_1(t)$ and $V_3(t)$ and the current signal $I(t)$ according to the experimental set-up sketched in Fig. 3.92

The simplest access to the prevailing spatio-temporal dynamics can be achieved by looking at the quasiperiodic state. As already mentioned in Sect. 3.4, its description requires at least three independent system variables. Then one may already observe some typical nonlinear effects. On the other hand, quasiperiodicity is very easily demonstrated by means of a spatial structure embracing two localized oscillators. Figure 3.94 shows the measurement for a specially chosen parameter set where the simultaneous presence of two competing oscillation centers inside one semiconductor sample was directly verified. The different time traces of the partial voltages V_1 and V_3 together with that of the integral current I are plotted in Fig. 3.94a. The power spectra corresponding to these signals can be seen in Fig. 3.94b. Both the time traces and the power spectra of the partial voltage drops bear witness to almost perfect periodic oscillations distinguished by different frequencies, whereas the current clearly displays quasiperiodic behavior. The arrows drawn between the power spectra clearly indicate that the current signal (reflecting two apparently incommensurate frequencies and their harmonics) is composed of two portions originating from the partial voltage drops. These experimental findings give rise to the interpretation that quasiperiodic current flow may result from the localized periodic oscillatory behavior of at least two spatially separated sample parts. Such spontaneous pattern-forming processes are commonly denoted as synergetic spatio-temporal dynamics that arise in a self-organized way inside the system considered (i.e., without being favored by external excitations or inhomogeneous boundary conditions).

Next we present further striking evidence that there are localized oscillation centers obtained by applying the voltage bias only over a part of the sample. A possible experimental set-up is sketched in Fig. 3.95. Provided that we have an oscillation center only due to the local conditions of the sample, its dynamical behavior should also be observable just when focusing our investigations solely on the corresponding semiconductor region. On the other hand, if the local dynamics results from an interaction of this part with the whole sample, a somewhat different behavior can be expected for the present experimental situation. In what follows, we discuss these ideas according to specific experimental findings.

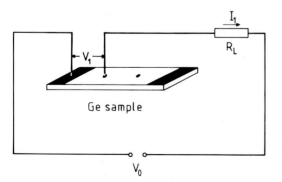

Fig. 3.95. Scheme of a particular experimental arrangement differing from that in Fig. 3.92 in that only part 1 of the sample is biased

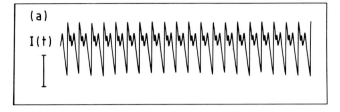

(a)

I(t)

(b)

V₁(t)

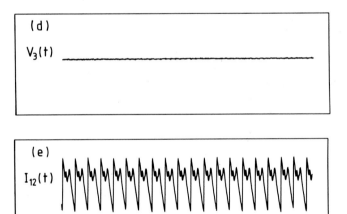

(c)

V₂(t)

(d)

V₃(t)

(e)

I₁₂(t)

Fig. 3.96a–h. Temporal current and voltage profiles for two corresponding periodic oscillatory states obtained from the experimental set-ups sketched in Figs. 3.92 and 3.95 by biasing either the whole **a–d** or different parts **e–h** of the sample arrangement, respectively, at the following constant parameters: bias current $I = 3.2004 \pm 0.0002$ mA, transverse magnetic field $B = 0.171$ mT, and temperature $T_b = 1.98$ K. All signals are plotted on the same vertical and horizontal scales. Their amplitude refers to the 2.5 mV voltage bar indicated in **a**. The current was measured as voltage drop at a 10 kΩ load resistor. The total time span amounts to 50 ms

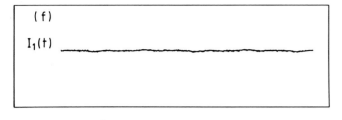

Fig. 3.96f–h

First, the problem arises as to what the appropriate control parameters are, in order to ensure that a specific part of the sample is forced to identical working conditions under the different experimental situations outlined in Figs. 3.92 and 3.95. Of course, the magnetic field and the temperature do not raise any difficulties. For the case of electric conditions like the bias voltage, the load resistance, and the current, it has turned out that the time-averaged current flow I (also called the bias current) represents the best parameter. Keep in mind that the current will be labelled by I_i or I_{ij} if application of the bias voltage V_0 is restricted to the sample part i or the sample parts i and j $(i, j = 1, 2, 3)$, respectively. The partial voltages V_{ij} corresponding to the latter case are specified accordingly. Note that the time-averaged current in any case must flow through the semiconductor sample, no matter how large the resistance of the ohmic contacts is, whereas the voltages can be affected sensitively by different contact resistances.

We start with the discussion of measurements performed while our experimental system is in a periodic state. Figure 3.96a–d provides the time signals $I(t)$, $V_1(t)$, $V_2(t)$, and $V_3(t)$, respectively, obtained when the bias voltage is over the whole sample (Fig. 3.92). It is clearly evident that the periodic oscillation can be detected mainly inside the sample parts 1 and 2. For comparison,

Fig. 3.96e–h presents the corresponding time signals $I_{12}(t)$, $I_1(t)$, $I_2(t)$, and $I_3(t)$, respectively, obtained when separately biasing different sample parts (Fig. 3.95). Only in the case of the arrangement investigating part 1 together with part 2 does there result a periodic signal $I_{12}(t)$ similar to $I(t)$, as expected. The almost identical shape of the corresponding phase portraits in Fig. 3.97 reconstructed from the time series of the currents I (a) and I_{12} (b) gives rise to the conclusion that in the present state being considered the total system dynamics is located inside the sample parts 1 and 2. Further support is provided by the magnetic field dependence of the frequency of those current signals, displayed in Fig. 3.98. Thus, we have found that the local dynamics of sample part 3 does not essentially contribute to the prevailing periodic state. In other words, it would be possible to cut off part 3 from the sample configuration without exerting any influence on the overall dynamics. However, dividing up the sample into further subregions (i.e., part 1 and part 2), one receives a new type of dynamics, obviously no longer correlated with the former one [compare the signals $I_2(t)$ and $V_2(t)$ in Fig. 3.96]. Not least this finding confirms our notion that the periodic current signal $I(t)$ in Fig. 3.96 results from the interaction of oscillatory subsystems that are located in part 1 and part 2 of the sample.

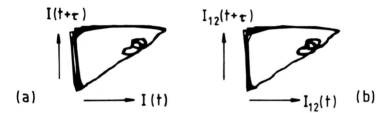

Fig. 3.97a, b. Phase portraits for the case of the two corresponding periodic oscillatory states in Fig. 3.96 (same sample and control parameters) obtained from the current signals $I(t)$ and $I_{12}(t)$ using a delay time of 100 μs

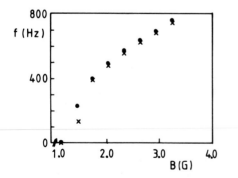

Fig. 3.98. Dependence of the frequency on the transverse magnetic field (1 G $\simeq$ 0.1 mT) for the case of the two corresponding periodic oscillatory states of the currents I and I_{12} (marked by *dots and crosses*, respectively) in Fig. 3.96 (same sample and constant parameters)

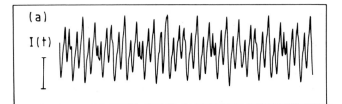

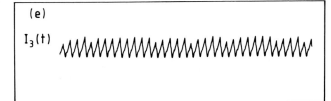

Fig. 3.99a–e. Temporal current and voltage profiles for two corresponding periodic and quasiperiodic oscillatory states obtained from the experimental set-ups sketched in Figs. 3.92 and 3.95 by biasing either the whole **a–c** or different parts **d, e** of the sample arrangement, respectively, at the following constant parameters: bias current $I = 2.284 \pm 0.001$ mA, transverse magnetic field $B = 0.506$ mT, and temperature $T_b = 2.10$ K. All signals are plotted on the same vertical and horizontal scales. Their amplitude refers to the 5 mV voltage bar indicated in **a**. The current was measured as voltage drop at a 10 kΩ load resistor. The total time span amounts to 33 ms

A second example of spatio-temporal behavior concerns the quasiperiodic state to be presented in the sequel. Analogous to the previous results of Fig. 3.94, quasiperiodicity in the current signal $I(t)$ of Fig. 3.99a can be attributed to the superposition of the two nearly periodic voltage traces $V_{12}(t)$ and $V_3(t)$ shown in Fig. 3.99b and c, respectively. Since all three time signals were digitized synchronously, the modulation of the amplitude of $V_{12}(t)$ is found to be highly correlated with the periodicity of $V_3(t)$. The current signals $I_{12}(t)$ and $I_3(t)$ corresponding to these partial voltages, measured accordingly at separate bias control of the sample parts under consideration, are plotted in Fig. 3.99d and e, respectively. Juxtaposition of $V_{12}(t)$ and $I_{12}(t)$ yields very similar time traces; even their frequencies coincide within an experimental accuracy of a few percent. If one compares $V_3(t)$ with $I_3(t)$, again reasonable agreement in the shape of both signals can be recognized, while the frequency of the latter is noticeably higher. So far, these findings prove the present quasiperiodic state to arise from two competing oscillation centers located in spatially separated parts of the semiconductor. One oscillator is localized inside the region that embraces parts 1 and 2, and the other, inside part 3 of the sample. An important result is that these localized oscillation centers could be activated independently from each other by addressing the voltage bias to the relevant sample part.

The characteristic magnetic field dependence of the frequency components involved gives prominence to the mutual relationship between one oscillation and the other. Figure 3.100a displays the successive development of two independent frequencies in the case of the experimental set-up of Fig. 3.92. The bifurcation to a quasiperiodic state takes place at a transverse magnetic field of slightly above 0.4 mT (compare with the similar situation in Fig. 3.42). The plot in the inset shows the square-root scaling of the second frequency with the magnetic field close to the bifurcation point. On the other hand, Fig. 3.100b and c unveils the relating frequency behavior for two particular arrangements of separately biasing different parts of the sample (Fig. 3.95). In accordance with the conclusions drawn from the time signals in Fig. 3.99, the graph of the first intrinsic frequency in Fig. 3.100a and that of the single frequency in Fig. 3.100b are almost identical. There is only little difference at the jump down to lower frequencies around 0.7 mT. Pronounced differences can be found if one compares the behavior of the second intrinsic frequency in Fig. 3.100a with that of the single frequency in Fig. 3.100c. Two features become immediately evident. First, the bifurcation point in situation (a) is shifted to higher magnetic fields such that all frequency values are generally smaller – with respect to the

Fig. 3.100a–c. Dependence of the intrinsic frequencies on the transverse magnetic field (1 G $\doteq$ 0.1 mT) for the case of the two corresponding periodic and quasiperiodic oscillatory states of the currents I **a**, I_{12} **b**, and I_3 **c** in Fig. 3.99 (same sample and constant parameters). The data of the two frequencies f_1 and f_2 existing simultaneously in the current signal $I(t)$ and the corresponding single frequencies of the current signals $I_{12}(t)$ and $I_3(t)$ are marked by crosses and circles, respectively. The insets in **a** and **c** demonstrate the scaling of the (circle) frequency data near the bifurcation point

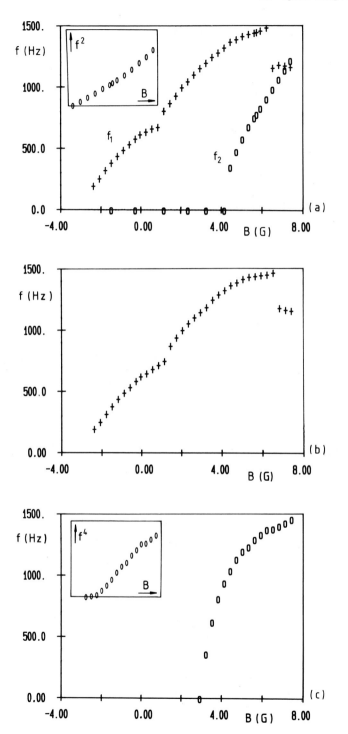

magnetic field applied – than the corresponding ones in situation (c). Second, the power law of the scaling in the vicinity of the bifurcation point changes from the exponent 1/2 in situation (a) to the exponent 1/4 in situation (c), as indicated by particular plots in the corresponding figure insets. Apparently, the interaction between the two oscillation centers manifests in the way that the presence of the first intrinsic system frequency (say f_1) can prevent the appearance of the second one (say f_2) over an extended control parameter interval.

Another experimental point of view concerns the spatio-temporal aspects underlying a quasiperiodic state whose frequency spectrum is dominated by protruding mixing components. As already shown in Sect. 3.4, they provide the first clue to the presence of a nonlinear interaction between the intrinsic oscillatory modes. From Sect. 3.5, we further know that such an interaction may lead to chaos. As a result of the prevailing informational content, the quasiperiodic state per se has the potential to establish a connection from pure quasiperiodicity to the more complex chaotic motion in phase space. Figure 3.101a–c gives the power spectrum of the integral current signal $I(t)$ together with that of the partial voltage drops $V_{12}(t)$ and $V_3(t)$, respectively, obtained by applying the bias voltage to the whole sample arrangement (Fig. 3.92). Following the notation previously used in Fig. 3.43, the different spectral peaks are distinguished by the indices (n, m) according to the frequency $f = nf_1 + mf_2$. There is no doubt that all three power spectra contain mixing components of the two fundamental frequencies f_1 and f_2. Upon separating the two oscillation centers via independent bias application to the relevant sample regions (Fig. 3.95), the power spectra of the resulting integral current flow $I_{12}(t)$ and $I_3(t)$ displayed in Fig. 3.101d and e, respectively, do not continue to exhibit any mixing component. Hence, the nonlinear coupling has been clearly switched off. Moreover, in both cases the remaining single frequencies are shifted to slightly different values. The lower frequency f_1 decreases further by 125 Hz, while the higher frequency f_2 increases by 62 Hz. We draw the conclusion that the interaction of the two oscillation centers (which at present can take place only when the experimental set-up shown in Fig. 3.92 is used) causes the two fundamental frequencies to attract each other.

Finally, it should be emphasized that our findings on the spatio-temporal dynamics of regular periodic and quasiperiodic states cannot be readily transferred to the more complex cases of ordinary chaos and hyperchaos, the phase portraits of which are outlined in Fig. 3.93d and e, respectively. As we have already mentioned above, there arises no problem in perceiving a consistent picture of the latter physical situation if one looks at the distinct chaotic properties of either the integral current flow or the partial voltage drops in the whole sample arrangement (Fig. 3.92), although one has to make use of a variety of independent numerical analysis procedures. However, when investigating only parts of the sample arrangement (Fig. 3.95), the two dynamical states reflect a totally different temporal behavior with respect to signal amplitudes and typical revolution times. Such a form of apparent inconsistency gives rise to the clarifying assumption that the dynamics of both the chaotic and the hyperchaotic state (and, of course, their higher-dimensional analogs) result from

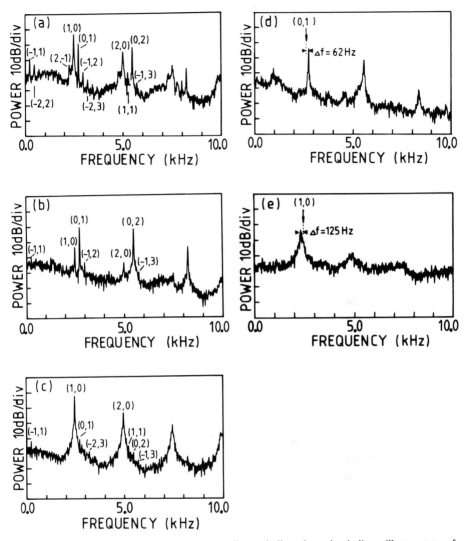

Fig. 3.101a–e. Power spectra for two corresponding periodic and quasiperiodic oscillatory states of the currents and voltages I **a**, V_{12} **b**, V_3 **c**, I_{12} **d**, and I_3 **e** obtained from the experimental set-ups sketched in Figs. 3.92 and 3.95 by biasing either the whole **a–c** or different parts **d, e** of the sample arrangement, respectively, at the following constant parameters: bias current $I = 1.774 \pm 0.001$ mA, transverse magnetic field $B = 0.225$ mT, and temperature $T_b = 2.10$ K. The spectral lines labelled by (n, m) are subject to the frequency $f = nf_1 + mf_2$. The differences of the two frequencies f_1 and f_2 existing simultaneously in the current and voltage signals $I(t)$, $V_{12}(t)$, and $V_3(t)$ to the corresponding single frequencies of the current signals $I_{12}(t)$ and $I_3(t)$ are indicated by Δf in **d** and **e**

a more refined partition into oscillatory subsystems than is provided by the present rather large-scaled sample configuration.

To conclude, we have disclosed several indications that the emergence of spontaneous oscillations in our semiconductor system is linked to the spatial formation of localized oscillation centers. The observation of different dynamical states obeying a hierarchical order of complexity is, therefore, purely a consequence of gradual rearrangements among distinct spatio-temporal transport structures in low-dimensional phase space. On the one hand, the spatially separated oscillatory subsystems develop in a self-organized way, i.e., without being favored by external excitations or boundary conditions. On the other hand, the mutual interaction between these oscillators can be sensitively influenced by means of appropriate control parameters. Such a picture of a coupled multicomponent synergetic system creating the measured spatio-temporal dynamics stands in contrast to a nonlocalized physical mechanism which seems to be relevant for the nonlinear dynamical behavior (in particular, the transition to chaos) of a somewhat different semiconductor experiment [3.92].

Problems

3.1 Study the bifurcation behavior of (3.4) by means of a linear stability analysis. Determine the parameter dependence of the eigenvalues and discuss the bifurcation.

3.2 A dynamics can also be expressed by a time-discrete process, such as $x_{n+1} = f(x_n)$. Here n denotes the discrete time given in multiples of an arbitrarily chosen time unit. Demonstrate numerically the sensitive dependence on initial conditions for $x_{n+1} = 4x_n(1 - x_n)$. Show the transient behavior of $x_{n+1} = 3.83x_n(1 - x_n)$ and, moreover, the existence of chaotic transients for some initial conditions. Always take $0 < x_n < 1$.

3.3 Verify the structural stability of the saddle node bifurcation $\dot{x} = \mu - x^2$. Use respectively a constant, a linear, and a quadratic perturbation term and show that the structure of the bifurcation is not changed. Check that a third-order perturbation term affects the structural form far away from the bifurcation point (reproduce the result in Fig. 3.10). Prove that, for any bifurcation with a term of the order n, an mth-order perturbation term with $m > n$ generally gives rise to structural changes far away from the bifurcation point.

3.4 Verify the structural instability of the transcritical bifurcation $\dot{x} = \mu x - x^2$. What is the influence of a constant perturbation term (with either positive or negative sign)? Determine the structure of the perturbed transcritical bifurcation by taking the perturbation term as a second independent control parameter. Construct the geometry of the manifold of the fixed points in analogy to Fig. 3.10.

3.5 Find the symmetry conditions for which perturbation terms leave the pitchfork bifurcation structurally stable. Determine the bifurcation diagram for the case of a structure-destroying perturbation. What does it look like for a fifth-order perturbation term?

3.6   Demonstrate by means of a coordinate transformation that the van der Waals equation of a real gas is nothing other than a cusp catastrophe.

3.7   Discuss the shape of the cusp catastrophe for the case of (3.14) on the basis of the physical variables V_0, R_L, and I. What is the influence of additional capacitances and inductances on the dynamical behavior? Show that oscillations can arise.

3.8   Look at the stability of the fixed points of the Hopf bifurcation described by (3.20). Prove that the control parameter μ_2 changes the bifurcation behavior of the radius variable in (3.24) from normal pitchfork to subcritical pitchfork. Determine the scaling of the hysteresis between the oscillatory state and the nonoscillatory state in control parameter space.

3.9   Show that for any return map $x_{n+1} = f(x_n)$ a fixed point obeying the condition $x_{n+1} = x_n$ is stable (unstable) if the magnitude of the derivative of the function f taken at the fixed point is smaller (larger) than unity. Calculate the fixed point of the logistic map $x_{n+1} = \mu x_n(1 - x_n)$ and demonstrate its stability. Classify the type of bifurcation that takes place at $\mu = 1$. Discuss it in the sense of a Hopf bifurcation by taking into account that a fixed point of the return map with nonzero value can be regarded as a periodic oscillation. Find the proper type of the bifurcation. Finally, calculate the period-two fixed point of the logistic map obeying the condition $x_{n+2} = x_n$ and discuss its stability.

3.10   Evaluate numerically the dynamics of the sine circle map close to a locking state, for example, at $K = 0.75$, as well as directly in the 1/1 locking state. Verify the scaling of the mean period of the resulting oscillation as a function of the distance of the control parameter from the critical value at the locking state. Verify the scaling of the corresponding frequency defined as the reciprocal mean period. Test the influence of slightly different initial conditions.

3.11   Solve (3.44) and determine the law governing the time separation of neighboring initial conditions.

3.12   Evaluate numerically the Lyapunov exponents for the case of the logistic map $x_{n+1} = \mu x_n(1 - x_n)$ with $0 < \mu < 4$. *Hint*: Iterate an initial value $0 < x_0 < 1$ several times. Calculate the absolute values of the slope at successive iterated points. The logarithm of the mean of these values gives the Lyapunov exponent.

3.13   The logistic map $x_{n+1} = \mu x_n(1 - x_n)$ exhibits type I intermittency close to the period-three window around $\mu = 3.83$. Evaluate numerically the characteristic features of type I intermittency in the parameter range $\mu < 3.83$. Find the scaling of the mean laminar length with the control parameter. Compare this scaling law with the one obtained for the saddle node bifurcation on a limit cycle and that of Problem 3.10. When working on this problem, keep in mind the results of Problem 3.2.

3.14   Visualize the characteristic winding-up dynamics of the maximum amplitudes for type II intermittency, as is given by (3.106) and (3.107). Think about the structural form of an experimental continuous dynamics having that property.

3.15 Discuss the geometry and, thus, the kind of bifurcation for the second iterate of type III intermittency in (3.108).

3.16 Calculate numerically the Lyapunov exponent for the case of the logistic map $x_{n+1} = \mu x_n(1 - x_n)$ with $\mu = 3.83$. In order to look at the transient behavior, determine the Lyapunov exponent for the first 10 iteration steps. For this purpose, use different initial conditions and compare the local (in the sense of initial conditions) Lyapunov exponents. Considering the fact that chaotic forcing of a bistability provides an essential mechanism for generating fractal boundaries, we next turn to the equation $x_{n+1} = F(x_n) + cy_n$, $y_{n+1} = G_\mu(y_n)$, where F is a bistable function like x_n^a and G_μ a chaotic map like the logistic one. Choosing the coupling constant $c = 0.1$, the bistability exponent $a = 2.0$, and the chaos control parameter $\mu = 3.98$, evaluate numerically the basin boundary of the above equation by iterating different initial values x_0, y_0 with $0 < x_0 < 1$ and $0 < y_0 < 1$. If during the first 50 iterations x_n becomes equal to or larger than unity, color the initial condition point x_0, y_0. For example, use white for points with $x_n \geq 1$, otherwise black. Calculate numerically the borderline where x_0 is a unique function of y_0. Magnify a small part of the borderline. Repeat this for $a > 2$ and $a < 2$. Show how, in the limit of larger magnifications, the borderline becomes striped in one case, and in the other, smooth. Finally, reveal an intermittent structure of the borderline by using the parameter values $a = 1.6$ and $\mu = 3.83$. (Hint, take advantage of the derivative of this borderline.) Explain such a spatially intermittent structure on the basis of transient chaos with transient positive Lyapunov exponents, as evaluated above.

4   Mathematical Background

The interest in nonlinear dynamical systems increased considerably when it was realized that very modest model cases could explain a whole range of dynamical behavior, from simple to very complex. In his pioneering work of 1963, E. Lorenz showed that his simple system of nonlinear differential equations describing a forced dissipative hydrodynamical flow was able to produce, in addition to regular behavior, aperiodic bounded solutions and discontinuous dependence on initial conditions. Earlier it had been believed that a description of what was called "turbulent motion" necessarily involved a large number of degrees of freedom, if a different interpretation from noise could be given at all. Now, suddenly, it was observed that a whole class (often called a "universality class") of apparently different theoretical and experimental systems behaved alike, when an "external" parameter changed in such a way that the system was led from simple to more complex behavior. The reason for this could be traced back, for both dissipative and nondissipative dynamical systems, from the differential equation to the Poincaré section and from there to a model map. This model map was shown to be characteristic for each universality class; it contains the whole information of how the change from simple to more complex behavior is achieved. In this way it became possible to describe two "generic" ways for this metamorphosis: period doubling and the quasiperiodic route to chaos. A representative collection of early works is furnished in [4.1].

Using ideas of renormalization theory, the values of the relevant parameters which describe the different universality classes of one-dimensional maps could be calculated. From that time on, the way to ordinary chaos was relatively well understood. A description of chaos itself seemed to be more involved at first. How could chaos be characterized? How could the different chaotic states be distinguished?

However, the first steps towards an answer to these questions had already been given. From 1960 onwards, in a famous series of papers, Kolmogorov, Sinai and, finally, Ornstein proved within the framework of ergodic theory that, for Bernoulli systems, the Kolmogorov–Sinai or metric entropy identifies such a system up to an isomorphism. In 1971, the concept of a strange attractor, which was characterized by means of sensitive dependence on initial conditions, was introduced. Mandelbrot was among the first to realize that strange attractors have fractal geometric properties [4.2]. It is through the influence of these people that tools known from long ago, like Lyapunov exponents and fractal dimensions, were brought back into the focus of interest [4.3]. In combination with ideas borrowed from statistical mechanics, they opened the road for a quantitative characterization of chaotic behavior.

Today many important points still remain to be clarified. What is the relation between the scaling properties of the measure and that of the support, between a temporal and a purely statistical description? How reliable are the different numerical calculations which can be used? What can they contribute to explaining the physics of a system?

The numerical characterization of the scaling behavior of chaotic dynamical systems is discussed in this chapter. A description of these effects involves spatial as well as temporal aspects. Emphasis is laid on the (temporal) description of the scaling behavior of the support and its relation to the (spatial) scaling behavior of the measure. For the former description various new or modified algorithms are introduced and applied; the algorithms and procedures belonging to the latter description are relatively well known. Here they are only discussed in a common context with the main topic. While the probabilistic description of effects in the phase space has been reported widely (fractals, fractal dimensions), not many publications deal with the numerically more involved dynamical description. Moreover, the relationship between these two scaling behaviors needed to be clarified.

In this chapter this gap is filled. By looking at simple examples, starting from a unified thermodynamical formalism for both ways of description, it is shown analytically what interrelations exist. It becomes clear that they describe fundamentally different aspects of the scaling behavior of a strange attractor and that only in special cases can the result of one description be deduced from the other. The extension of the concepts developed for model cases to generic systems is outlined. It is discussed how the scaling behavior of a strange attractor can be characterized by the use of Lyapunov exponents calculated from time series. To this end, the way in which the associated distribution of effective Lyapunov exponents (scaling function for the scaling of the support) has to be interpreted is shown. While the theoretical basis needed for such an interpretation has been developed and published during the time this chapter was being written, here the first examples of phase-transition-like behavior of scaling functions of Lyapunov exponents calculated from experimental data are shown. For the evaluation of Lyapunov exponents and associated distributions of effective Lyapunov exponents, various specific numerical algorithms were developed. To test their reliability, they were applied to data obtained from different nonlinear model systems, such as the Lorenz system, Hénon's map and the NMR laser model equations; different issues resulting from these investigations are discussed.

Of special interest is the question of how the above-mentioned concepts can be applied to experimental systems and what information can be gained about the physics or the modelling of the system under investigation. Great efforts in experimental research have to be made to provide us with the data to permit an analysis along the lines outlined above. Activities in the fields of fluid dynamics, solid state physics, chemical reaction processes, and laser physics have been most successful. In this work, two experimental systems are investigated in detail. One is the NMR laser system [4.4], which shows a strongly nonlinear reaction to weak external modulation of parameters at low frequencies (~ 100 Hz). The behavior of this system can be modelled for these conditions with high accuracy by a low-dimensional differential equation system derived from the Maxwell–Bloch equations. As to the experimental germanium semiconductor system [4.5], the nonlinear behavior results from the auto-catalytic nature of the avalanche breakthrough at low temperatures (the "natural" frequency is around 1000 Hz). Full details about the ingredients have been given already in Chaps. 2 and 3. This system can best be modelled by a Belousov–Zhabotinsky type reaction–diffusion model, consistent with the fact that many degrees of freedom can be involved (but need not all be active). For both experimental systems, the different aspects of the scaling behavior result in consistent pictures.

Finally, the problem of the evaluation of Lyapunov exponents from high-dimensional embedding spaces is addressed. A conceptually new algorithm, which is superior to the algorithms known so far for these cases, is introduced. With the help of this algorithm a whole class of systems which are able to produce only short time series can be treated.

In this chapter an attempt has been made to treat in a coherent way the complete temporal and spatial scaling behavior of a dynamical system as far as the evaluation of scaling functions of generic systems is concerned. The invest-igations on the Lyapunov exponents upon a variation of noise amplitude or sampling time show relevant new aspects. The characterization of experimental systems by means of spectra and scaling functions of Lyapunov exponents, as performed in this work, will, sooner or later, become one of the central points in the numerical investigation of experimental systems.

We proceed as follows in the remaining part of the work. In Sect. 4.1 an outline of the theory of dynamical systems is given. We discuss how complex behavior is created and observed. Section 4.2 deals with the scaling behavior of an attractor arising from a dynamical system. Useful tools borrowed from statistical mechanics are outlined. For the attractor, in analogy to the thermo-dynamical formalism, a generalized free energy is defined and associated entropies are derived. Different, important model cases are then discussed in a coherent way to work out the connection between the scaling of the support and the scaling of the measure. In the first half of Sect. 4.3 the general concepts of the preceding sections are made suitable for generic systems. Scaling functions for the dynamical (scaling properties of the support) and for the probabilistic (scaling of the measure) points of view are introduced and their relationship investigated. In the second half of Sect. 4.3 the occurrence of nonanalytic behavior of the thermodynamical functions is discussed. Examples of such a behavior for nonhyperbolic maps are given together with the theoretical derivation. Evidence of phase-transition-like behavior in experimental systems is shown and interpreted. In Sect. 4.4 a closer look is taken at the numerical procedures which are used to investigate the scaling behavior of experimental systems. Emphasis is laid on the evaluation of the Lyapunov exponents from time series. Well-known differential equation systems and model maps, on the one hand, and experimental systems, on the other hand, serve as a testing ground for the different numerical procedures. The results obtained from different algorithms are compared and the origins of deviations explained. Along with these tests the behavior of the two experimental systems is analyzed and discussed. A conceptually new variant of the algorithm to evaluate Lyapunov exponents from time series discussed in Sect. 4.4 is proposed in Sect. 4.5. It is shown that this algorithm is favorable for the extraction of Lyapunov exponents in high-dimensional embedding spaces or for small data bases. Furthermore, it is outlined how this algorithm can be used as a severe test for the occurrence of hyperchaos. In Sect. 4.6 the degree of mapping, which is a common tool in the theory of nonlinear equations, is considered in a general survey. The connection between the degree of mapping and the concept of Lyapunov exponents is worked out and possible applications are sketched. Finally, a summary of the relevant exact results on the connection between Lyapunov exponents and curvature is given.

4.1   Basic Concepts in the Theory of Dynamical Systems

4.1.1   Dissipative Dynamical Systems and Attractors

This section contains the main definitions and concepts used in the theory of dynamical systems. Although it has been attempted to argue in an intuitive way, this section is not designed as an elementary introduction to the topic. Instead, we have tried to work out the main concepts used for the analysis of the various systems later on and present these concepts in a not-oversimplified way. Where it was not possible to achieve this goal in a satisfactory way, references for further reading are given. No care, however, has been taken to review all the references and to give due credit to all authors; the references given are selected mainly in an attempt to cover with a few references the widest possible terrain.

For a compact manifold M, a continuous-time *dynamical system* is a vector field X on M. By m we denote the algebraic dimension of M; time will always be measured in units of seconds. The corresponding evolution equation has the form

$$X = Fa(x) \,, \tag{4.1}$$

where $x \in M$ and a denotes a list of relevant external parameters. Fa will be called the *dynamical map*. The dynamical map Fa is supposed to depend differentiably on a and differentiably on x almost everywhere. Once values of the external parameters and initial conditions x_0 are chosen, the system will evolve according to (4.1) such that, at any point x of the evolution curve $x(t)$ of x_0 in M, the vector field X is tangential to $x(t)$. After a possible transient behavior, the system is assumed to settle on a well-defined asymptotic dynamics. If a or x_0 is changed, the resulting asymptotic behavior may change too.

An important class of dynamical systems is furnished by the *dissipative systems*. In many physical systems energy is lost due to friction, and an ever smaller part of the associated phase space can be reached. In this case the dynamical map contracts the phase space volume (at least in the time average), a property which can be formulated with the help of the divergence ∇ as

$$\nabla Fa < 0 . \tag{4.2}$$

For discrete-time systems which will be written for convenience also in the form

$$x_{n+1} = Fa(x_n) \tag{4.3}$$

(whether the generator of a flow or a map is considered will be clear from the context), the analogous property is

$$\| DFa(x) \| < 1 , \tag{4.4}$$

where DFa denotes the linear approximation of the map Fa. Since any continuous-time system can be converted into a discrete-time system by means of a Poincaré section, we will, for convenience, change freely between the two possible descriptions.

As a simple example of a dissipative system we take the Lorenz model [4.1, 6]

$$\dot{x}_1 = -\sigma x_1 + \sigma x_2 , \tag{4.5}$$

$$\dot{x}_2 = -x_1 x_3 + r x_1 - x_2 , \tag{4.6}$$

$$\dot{x}_3 = x_1 x_2 - b x_3 . \tag{4.7}$$

With the help of this example, a deeper insight into the asymptotic behavior of a general dissipative system can be gained. If a phase space element is denoted by U and its volume by V, we find that V changes in time as

$$\frac{dV(t)}{dt} = \int_{U(t)} dx^3 \, \nabla Fa . \tag{4.8}$$

Since the system has a constant divergence ∇Fa, it follows that

$$\frac{dV(t)}{dt} = (-\sigma - 1 - b) V(t) < 0 \tag{4.9}$$

if $\sigma, b > 0$. The latter equation indicates that the volume V of the phase space element U shrinks in time as

$$V(t) = V(0) e^{t(\nabla Fa)} . \tag{4.10}$$

After the transient behavior for dissipative systems the accessible region in phase space will, therefore, have zero volume. Contrary to one's intuitive feeling, this only means that the motion is restricted to a submanifold of the original, underlying manifold. For obvious reasons, for the resulting limiting set the term *attractor* has been coined. More precisely, an attractor A can be defined by the following requirements [4.7]:

1. A bounded neighborhood N of A exists, such that $Fa((\bar{N})) \subset Interior(N)$ and $A = \bigcap_{n > 0} Fa^n(N)$.
2. There is an orbit of Fa which is dense on A.

For flows, familiar examples of attractors are attracting fixed points, limit cycles and tori. These cases are naturally associated with steady, periodic and quasiperiodic states of the time evolution. They can be classified according to the number of contracting directions in relation to the phase space dimension. For example, a fixed point is characterized by contraction in all directions, whereas for limit cycles and tori the contraction takes place transversely to a submanifold of dimension 1 and of dimension $m \geq 2$, respectively. However, a more interesting case is provided for flows in dimension at least 3 (or for maps in dimension at least 2) in the following way.

Assume that a direction can be found along the flow in which distances are being stretched, at least on the average. In view of requirement (4.2), contraction has to be present in perpendicular directions. If, in addition, boundedness of the phase space region is assumed, then a structure reminiscent of a *Cantor set* [4.7] will evolve. In a schematic picture this is shown in Fig. 4.1. For the precise definition of a Cantor set we refer to Sect. 4.2.

An attractor which has been generated by a folding process of the kind considered can, therefore, at least for most of its points, locally be described

(a)

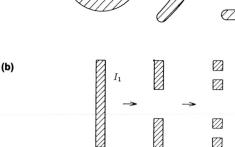

(b)

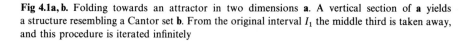

Fig 4.1a, b. Folding towards an attractor in two dimensions **a**. A vertical section of **a** yields a structure resembling a Cantor set **b**. From the original interval I_1 the middle third is taken away, and this procedure is iterated infinitely

by a direct product of a Cantor set with a continuum. This fact then suggests a *renormalization treatment* and a characterization via fractal dimensions. Sections 4.2 and 4.3 will deal with these topics. To add a realistic and instructive example let us follow in Fig. 4.2 the folding process generated by the iteration of Hénon's map [4.8]. As can be seen in Fig. 4.3, the final attractor displays a rather complicated structure which shows self-similarity. The lines which constitute the final attractor are called the *unstable manifolds* of the (unstable) fixed points of the attractor. The related concepts will be outlined in Sect. 4.1.3. Note that the plane is folded onto itself exactly once.

4.1.2   Invariant Probability Measures

To work out the connection between the process of successive mapping and the final appearance of the attractor, we need the notion of a *probability measure* ρ on A [4.3]. ρ is defined in the following way:

$$\rho(x) = \lim_{T \to \infty} \frac{1}{T} \int_0^T dt \, \delta(x - x(x_0, t)) , \tag{4.11}$$

where $x(x_0, t)$ denotes the evolution curve starting from point x_0 and $\delta_x(t)$

Fig. 4.2. Three rows showing the folding of a circular disc towards the attractor for successive iterations of Hénon's map $(x_{n+1}, y_{n+1}) = (1 - ax_n^2 + y_n, bx_n)$, where $a = 1.4$ and $b = 0.3$. The circular disc from which it started is not shown. For convenience all figures have been compressed in the vertical direction

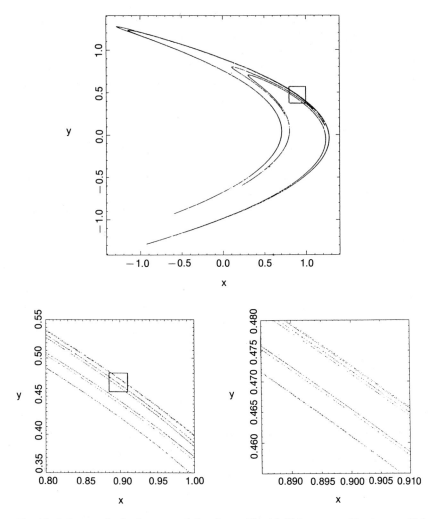

Fig. 4.3. Points on the final attractor belonging to Fig. 4.2. Using a magnification, a self-similar structure can be discovered

denotes the delta function at point x. Similarly, we define for an observable $\varphi : \mathbb{R}^m \to \mathbb{R}$

$$\rho(\varphi) = \lim_{T \to \infty} \frac{1}{T} \int_0^T dt \, \varphi(x(x_0, t)) \, . \tag{4.12}$$

The part of $\mathbb{R}^m$ on which ρ is nonzero is the *support* of the measure. ρ is called an *invariant* probability measure if ρ is invariant under time evolution:

$$\rho(\varphi \circ Fa) = \rho(\varphi) \, . \tag{4.13}$$

Note that in Fig. 4.3 we have shown a finite-n approximation to the density of the invariant probability measure of the Hénon map, where n denotes the number of iterations. An invariant measure satisfies $\rho(Fa^{-n}(E)) = \rho(E), n > 0$, $E \subset \mathbb{R}^m$. Due to the fact that we have defined the final attractor as a compact set, which is invariant under the application of Fa, the Markov–Kakutani fixed-point theorem can be applied and $\rho(x)$ can be chosen to be *ergodic* [4.9]. In this case, temporal and spatial averages are equal in the sense that $\rho(\varphi) = \int_A \varphi(x)\rho(dx^m) \equiv \langle \varphi \rangle$ for almost all initial conditions, where the notation of [4.9e] was used.

For a generic [4.10] system uncountably many invariant measures exist. From a mathematical point of view then the question arises which one should be chosen in order to be able to reflect the physical properties of the system.

The following selection process goes back to Smale axiom A systems and was generalized by Sinai, Ruelle, and Bowen to the so-called *SRB measures* [4.3]. Consider an invariant measure, perturbed stochastically. In the limit of the perturbation amplitude going to zero, it will converge to some distinct measure, which will be called natural [4.11] or physical [4.3]. Under certain conditions it will fulfill the SRB conditions: the measure will be singular transverse to the unstable manifolds, of which the attractor is essentially composed (Fig. 4.2), while it is assumed to be smooth (to be more precise, absolutely continuous [4.3, 12] with respect to the Lebesque measure) along these manifolds. In that way the measurable sets are composed as unions of open, disjoint pieces of (local) unstable manifolds, $\bigcup_{\alpha \in \mathscr{A}} S_\alpha$. The measure carried by these pieces can be expressed as $\rho_\alpha(d\xi) = \varphi_\alpha(\xi)d\xi$. Here φ_α is an integrable function, $d\xi$ denotes the volume element when S_α is smoothly parametrized by a piece of $\mathbb{R}^{m^+}$, and m^+ denotes the number of unstable directions. The SRB conditions are important for the mathematical treatment of model systems from a technical point of view because of this absolute continuity property. However, it is generally difficult to establish the SRB properties for nontrivial model systems.

Let us now turn to a more explicit and less technical approach to the invariant probability measure, which will be used explicitly in Sect. 4.2.5. Denote by DFa^+ the product of the eigenvalues of DFa which are larger than 1, by β, a real number, and by $R(\beta)$, a real function of β. For the present purpose, where not the most general formulation is needed, β can be thought of as being equal to 1. Then the invariant probability measure ρ is seen to evolve as the asymptotic solution of the iterative *generalized Frobenius–Perron equation* (for an explicit discussion of the one-dimensional case see [4.13]):

$$\rho_{k+1,\beta}(y) = R(\beta) \sum_{x \in Fa^{-1}(y)} \frac{\rho_{k,\beta}(x)}{|DFa^+(x)|^\beta} \,. \tag{4.14}$$

If β is chosen equal to one, it is easily seen that this iterative process describes in a very simple mathematical form the effect of the folding and the stretching that we have been observing in the last section. Only for a distinct value of $R(\beta)$ does an arbitrary starting density ρ_0 converge to a finite, positive $\rho_{\infty,\beta}$. The value

$$\lim_{n \to \infty} \frac{1}{n} \log R(\beta) \tag{4.15}$$

is called the *generalized escape rate* and for $\beta = 1$ simply the escape rate. For attractors, the generalized escape rate is equal to zero. Starting from uniform densities,

$$Q_{\beta,n}(y) = R^n(\beta) \sum_{x \in Fa^{-n}(y)} e^{-\beta \log |DFa^{n,+}(x)|} \tag{4.16}$$

is obtained, and in the limit $n \to \infty$ eigenfunctions $Q_\beta(x)$ of the iterative process are found. If they can be calculated, they serve as an important tool for the characterization of phase-transition-like behavior of dynamical systems.

Note that for the formulation of the invariant measure with the help of the Frobenius–Perron equation the absolute continuity of the dynamical map along the unstable manifold is implicit. This property of the dynamical map is, in turn, intimately related to the notion of hyperbolicity, which provides another important characterization of dynamical systems. In loose terms, a dynamical system is often called *hyperbolic* if none of the points of its attractor has as largest eigenvalue an eigenvalue of modulus less than or equal to one. For a more precise definition and discussion of hyperbolicity we must refer to the literature (e.g. [4.7]). This characterization applies for maps; if flows are considered then the direction of the flow, where the average distance between points is not changed, leads, of course, to an eigenvalue equal to one, as expected. For one-dimensional hyperbolic attractors it can rigorously be shown that the SRB measure is the natural measure and $Q_{\beta=1}(x)$ is its density. In this case, and also for some nonhyperbolic, completely chaotic maps of the interval, the SRB approach and the Frobenius–Perron approach are equivalent and $-\log[R(\beta)]$ is equal to the SRB pressure [4.13]. Nonhyperbolic systems may have singularities on the unstable manifolds, which complicate the use of SRB arguments [4.12].

4.1.3   Invariant Manifolds

So far the natural measure has been established as the central quantity for the observation of an attractor. In this subsection the concept of stable/unstable manifolds is introduced and its relationship to the invariant probability measure is elucidated. In this way it will be possible to give the simplest example of how a highly irregular dynamical behavior, as observed in chaotic systems, can be achieved.

The link between the two concepts can be found starting from a local analysis. For reasons that will become evident, let us consider a hyperbolic periodic point x_0 on the attractor. x_0 is called a hyperbolic fixed point if x_0 is provided with a tangent space T_{x_0} that can be written as a (possibly trivial) direct sum $T_{x_0} = E^u_{x_0} \oplus E^s_{x_0}$ such that (a) $DFa(E^s_{x_0}) = E^s_{Fa(x_0)}$, $DFa(E^u_{x_0}) = E^u_{Fa(x_0)}$, respectively, and (b) there are constants $c > 0$, $\lambda \in (0, 1)$ such that $\|DFa^n(x_0)v\| \le c\lambda^n\|v\|$ whenever $v \in E^s_{x_0}, n \ge 0$ and $\|DFa^{-n}(x_0)v\| \le c\lambda^n\|v\|$ whenever $v \in E^u_{x_0}, n \ge 0$. Now let the external parameter change continuously in such a way that the system is led from an ordered to a more complex behavior. During this process more and more attracting hyperbolic periodic points lose their stability and many of them become saddle points. We will explain how the genesis of an attractor by iteration can finally be visualized as a random walk on the union of all (unstable) periodic points.

The appropriate mathematical tool is provided by the stable/unstable manifold theorem. Suppose that the dynamical map Fa is of differentiability C^r. Then the stable/unstable manifold theorem [4.10] asserts the existence of locally defined, nonlinear C^r-differentiable manifolds, called local stable/unstable manifolds $W^s_{x_0}/W^u_{x_0}$, which extend through x_0 and lie tangentially to the eigendirections of DFa at x_0. By definition all points on $W^s_{x_0}/W^u_{x_0}$ tend to x_0 under forward/backward iteration of the dynamical map. For an additional point x in the vicinity of x_0 the fate under the action of the flow can now be predicted: its trajectory is attracted towards the nearby periodic point x_0 along the stable manifold and then ejected along the unstable manifold of point x_0. A picture as shown schematically in Fig. 4.4 results. Locally defined manifolds $W^s_{x_0}/W^u_{x_0}$ of different hyperbolic periodic points are connected in a natural way by iteration. They form global objects called *stable/unstable* manifolds Ws/Wu:

$$Ws(x_0) = \overset{n > 0}{\bigcup} Fa^{-n}(W^s_{x_0}) = \{y \backslash Fa^n(y) \to x_0\}, \qquad (4.17)$$

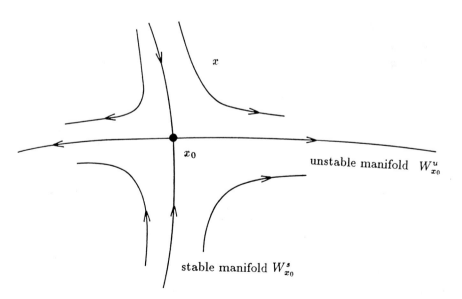

Fig. 4.4. Stable and unstable manifold of the hyperbolic point x_0 (see text)

$$Wu(x_0) = \overset{n>0}{\bigcup} Fa^{+n}(W_{x_0}^{u}) = \{y \backslash Fa^{-n}(y) \to x_0\} \ . \tag{4.18}$$

As can easily be seen, an attractor must contain the unstable manifolds of all points. Whether one can go by iteration from one periodic point x_0 to another periodic point x_0' on the attractor depends upon whether the unstable manifold of x_0 and the stable manifolds of point x_0' intersect. It will be one component of the definition of chaos (Sect. 4.1.4) to ensure that from one point of the attractor all other points of the attractor can be reached. As observed long ago by Poincaré in the context of the three-body problem, the global stable/unstable manifolds provide the key to understanding the simplest case of irregular dynamical behavior. Suppose there is an intersection of the stable and the unstable manifold of a hyperbolic fixed point x_0 (*homoclinic intersection*). Since the stable and the unstable manifold of x_0 are both invariant, the existence of one such point implies an infinity of such points. As can be seen in Fig. 4.5, this forces $Wu(x_0)$ to oscillate wildly as it approaches x_0, and similarly for $Ws(x_0)$ [4.14]. In this way even one single hyperbolic periodic point can lead to a wildly erratic orbit structure.

Poincaré writes (*Celestial Mechanics*, 1892): "The complexity of this figure will be striking, and I shall not even try to draw it". For obvious reasons, such a point x_0 is often called "snap-back repeller". Heteroclinic intersections (here the stable and the unstable manifolds of different points intersect) can lead to similar effects. Other types of intersections (self-intersections, intersections of manifolds of the same type of different points) are forbidden by the uniqueness property of the solutions of ordinary differential equations. Two manifolds can, however, intersect essentially in two different ways: transversely and tangen-

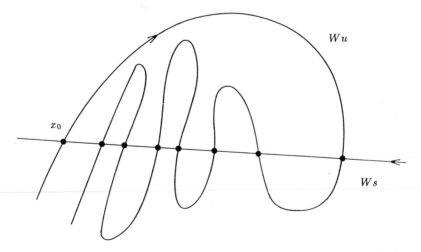

Fig. 4.5. Oscillation of manifolds due to the homoclinic orbit belonging to the saddle point x_0. Homoclinic points are marked by a full dot

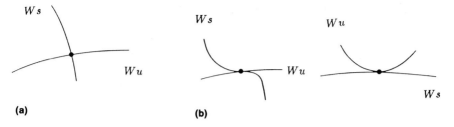

Fig. 4.6a, b. Transverse **a** and tangential **b** intersection of manifolds

tially. An illustration of these types of intersections is given in Fig. 4.6. Homoclinic tangency points account for an important phase-transition-like effect which will be discussed in Sects. 4.3.4 and 4.3.5.

We argued earlier that through the folding process Cantor-like structures will develop. From a rigorous mathematical point of view, it can, in fact, be shown that the existence of a transverse homoclinic point alone is sufficient for the existence of a Cantor structure. However, one may therefore ask oneself: what importance for "real-life" situations can an argumentation which is mainly based on model maps or model properties have? The question can partially be settled in a firm mathematical context by considering only those models as relevant whose topological properties can persist under a small enough C^k-perturbation of the system, $k \geq 1$. Only those systems, which are then said to be *structurally stable* [4.10], guarantee that all their dynamical and topological properties can be observed in reality or in computer experiments. For example, the famous "horseshoe map" has been proven to be both chaotic (in the strict sense which will be given in Sect. 4.1.4) and structurally stable. Since any system with a transverse homoclinic orbit must contain a horseshoe as a subsystem (e.g. [4.10]), all such systems are therefore proven to be chaotic. It can be shown, furthermore, that horseshoes exist in all dimensions greater or equal to two; chaotic systems therefore exist in great abundance. Note that an attractor which contains a transverse homoclinic orbit is sometimes called *strange* [4.15], but this term is also used in a less restrictive way [4.3]. Unfortunately, many of the so-called strange attractors investigated in this work cannot be interpreted in terms of structural stability (due to their nonhyperbolic structure) and the existence of a horseshoe cannot be proven. But, on the other hand, this might not be necessary since structural stability has turned out to be a nongeneric property anyway. The easiest way out of this dilemma is possibly to focus on the properties rather than on the systems, and to call those properties which can withstand such a perturbation of the model physically relevant. In this sense also a study of non-structurally stable systems can be helpful because they exhibit features common to all.

To give an example we consider the hyperbolic analog of the Hénon map. In many respects the *Lozi map* [4.16]

$$(x_{n+1}, y_{n+1}) = (1 - a|x_n| + by_n, x_n) \tag{4.19}$$

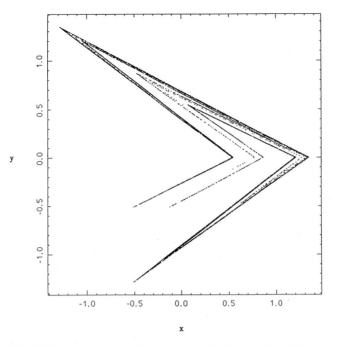

Fig. 4.7. Strange attractor of the Lozi map. For the equation of the map see text

provides a good approach to the former. For instance, the similarity between the strange attractor of the Lozi map obtained for $a = 1.7, b = 0.5$ and the attractor of the Hénon map is evident (compare Figs. 4.3 and 4.7). The unstable manifold can be made visible in the following way [4.15]. Choose a point x near a hyperbolic periodic point, draw a line segment between that point and its image point T and let the map iterate for a number of times. The picture which is obtained corroborates the observation that a strange attractor consists essentially of the unstable manifolds of its unstable periodic points.

4.1.4   Chaos

In the preceding subsections different aspects of highly irregular behavior in dynamical systems have been worked out. As a summary, highly irregular behavior has been seen to be related with the following features:

1. Loss of stability of the dynamical map and as a consequence amplification of perturbations.
2. Cantor-like structures develop for the support of the natural measure.

A simple necessary condition for this to happen may be formulated as follows. Suppose that a covering of the phase space by solid spheres $B(x_0, \varepsilon)$ with centers x_0 and, for convenience, fixed infinitesimal radius ε is given. Under k-fold iteration these balls are then mapped into ellipsoids with axes ε_i', $i = 1, \ldots, m$.

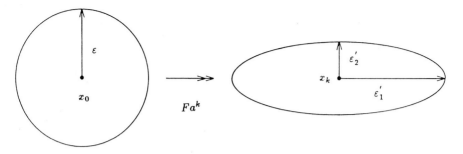

Fig. 4.8. A ball of radius ε is mapped onto an ellipsoid with semiaxes ε_i' (see text)

Using a short notation this can be written as

$$\varepsilon_i' \sim \varepsilon\, e^{k\lambda_i(x_0,\,k)} \ . \tag{4.20}$$

This kind of notation is further explained in Sect. 4.2.1. The situation is shown in a schematic picture in Fig. 4.8.

The logarithmic stretching or shrinking rates $\lambda_i(x_0, k)$ are called the effective Lyapunov exponents of k steps of the initial point x_0 [4.17, 18]. In Sect. 4.4 it will be discussed in more detail that in the limit $k \to \infty$ these exponents converge ρ-almost everywhere on the attractor towards fixed numbers λ_i called *Lyapunov exponents*. In what follows let us assume that the Lyapunov exponents λ_i are ordered according to magnitude such that $\lambda_1 \geq \lambda_2 \geq \ldots \geq \lambda_m$ and that at least one exponent is positive. It is then evident that small deviations or perturbations are exponentially amplified in the directions corresponding to positive Lyapunov exponents, and the system is said to display a sensitive dependence on initial conditions. At the same time, due to the boundedness of the phase space region, Cantor-like structures are generated.

As a consequence, the most common definition of chaos is given as follows. A dynamical system on an attractor A is called *chaotic* [4.7] if it has a sensitive dependence on initial conditions, if it is topologically transitive (this means roughly that it cannot be broken into two parts which do not interact under Fa), and if the periodic points are dense in A. For both experimental and model systems an analysis of the Lyapunov exponents is therefore of primary interest. In this work this analysis is provided for easy model systems in Sect. 4.2 while the more involved generic case is treated in Sects. 4.3–4.5.

4.2   Scaling Behavior of Attractors of Dissipative Dynamical Systems

4.2.1   Scale Invariance

In the previous section it has been argued that through the folding process self-similar structures will develop. In simple words an object can be called self-similar if it does not change its appearance under magnification. A precise

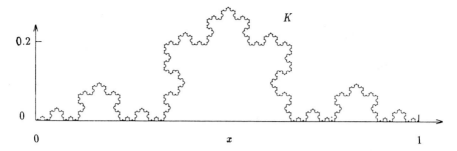

Fig. 4.9. Approximated "triadic" Koch curve (k = 5th generation). The final self-similar curve evolves for $k \to \infty$. Starting from the interval $[0, 1]$, the middle third is removed and replaced by two sides of an equilateral triangle, pointing outwards. In the next step this procedure is repeated for each of the new intervals of length $1/3$ with respect to the previous generation, and so on

definition of self-similarity is presently not needed; it will be given in Sect. 4.2.5. Consider in Fig. 4.9 the simple self-similar object of a Koch curve K. K can be interpreted as the graph of a relation $[x = t, y = F_{\text{Koch}}(t)]$, $t \in [0, 1]$.

Although F_{Koch} is not single-valued, if corresponding points are considered, it is easy to see that with $\mu_0 = 3^{-n}$, $n = 0, 1, 2, \ldots$, $F_{\text{Koch}}(t)$ satisfies the homogeneity condition

$$F_{\text{Koch}}(\mu_0 t) = \mu_0^\alpha F_{\text{Koch}}(t) , \tag{4.21}$$

with the scaling exponent $\alpha = 1$, at each point of the graph. More generally, a function F is said to be scaling or scale invariant [4.19] if $F(\mu t) = \mu^\alpha F(t)$, $\forall \mu > 0$. Functions with that property are called homogeneous functions (the power-law functions, for example). This class of functions has been most useful in describing critical phenomena near second-order phase transitions since renormalization group theory showed that the associated free energy has a scaling form. Necessary conditions to be satisfied by a scaling function F are positiveness of $F(\varepsilon)$ and the existence of the limit

$$\alpha = \lim_{\varepsilon \to 0} \frac{\log F(\varepsilon)}{\log \varepsilon} . \tag{4.22}$$

Equivalently, the last property is written as $F(\varepsilon) \sim \varepsilon^\alpha$. Note, however, that such an expression does not imply $F(\varepsilon) = C_1 \varepsilon^\alpha$, which is seldom realized in real physical situations, but, more generally, $F(\varepsilon) = c_1 \varepsilon (1 + c_2 \varepsilon^\alpha + \ldots)$, $\alpha > 0$. In two dimensions we note that the corresponding condition $F(\mu x, \mu y) = \mu^\alpha F(x, y)$ can be written as $F(x, y) = y^\alpha \tilde{F}(x/y)$, with $\tilde{F}(z) = F(z, 1)$. For a discussion of generalized homogeneous functions see [4.20].

For dynamical systems two kinds of scaling behavior with regard to the dynamical map have to be distinguished. To have a simple example we consider the logistic map $x_{n+1} = a x_n(1 - x_n)$ [4.7, 21]. As pointed out before, a continuous change in the external parameter (in our case a) can lead the system, e.g., on the period doubling route, to chaos. It can be seen that in this way various

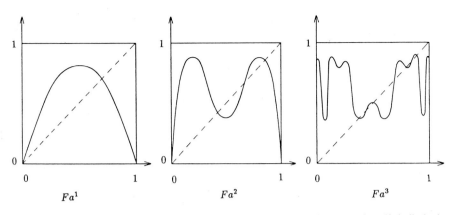

Fa^1 Fa^2 Fa^3

Fig. 4.10. Scaling property of the dynamical map with respect to iteration. Note the self-similarity in the graph for $n \to \infty$. The way each graph evolves from the previous graph resembles the creation process of the Koch curve in Fig. 4.9, although the present situation is somewhat more complicated

distinct scaling laws are obeyed (exponents: Feigenbaum's delta and alpha [4.1, 7, 21, 22]). For a fixed external parameter, however, the scaling behavior with respect to iteration of the dynamical map is important (Fig. 4.10).

The observed scaling property can be shown to be universal in the sense that it depends only on the order of the maximum. Returning to the two-dimensional case we see that in an analogous way the dynamical map will develop self-similar structures of the invariant probability measure in the plane, as has been observed already for the Hénon map. These structures can best be described with the help of a logarithmic scaling exponent of the measure. The *local crowding index* $\alpha(x_0, \varepsilon)$ is then introduced as

$$\alpha(x_0, \varepsilon) = \frac{\log P(B(x_0, \varepsilon))}{\log \varepsilon} , \tag{4.23}$$

where $P(B(x_0, \varepsilon))$ denotes the probability of the ball $B(x_0, \varepsilon)$ of radius ε and center x_0 with respect to the natural measure $P(B(x_0, \varepsilon)) = \int_{B(x_0, \varepsilon)} d\rho(x)$. In the limit $\varepsilon \to 0$ the quantity is called the *local pointwise dimension* [4.11] around x_0. This can be seen in complete analogy with the definition of the *effective Lyapunov exponents* of k steps as the logarithmic stretching rates $(1/k) \log[DFa^k(x_0)]$ which describe locally the dynamical scaling properties of the support.

At this point it might be asked whether there is a relation simpler than (4.14) between the scaling of the support and the probability measure of the attractor. It could be presumed that the distribution of the positive effective Lyapunov exponents and the distribution of the local crowding indices might contain identical information. This, however, is not the case generically, although important exceptional cases will be considered in the sequel. A distinct natural measure on the attractor can be created by different dynamical processes; on the other hand, the information about the local stretching rates alone cannot specify the "shape" of an attractor.

4.2.2 Symbolic Dynamics

Different regions of the phase space with the same scaling behavior with respect to the scaling of the support (Lyapunov exponents) or with respect to the scaling of the measure (local dimensions) can be considered to constitute classes with identical scaling properties. The number of classes needed for this description is one of the relevant parameters which determine the complexity of the system (for a precise definition of complexity see, e.g., [4.23]).

Such a *partition* of the phase space can be seen to be related to a finite precision measurement of a system at successive times and the information gain contained in one single measurement. The measurement process is then most successful if an infinite number of measurements can individuate one single initial condition; in this case the partition is called *generating*. Let us denote the different regions of a chosen partition by B_j, $j = 1, \ldots, M$ and associate with each region a symbol s_j, $j = 1, \ldots, M$. The result of a continued measurement can then be coded as a string of symbols $\{\ldots, s_i, s_{i+1}, \ldots\}$. Moreover, this string is seen to describe the dynamics of the system equivalently in the class of homeomorphisms if the partition is generating; the metric in the space of bi-infinite symbolic sequences is given by $d(a, b) = \sum_{i=-\infty}^{\infty} |a_i - b_i|/M^{|i|}$. This leads to the description of a dynamical system via its *symbolic dynamics* [4.7]. For the simple example of the tent map the procedure is shown in Fig. 4.11.

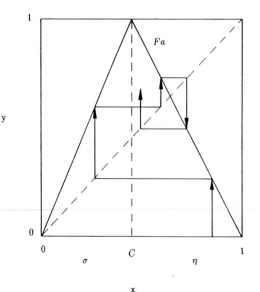

Fig. 4.11. Explanation of the idea of symbolic dynamics. As an example the tent map Fa is taken: $Fa: x \to ax/C$, for $x \in [0, C]$; otherwise $x \to a(1 - x)/(1 - C)$, $a \in \mathbb{R}$. The tent map is considered in the case of fully developed chaos [$Fa(C) = 1$]. The symbolic sequence $\{\ldots, \eta, \sigma, \eta, \eta, \ldots\}$ is produced. The partition requires 2 classes

In the space of symbols the dynamical map Fa is represented as a shift map

$$F_s : \{ \ldots, s_i, s_{i+1}, \ldots \} \rightarrow \{ \ldots, s_{i+1}, s_{i+2}, \ldots \} \, . \tag{4.24}$$

Let us suppose we have a generating partition. If the dynamics considered are chaotic then the image of one element of the partition B_j under the n-fold action of the dynamical map will be far-spread over the attractor. The information gain contained in one additional measurement of the system can then be measured in the following way. With each part $S_n = \{ s_0, s_1, \ldots, s_{n-1} \}$ of length n of a string of symbols (called n-cylinder [4.13]) a probability is associated:

$$P(S_n) = \rho^*(B) \, , \tag{4.25}$$

where $B = \bigcap_{k=0}^{n-1} Fa^{-k}(B_{s_k})$, because only the initial conditions contained in B can lead to this n-cylinder. The probability measure ρ^* can be chosen, e.g., as the probability of observing or measuring this specific substring among all substrings of length n. In this case ρ^* coincides with the natural measure $P(S_n) = \int_{G(S_n)} d\rho(x)$, where $G(S_n)$ denotes the area in phase space which leads to the same symbolic sequence S_n. For the moment, however, let us choose ρ^* differently. The roughest approach to estimate the complex behavior of a system is made by letting $\rho^*(B) = 0$ for B empty and $\rho^*(B) = \text{const} > 0$ for B nonempty. To characterize the system as a whole, this quantity must be averaged or summed over the system. Since the total number $N(n)$ of all possible n-cylinders will grow exponentially in n, the quantity

$$h_{\text{top}} = \lim_{n \to \infty} \frac{\log N(n)}{n} \, , \tag{4.26}$$

can be considered. h_{top} is called the *topological entropy* of the system. If this quantity is larger than 0, this means that orbits with nearby starting points have to separate under iteration of the dynamical map. The topological entropy has therefore been used as an indicator of chaotic behavior ($\infty > h_{\text{top}} > 0$: chaotic behavior). If a generating partition is not available, the above definition is replaced by the supremum over all possible partitions $\mathscr{P}$: $h_{\text{top}} = \sup_{\mathscr{P}} h_{\text{top}}(\mathscr{P})$. The supremum can always be reached if ever finer partitions are chosen, which forces the diameters of the sets B_j to go to zero. If $fix(Fa^n)$ denotes the set of fixed points of Fa^n then it can be shown that for a large class of dynamical systems $h_{\text{top}} = \lim_{n \to \infty} (1/n) \log[\text{card}(fix(Fa^n))]$ [4.9d]. This fact will be used in Sect. 4.2.5. The use of the natural measure ρ leads to the definition

$$K(q) = - \sup_{\mathscr{P}} \lim_{n \to \infty} \frac{1}{n} \frac{1}{q-1} \log \sum_{S_n} P^q(S_n) \, , \tag{4.27}$$

which coincides for $q = 0$ with the topological entropy h_{top}. For $q \to 1$ another important quantity is obtained, the Kolmogorov–Sinai or metric entropy, which can also be written as

$$K(1) = - \sup_{\mathscr{P}} \lim_{n \to \infty} \frac{1}{n} \sum_{S_n} P(S_n) \log P(S_n) \, . \tag{4.28}$$

It is easily seen that $K(1)$ is related with Shannon's information entropy; a different interpretation can be given as the average spread rate of the mass initially contained in unit boxes covering the attractor. In contradistinction to other entropy-like quantities we will refer to $K(q)$ as *information-theoretical entropies*. For the case of the fully chaotic tent map considered in Fig. 4.11 one obtains $K(0) = \log 2$ and $K(q) = [-1/(q-1)] \log [C^q + (1-C)^q]$.

The borders of the different regions of a generating partition are generally chosen as $(n-1)$-dimensional submanifolds of the underlying space. Unfortunately, for general dynamical systems there is no general recipe known for the construction of a generating partition. However, for two-dimensional nonhyperbolic systems it has been conjectured that by a line connecting "principal" tangency points such a partition could be obtained [4.24–26]. In this way, interesting progress using symbolic dynamics has been made for low-dimensional systems (Hénon and Lozi map, circle map, [4.24]). For higher-dimensional systems the latter approach seems to be very difficult. Since the correspondence between the original system and the symbolic representation is only via a homeomorphism, a symbolic string itself does not contain the whole information about the dynamics of the system. The missing information about the size of Lyapunov exponents has to be provided explicitly (e.g., by exploration of periodic orbits).

A description corresponding to the approach by the Frobenius–Perron equation can now be given for the scaling behavior around a point y on the attractor by means of a scaling function $\breve{\sigma}$ [4.27] in the symbolic space. $\breve{\sigma}$ is obtained with the help of successive approximations of order n and $n+1$ to the chaotic system by symbolic sequences of length n and $n+1$, respectively. Let ε_n and ε_{n+1} denote the maximal radius of balls around the $(n-1)$-times and n-times backward iterated points with associated symbols s_{-1} and s_0, respectively, such that the $(n-1)$-times and n-times forward iterations of that ball yield the same symbolic sequence for the whole ball. In this way, with each S_n a length ε has been associated. The "daughter/mother" scaling function $\breve{\sigma}$ can then be defined as

$$\breve{\sigma}(s_{-n}, \ldots, s_0) = \varepsilon_{n+1}(s_{-n}, \ldots, s_0)/\varepsilon_n(s_{-n}, \ldots, s_{-1}) \; ; \quad \cdot \tag{4.29}$$

it can, thus, be identified with DFa^{-1} in the previous approach.

In the next sections we follow a general statistical approach to describe the scaling behavior of a dynamical system, partly making use of the concept of symbolic dynamics.

4.2.3 Analogy with Statistical Mechanics

After having defined local scaling exponents (the effective Lyapunov exponents and the crowding index) we saw that it is natural to consider a partition of the phase space into classes of identical scaling behavior. The present situation resembles the starting point of statistical mechanics. Statistical mechanics deals with dominant properties (properties of a certain average) of a mechanical system. From *microscopic* knowledge, *macroscopic behavior* should be predicted. In the present case we are interested in the calculation of average scaling from individual, local scaling behavior. This is the origin from which an analogy between our further development and the formalism of statistical mechanics arises. We will, therefore, proceed by first recalling the relevant facts and notation of statistical mechanics before applying related concepts to the scaling behavior of dynamical systems.

In statistical mechanics, the partition function for constant pressure and temperature

$$Z_G(p, T) = \sum_s e^{-[H(s) + pV(s)]/k_B T} \tag{4.30}$$

is often considered, where s runs over the states of the system, $H(s)$ denotes the energy of the states, $V(s)$, its volume, T, the temperature and k_B, Boltzmann's constant. Then the function

$$F_G(p, T) = -k_B T \log Z_G \tag{4.31}$$

is derived (F_G is often called Gibbs free energy or enthalpy, or Gibbs potential). The probability P for the system to be in state s is

$$P(s) = Z_G^{-1} e^{-[H(s) + pV(s)]/k_B T}. \tag{4.32}$$

If X is an observable, then the expectation value is given by

$$\langle X \rangle = Z_G^{-1} \sum_s X(s) e^{-[H(s) + pV(s)]/k_B T}. \tag{4.33}$$

From the Gibbs free energy, all the other thermodynamical quantities can be calculated. Other partition functions can be used; for our problem the use of this specific partition is most appropriate.

An important example of macroscopic behavior to be predicted is the phenomenon of a *phase transition* [4.19, 28]. The classical situation is recalled in Fig. 4.12 as the transition between liquid and gaseous phases. In this figure the constant-temperature curves for an equation of state are shown, where T denotes the temperature, V, the volume and p, the pressure. The broken curve is called the "coexistence curve" for the two phases. Beneath this curve, both phases can coexist along the horizontal lines. The endpoints on the coexistence curve correspond to pure phases. They are determined by Maxwell's rule of equal areas.

For $T < T_c$, V changes discontinuously with p; it changes in a C^∞ way for $T > T_c$. For $T = T_c$, V changes continuously but

$$\left(\frac{\partial V}{\partial p}\right)_T \tag{4.34}$$

diverges. Therefore, for $T < T_c$, the "phase transition" is said to be of first order, otherwise the phase transition is said to be a continuous transition or of higher

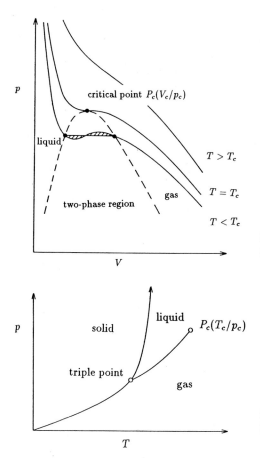

Fig. 4.12. V–p diagram for two phases (liquid/gas). See text

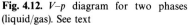

Fig. 4.13. Domain of nonanalytical behavior of V in the T–p plane

than first order. In the latter case "universal behavior" is displayed: all critical phenomena do not depend on details of the local forces but only depend on global characteristics such as the dimension of the system, the number of components, or the range of forces (see renormalization theory). A different aspect of the same phenomenon can be given in the T–p plane (Fig. 4.13). V is an analytic function of p and T except on the lines drawn, where it is discontinuous, except at the critical point $P_c(T_c/p_c)$.

For a general system consider again the Gibbs potential $F_G(p, T)$. In order for the system to have a phase transition of order n, it is required that one of the terms

$$\left(\frac{\partial^n F_G}{\partial^n p}\right)_T \tag{4.35}$$

or

$$-\left(\frac{\partial^n F_G}{\partial^n T}\right)_p \tag{4.36}$$

diverges or jumps. Unfortunately, this classification of phase transitions is not yet exhaustive. Therefore, a physical phase transition has to be defined as the subset of mathematical points of nonanalytical behavior of the thermo-dynamical function F_G to which a physical relevance can be attributed.

As a simple example we consider a partition function

$$Z_G(p, T) = \sum_s^N e^{-[H(s) + pV(s)]/k_B T} , \tag{4.37}$$

where each term to be summed is smooth in T and p and infinitely differentiable $(T \neq 0)$. F_G can become nonanalytical only in the thermodynamical limit $N \to \infty$. On the other hand, analytical dependence means that there is a conver-gent series expansion

$$F_G(p, T) = \sum_{n=0}^{\infty} g_n(T) p_i^n . \tag{4.38}$$

A system with phase transitions can, therefore, be constructed as follows: for $T < T_c$ a singular term is added, for example, $C(T)|p_i|^k$, $k \in R$, yielding for $p_i = 0$, if k is not even, a phase transition which for $k = 1$ is of first order, for $1 < k < 2$, of second order and generally for $n - 1 < k \leq n$, of nth order.

In order to characterize the scaling behavior of a dynamical system let us proceed analogously. First we define a partition function from which a free energy can be derived as the fundamental function which describes the scaling behavior of a dynamical system. The notions of different entropy-like quantities used in the literature are interpreted in terms of that free energy. Then the points of nonanalytical behavior are investigated and their meaning is discussed. It should be pointed out that an exact evaluation of Z_G [given in (4.30)] is generally only possible for simple model systems. Furthermore, most real phase transitions arise as a result of two competing microscopic effects, where one effect tends to order the system and the other tends to randomize it. We will see that analogous statements are true for dynamical systems.

4.2.4 Partition Function for Chaotic Attractors of Dissipative Dynamical Systems

For the attractor of a dissipative dynamical system, the partition creates classes of points with identical scaling behavior, corresponding to the classes given by $H(s)$ in (4.30). In contrast to the case just considered, the present scaling behavior can no longer be described with the help of one single variable, since the scaling of the support as well as the scaling of the measure should be taken into account. Nevertheless, this procedure leads us to a partition, which can be seen as the analog of an isotherm–isobar ensemble [4.28c] in statistical mechanics.

A *generalized partition function* for the attractor A is then defined in the following way [4.29]:

$$GZ(q, \beta, n) = \sum_{j \in (1, \ldots, M)^n} l_j^\beta p_j^q. \tag{4.39}$$

Here p_j is the probability of the system falling in the jth region of the partition (depending on the iteration n), l_j describes the size of the region and β and q can be called "filtering exponents". To account for the anisotropy of the attractor, they can be thought of as vectors where necessary. For simplicity it is assumed above that a generating partition of M elements has been found. The evaluation of the average of the scaling quantities over the attractor then becomes simple, since only the scaling properties of the finite number of elements in the generating partition have to be evaluated. Although the starting point is the same as in [4.29], a more general formalism is now developed and a different interpretation is given. In what follows we owe much to the arguments given in [4.30], where, however, the connection of the present formalism with neither the scaling function of Lyapunov exponents nor the information-theoretical entropies was established.

Let us introduce in GZ a "local scaling assumption" with respect to iteration. It is assumed that l and p scale with n (and the scaling exponents ε and α, respectively) in the following way:

$$l_j = e^{-n\varepsilon_j}, \tag{4.40}$$

$$p_i = l_j^{\alpha_j}. \tag{4.41}$$

The generalized partition function is then obtained as [4.30]

$$GZ(q, \beta, n) = \sum_{j \in (1, \ldots, M)^n} e^{-n\varepsilon_j(\alpha_j q + \beta)} \tag{4.42}$$

and the *generalized free energy*

$$GF(q, \beta) = \lim_{n \to \infty} \frac{1}{n} \log GZ(q, \beta, n) \tag{4.43}$$

can be derived.

A *generalized entropy* $GS(\alpha, \varepsilon)$ is related to a "global scaling assumption": it is assumed that the number of regions whose scaling exponents are between (α, ε) and $(\alpha + d\alpha, \varepsilon + d\varepsilon)$ can be written as

$$N(\alpha, \varepsilon) \, d\alpha \, d\varepsilon \sim e^{n GS(\alpha, \varepsilon)} \, d\alpha \, d\varepsilon \tag{4.44}$$

in the limit $n \to \infty$.

With the help of the generalized entropy the generalized partition function can be written as an integral in the following way:

$$GZ(q, \beta, n) \sim \int d\varepsilon \int d\alpha \, e^{n[GS(\alpha, \varepsilon) - (\alpha q + \beta)\varepsilon]}. \tag{4.45}$$

Using a saddle point approach in the limit $n \to \infty$ the relation

$$GF(q, \beta) = GS(\langle \alpha \rangle, \langle \varepsilon \rangle) - (\langle \alpha \rangle q + \beta) \langle \varepsilon \rangle \tag{4.46}$$

is obtained, where $\langle \varepsilon \rangle$, $\langle \alpha \rangle$ lead to the maximum of the bracket [   ] above.

This shows that $GF(q, \beta)$ and $GS(\alpha, \varepsilon)$ are connected via an (unusual) generalization of a two-dimensional *Legendre transformation*. $\langle \varepsilon \rangle$ and $\langle \alpha \rangle$ can be calculated from the free energy $GF(q, \beta)$ as

$$\langle \varepsilon \rangle = -\frac{\partial}{\partial \beta} GF(q, \beta) , \tag{4.47}$$

$$\langle \alpha \rangle = -\frac{\partial}{\partial q} \frac{GF(q, \beta)}{\langle \varepsilon \rangle} ; \tag{4.48}$$

the generalized entropy GS can be derived in a similar way as

$$GS(\langle \alpha \rangle, \langle \varepsilon \rangle) = GF(q, \beta) - q\frac{\partial GF(q, \beta)}{\partial q} - \beta\frac{\partial GF(q, \beta)}{\partial \beta} . \tag{4.49}$$

To focus on the dependence of the entropy on one of the two scaling exponents α and ε alone, two additional entropies can be introduced [4.30]. The entropy function for the characterization of the scaling behavior of the support, denoted by $GS(\varepsilon)$, is introduced according to

$$e^{nGS(\varepsilon)} = \int d\alpha\, e^{n[GS(\alpha, \varepsilon)]} . \tag{4.50}$$

In complete analogy, also an entropy function for the scaling of the measure, which is denoted by $GS(\alpha)$, can be introduced by the relation

$$e^{nGS(\alpha)} = \int d\varepsilon\, e^{n[GS(\alpha, \varepsilon)]} . \tag{4.51}$$

The understanding in these definitions is that $GS(\alpha)$ is given by the maximal value of $GS(\alpha, \varepsilon)$ with respect to variation of ε alone and likewise for $GS(\varepsilon)$. Therefore, the argument of $GS(\alpha)$ and q, β are related by the equation $\alpha q + \beta = 0$ and $GS(\alpha)$ assumes the same values as $GF(q, \beta)$ [see (4.49)]. $GS(\varepsilon)$, on the other hand, is given by (4.49) with the help of the condition $q = 0$. We note that this entropy function coincides with the entropy function defined by *Oono* and *Takahashi* [4.31] for the thermodynamical formalism for dynamical systems.

In the next section we apply the concepts developed so far. By inspection of a few model cases, we investigate characteristic forms that are assumed by the different entropy functions and establish the relationship with the concept of generalized fractal dimensions. The connection to the $f(\alpha)$ formalism developed in [4.29] is worked out in the next section.

4.2.5   Discussion of the Partition Function

An attractor is called *self-similar* [4.2] if a partition can be found such that in any step i of the n steps, the scaling region can be divided into the M subregions of the same scaling behavior as the old region. More explicitly, it is required that the partition function GZ can be factorized into an n-fold product of the form

$$GZ(q, \beta, n) = \sum_{j \in (1, \ldots, M)} l_j^\beta p_j^\alpha \cdot \sum_{j \in (1, \ldots, M)} l_j^\beta p_j^\alpha \cdot \ldots \cdot \sum_{j \in (1, \ldots, M)} l_j^\beta p_j^\alpha, \quad (4.52)$$

(n times) such that $GZ(q, \beta, n)$ can be written in a simpler way as

$$GZ(q, \beta, n) = GZ(q, \beta, 1)^n . \tag{4.53}$$

Thus, the requirement of self-similarity is equivalent to the existence of a generating partition. The formalism above can, however, be generalized for systems and partitions for which

$$GZ(q, \beta, n) \to \widetilde{GZ}(q, \beta, n) \tag{4.54}$$

asymptotically, meaning that the expression (4.39), if the actual partition is used, converges. In this case the attractor is called asymptotically self-similar. Furthermore, if the scaling exponents of a system are not identical with respect to the different directions in the phase space, the attractor is called *self-affine* [4.2].

The analogy with statistical mechanics can now be made explicit with the following identification:

$$-n\varepsilon = \text{energy} = E,$$

$$n\varepsilon\alpha = \text{volume/temperature} = V/k_B T,$$

$$-\beta = \text{inverse temperature} = 1/k_B T,$$

$$q = \text{pressure} = p,$$

where k_B denotes Boltzmann's constant. The ensemble chosen can be interpreted as an isotherm–isobar partition. n can be identified with the number of spins with as many possible states as given by the associated symbolic dynamics. It can be shown that all thermodynamic relations follow. Note that E and V are not independent and that the sum extends over the symbolic sequences S_n, where a covering of the attractor with finite diameters has been chosen, irrespective of the probability of S_n itself. A more detailed interpretation from the point of view of random sampling will be given in the next section.

Model Cases

To show the different characteristic realizations of the formalism developed we investigate three models: the "Cantor construction" and the escape from a strange repeller as one-dimensional examples, and the two-dimensional baker map.

Let us discuss first the Cantor construction. Before turning to more refined examples of this construction, the uniform Cantor set

$$C := \{x \in \mathbb{R} \mid x = 2 \sum_{i=1}^{\infty} s_i 3^{-i}, s_i \in \{0, 1\}\} \tag{4.55}$$

is considered. C can be taken as a simple model of a dynamical system (Fig. 4.1).

For $n \to \infty$ a closed, totally disconnected, perfect set is obtained (totally disconnected: contains no intervals; perfect: every point in the set is an accumulation
point or limit point of other points in the set). These properties serve as a more
general definition for a *Cantor set* [4.7]. In the multidimensional, nonisotropic
case this definition generalizes to the notion of a Cantor direction [4.32].

Let us first assume that the measure of the middle third is distributed to the
remaining two thirds. Then it is easily found that the partition function GZ has
the form

$$GZ(q,\beta,n) = \sum_{k=1}^{2^n} (3^{-n\beta}(1/2)^{nq}) , \tag{4.56}$$

and the free energy GF can be expressed as

$$GF(q,\beta) = \log(3^{-\beta}2^{-q+1}) . \tag{4.57}$$

A particularly interesting case is the case of a *zero of* $GF(q,\beta)$. If q is
interpreted as the independent variable and β as the dependent variable then
let us denote this zero by $\beta_0(q)$. For $n \to \infty$, $\beta_0(q)$ leads to a maximum $\neq \infty$
of the partition function. Hence, it can be taken as a generalization for $q \neq 0$
of the Hausdorff dimension [4.33]. As a different generalization of the
Hausdorff dimension, also the quantity $-\beta_0/(q-1)$ has been considered
[4.37]. Figure 4.14 shows a plot of both quantities. Note that for the present
case $-\beta_0/(q-1)$ is constant. Fractals with this property are called *uniform*. We

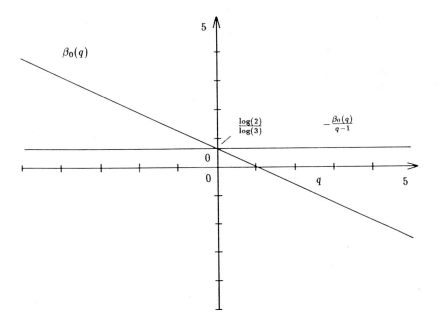

Fig. 4.14. Functions β_0 and $-\beta_0/(q-1)$ for the uniform Cantor set (see text)

note that $-\beta_0(q)$ has been denoted in [4.29] by $\tau(q)$. Observe, however, that in an earlier definition [4.34] $\tau(q)$ was defined using a fixed-length scale ε, contrary to the present procedure.

The derivative of the free energy with respect to β,

$$-\frac{\partial GF(q,\beta)}{\partial \beta}, \tag{4.58}$$

evaluated at $q = 1$ and $\beta = 0$, gives the time average of the logarithmic stretching rates, the Lyapunov exponent of the system (log 3), where some ambiguity with respect to the sign of the Lyapunov exponent seems to be implicit. This sign can, however, be understood by observing that for an ordinary dynamical system the contracting direction corresponds to attraction under backward iteration in time. The escape rate κ is zero. If the measure of the middle third is not redistributed, it is seen that the escape rate has the value

$$\kappa = \log 3 - \log 2 . \tag{4.59}$$

The generalized entropy is different from zero only for $\varepsilon = \log 3$ and $\alpha = \log 2/\log 3$. Here the value

$$GS(\alpha, \varepsilon) = \log 2 \tag{4.60}$$

is assumed.

Of course, the model of a uniform Cantor set is not very satisfactory for the approximation of the Cantor structure observed, e.g., for the Hénon map. A better approximation can be given if two different stretching factors are present, e.g., $l_1 = 1/4$, $l_2 = 2/5$. In this case a *nonuniform Cantor set* [4.2] is obtained. The partition function GZ can be written as

$$GZ(q,\beta,n) = \sum_{k=1}^{n} (b_{n,k})(l_1^{\beta+q})^{n-k}(l_2^{\beta+q})^k , \tag{4.61}$$

where $b_{n,k}$ is used to denote the binomial coefficients, whereas the free energy GF assumes the simple form

$$GF(q,\beta) = \log(l_1^{\beta+q} + l_2^{\beta+q}) . \tag{4.62}$$

The zero of $GF(q,\beta)$ is then obtained by solving

$$l_1^{\beta_0+q} + l_2^{\beta_0+q} = 1 \tag{4.63}$$

for $\beta_0(q)$.

For $q = 0$, again, the Hausdorff dimension is obtained from $\beta_0(q)$ or $-\beta_0/(q-1)$, which in the present case are nonlinear functions of q, in contrast to the uniform case. As can easily be calculated, $GS(\alpha, \varepsilon)$ is now nonzero on a segment of a straight line in the α–ε plane given by $\alpha = 1$, and the simple equality $GS(\alpha, \varepsilon) = GS(\varepsilon)$ is found to hold. In Fig. 4.15 the graph of $GS(\varepsilon)$ is shown for $l_1 = 1/4$ and $l_2 = 2/5$. Note that in the present model the probability p_i has been chosen equal to the length l_i, $i = 1, 2$. In general, however, this will

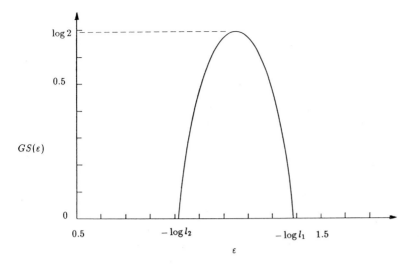

Fig. 4.15. $GS(\alpha, \varepsilon)$ for the nonuniform Cantor set. For convenience, only the ε-axis is shown since $\alpha = 1$. Therefore, the support of $GS(\alpha, \varepsilon)$ consists of a straight line. It is found that $GS(\alpha, \varepsilon) = GS(\varepsilon)$, and $GS(\alpha)$ is nonzero only for $\alpha = 1$, where the value $\log 2$ is taken

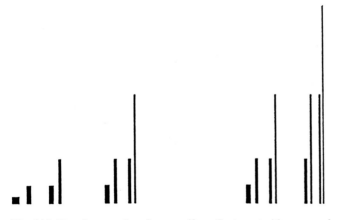

Fig. 4.16. Fourth generation of a nonuniform Cantor set with measure, $l_1 = 1/4$, $l_2 = 2/5$, $p_1 = 3/5$, $p_2 = 2/5$

not be the case. Note, furthermore, that by the above procedure no probability measure is obtained since the measure is not normalized.

If the probabilities are normalized (they need not, however, be proportional to the length of the pieces, e.g., $l_1 = 1/4$, $l_2 = 2/5$, $p_1 = 2/5$, $p_2 = 3/5$), then the limiting set is called a *nonuniform Cantor set with measure*. As an illustration, an intermediate step of this construction is shown in Fig. 4.16. It is seen that for the partition function and the free energy the expressions

$$GZ(q, \beta, n) = \sum_{k=1}^{n} (b_{n,k})(l_1^\beta p_1^q)^{n-k}(l_2^\beta p_2^q)^k ,$$ (4.64)

and

$$GF(q, \beta) = \log(l_1^\beta p_1^q + l_2^\beta p_2^q)$$ (4.65)

are obtained. The zero of $GF(q, \beta)$ is evaluated from the solution of the equation

$$l_1^{\beta_0} p_1^q + l_2^{\beta_0} p_2^q = 1$$ (4.66)

for $\beta_0(q)$. The function $-\beta_0/(q-1)$ for this Cantor set was first calculated in [4.29]. There it was called again the Renyi dimension $D(q)$, although the original definition of $D(q)$ [4.34] applies only for a fixed-length scale ε (see also Sect. 4.3.1). Again, $\beta_0(q)$ is a nonlinear function of q. The difference in comparison with the previously considered nonuniform Cantor set can better be worked out with the help of the entropy function $GS(\alpha, \varepsilon)$. For the nonuniform Cantor set with measure this function is displayed in Fig. 4.17. Note that the support of $GS(\alpha, \varepsilon)$ is now no longer a linear manifold, in contrast to Fig. 4.15.

Let us now briefly interpret the two-scale Cantor set in the framework of symbolic dynamics. The two-scale Cantor set model corresponds to a system which can be described by a binary symbolic string. The system has a *complete* symbolic tree, i.e., all sequences of the symbols are allowed, they all have nonzero probability. Furthermore, the probabilities factorize $[P(S_{n+n'})$ $= P(S_n) * P(S_{n'})]$. Such a system, although having maximal metric entropy $(\log 2)$, is nevertheless a rather uninteresting one, as it is not able to produce

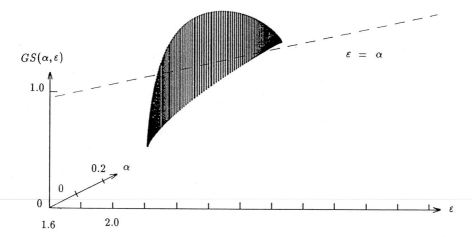

Fig. 4.17. $GS(\alpha, \varepsilon)$ for the nonuniform Cantor set with measure described in Fig. 4.16. $GS(\alpha, \varepsilon)$ is shown on the (ε, α) plane, viewed from an angle of 45°. The support of $GS(\alpha, \varepsilon)$ is still one-dimensional, but no longer a linear manifold

surprises. In other words, it is not as complex as it could be, although several (in our case two) alternatives exist. For that reason, in [4.23] the complexity of a system has been measured by a hierarchy of complexity exponents, the first being defined as $\lim_{n \to \infty} \log[N_a(n)]/n$, where $N_a(n)$ is the number of admissible subsequences. The second-order complexity is evaluated as $\lim_{n \to \infty} \log[N_f(n)]/n$, with N_f denoting the number of irreducible forbidden finite substrings (called "words"), where irreducible means that this substring contains no smaller forbidden substring.

As it is assumed that a generating partition is used, with any symbolic string S_n a length $\varepsilon(S_n)$ and a probability $P(S_n)$ can be associated (compare Sect. 4.2.2). The scaling exponent for the scaling of the measure in the symbolic space can then be introduced as $\alpha(S_n) = \log[P(S_n)]/\log[\varepsilon(S_n)]$, $n \to \infty$. Of course, this exponent has the same value as the corresponding exponent in the phase space. Observe that for composed strings of length $n + n'$, it is generally obtained that $P(S_{n+n'}) \neq P(S_n) * P(S_{n'})$. The scaling exponent α, however, is still found through the expression $\alpha(S_{n+n'}) = \log[P(S_{n+n'})]/\log[\varepsilon(S_n) * \varepsilon(S_{n'})]$. Even if the probabilities factorize, the former relation is usually nonlinear. This fact, of course, is responsible for the curvilinear nature of the support of GS in Fig. 4.17.

An important and yet simple example of a one-dimensional Cantor set is furnished by the escape from a *strange repeller* [4.13, 35]. To have the simplest example in mind we could choose the symmetric tent map (Fig. 4.11), modified such that $Fa(C) > 1$, but also the logistic map with $a > 4$ belongs to the class to be considered. For these maps on the interval the partition function can be expressed as

$$GZ(q, \beta, n) = \sum_{x \in Fa^{-n}(y)} e^{-\beta \log |DFa^n(x)|} \left(\frac{e^{-\log |DFa^n(x)|}}{\sum_{i=1}^{} |Fa^{-n}(y)| \, l_i} \right)^q, \quad l_i := e^{-\log |DFa^n(x_i)|} ,$$

$$(4.67)$$

where the first factor on the right-hand side represents the scaling of the support and the second, the scaling of the associated measure, which is supposed to be proportional to the length. Now the connection with the Frobenius–Perron equation of the form given in (4.16) can be established as follows:

$$GZ(q, \beta, n) = \left(\frac{1}{\sum_i l_i} \right)^q \sum_{x \in Fa^{-n}(y)} e^{-(\beta+q) \log |DFa^n(x)|} . \qquad (4.68)$$

In view of (4.16), the factor in front of the summation can be seen to be responsible for the escape rate κ. Now a relation between the information-theoretical entropy $K(1)$, the Lyapunov exponent λ, the pointwise dimension α and the escape rate κ and the generalized free energy GF can be worked out. It is easily seen that GF may be written in the form

$$GF(q, \beta) = GF(q = 0, \beta + q) + \kappa q , \qquad (4.69)$$

and, therefore, the equality

$$GF(q = 1, \beta = 0) = GF(q = 0, \beta = 0 + 1) + \kappa(q = 1) \qquad (4.70)$$

holds. Knowing, furthermore, that also the Lyapunov exponent can be calculated from the free energy using

$$\lambda = -\frac{\partial GF}{\partial \beta} \qquad (4.71)$$

for $q = 1$, $\beta = 0$ [see (4.58)], and since $GS(q = 1, \beta = 0)$ is identical with the Kolmogorov–Sinai entropy $K(1)$, we obtain the equation

$$GF(q = 1, \beta = 0) = K(1) - \lambda + \kappa . \qquad (4.72)$$

Because both $\langle \alpha \rangle$ and $\langle \varepsilon \rangle$ are functions of q, β, the meaning of $GS(q, \beta)$ is obvious. *Bohr* and *Tél* [4.13] showed, and it can easily be derived from the present formalism, that for hyperbolic systems the relation $\lambda \alpha = \lambda - \kappa$ holds (where α is the pointwise dimension). This fact leads us to the desired relation between the free energy, Kolmogorov–Sinai entropy, Lyapunov exponents and pointwise dimension: $-\lambda \alpha = GF(q = 1, \beta = 0) - K(1)$. This equation demonstrates explicitly that not even for these systems it can be assumed that $\lambda \alpha = K(1)$.

As pointed out before, the *tent map* $x \to ax/C$ for $x \in [0, C]$, $x \to a(1 - x)/(1 - C)$ for $x \in [C, 1]$ serves as an explicit example for this class if C is mapped onto a value larger than one. It is then convenient to replace the parameters a and C by two parameters l_1, l_2 chosen such that $Fa(l_1) = 1$ and $Fa(1 - l_2) = 1$. With the help of these parameters the map is written as $x \to x/l_1$ for $x \in [0, l_1/(l_1 + l_2)]$, $x \to (1 - x)/l_2$ for $x \in [l_2/(l_1 + l_2), 1]$. The fully chaotic case is obtained by the condition $l_1 + l_2 = 1$, whereas the strange repeller is characterized by $l_1 + l_2 < 1$. For both cases the partition function can be written as

$$GZ(q, \beta, n) = (l_1 + l_2)^{-nq}(l_1^{\beta+q} + l_2^{\beta+q})^n , \qquad (4.73)$$

from which the form of the free energy $GF(q, \beta) = q\kappa + \log(l_1^{\beta+q} + l_2^{\beta+q})$ and the fundamental relation for strange repellers $\alpha = 1 - \kappa/\varepsilon$ is easily derived.

For the fully chaotic case, at variance with the usual Cantor construction, a uniform density $\rho(x)$ is obtained:

$$\beta_0 = -(q - 1) , \qquad (4.74)$$

such that the Hausdorff dimension is equal to 1.

The Lyapunov exponent is easily calculated to have the value

$$\lambda = -(l_1 \log l_1 + l_2 \log l_2) ; \qquad (4.75)$$

it can be seen that λ is equal to the Kolmogorov–Sinai entropy in the fully chaotic case. For later purposes we note that the entropy function can be expressed as

$$GS(\alpha, \varepsilon) = \log(l_1^{\beta+q} + l_2^{\beta+q}) - \frac{(\beta + q)(l_1^{\beta+q} \log l_1 + l_2^{\beta+q} \log l_2)}{(l_1^{q+\beta} + l_2^{q+\beta})} . \qquad (4.76)$$

Again, q, β on the right-hand side of (4.76) denote those values which lead to the corresponding values of α, ε considered on the left-hand side.

For this simple model system also the entropy functions $GS(\varepsilon)$ and $GS(\alpha)$ can easily be calculated. Particularly simple relations are obtained for the fully chaotic case. For $q = 0$ the equality of $GS(\varepsilon)$ and $GS(\alpha, \varepsilon)$ can be verified. Since, as is immediately seen, we have $\langle \varepsilon \rangle = \langle \alpha \rangle \langle \varepsilon \rangle$, where $\langle \varepsilon \rangle \neq 0$, it follows that $\langle \alpha \rangle = 1$ and $\beta = -q$. This, finally, implies that the entropy function $GS(\alpha)$ is of the simple form $GS(\alpha) = 1$ for $\alpha = 1$. Note that this explicit discussion of the tent map has been given because this example will be reconsidered in Sect. 4.3.1.

In the previously considered examples the scaling exponent α for the scaling of the measure was either constant or there was a bijective relationship between α and ε. In the most general case, however, this is no longer true. A typical example is furnished by the *three-scale Cantor set with measure*. The case of a "ternary" Cantor set has been shown to be of relevance for the scaling behavior of the postcritical regime of the circle map [4.24b], where the tree is not complete.

In Figs. 4.18–21 we give a full view of the scaling properties of this system (see the figure caption for the chosen parameters). As can be seen, the support of GS is no longer a one-dimensional manifold. The borders of GS are curved and so are the lines along which the different functions, which describe the various

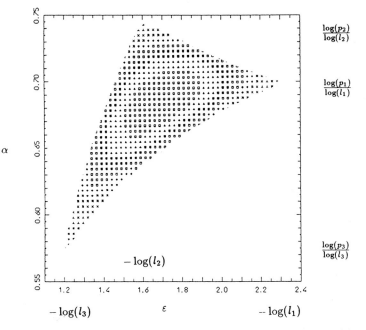

Fig. 4.18. Support of $GS(\alpha, \varepsilon)$ in the ε–α plane for the three-scale Cantor set with $p_1 = 0.2$, $p_2 = 0.3$, $p_3 = 0.5$, $l_1 = 0.1$, $l_2 = 0.2$, $l_3 = 0.3$. Increasing from 0 in steps of 0.1, different symbols have been associated with the different values assumed by GS

aspects of the scaling behavior of the system, are evaluated. This means that for a full characterization of the scaling behavior of a system both aspects – the scaling of the measure and the scaling of the support – have to be investigated.

As a first summary concerning one-dimensional examples we note that starting from a generalized entropy function built on both distributions of α and ε the calculation of model cases has been shown to turn out in a straightforward way, in analogy with the procedures known from statistical mechanics. Furthermore, on this basis explicit relationships between the scaling of the measure and the scaling of the support have been shown. Only in special cases is $GS(\alpha)$ sufficient (if there is no distribution of ε) to give a complete characterization of the scaling behavior of the system. $GS(\varepsilon)$ is sufficient if a generating partition with constant mass can be found.

The dependence of $GS(\alpha, \varepsilon)$ on α and ε can, however, better be read off from Figs. 4.19–21. As pointed out before, more precise information about the values of $GS(\alpha, \varepsilon)$ can now be obtained from the graphs of the functions $GS(\alpha)$ and $GS(\varepsilon)$. These graphs are displayed in Figs. 4.20 and 4.21.

As a final example let us look at the scaling behavior of the two-dimensional *baker map* [4.27] as the prototype of a two-dimensional hyperbolic map. For the equation and the attractor we refer to Fig. 4.22.

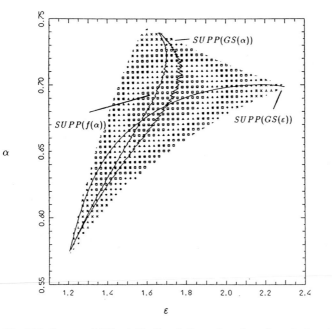

Fig. 4.19. Support of $GS(\alpha, \varepsilon)$. The lines indicate the values of ε and α for which the functions $GS(\alpha)$, $GS(\varepsilon)$ and $f(\alpha)$ are evaluated. For the definition and the meaning of the function $f(\alpha)$ we refer to Sect. 4.3.1, where in Fig. 4.23 the graph of $f(\alpha)$ associated with the present example is shown

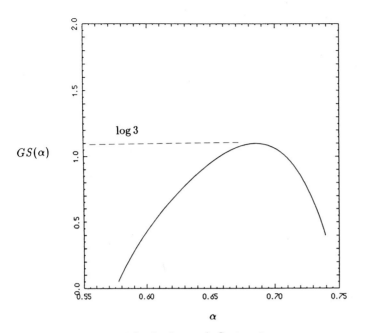

Fig. 4.20. Function $GS(\alpha)$ for the three-scale Cantor set

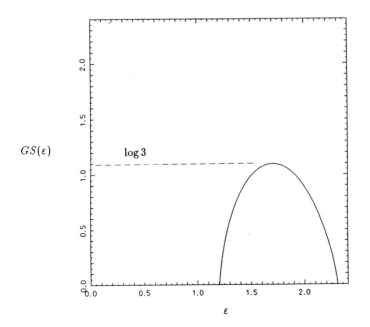

Fig. 4.21. Function $GS(\varepsilon)$ for the three-scale Cantor set

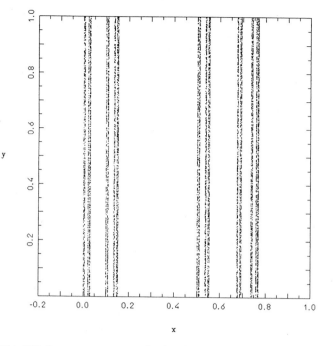

Fig. 4.22. Strange attractor of the baker map: $(x_{n+1}, y_{n+1}) = (\xi_a x_n, y_n/a)$ for $y_n \leq a$; $(x_{n+1}, y_{n+1}) = (1/2 + \xi_b x_n, (y_n - a)/b)$ otherwise; $a = 2/5$, $b = 3/5$, $\xi_a = 1/5$, $\xi_b = 7/20$. Note that $\xi_a + \xi_b < 1$

Due to the simple structure of the attractor, in accordance with the formalism developed, the two-dimensional space can be factorized into the directions of the x- and y-axes. Then individual "partial" free energy functions, entropies, Lyapunov exponents and dimensions are separately calculated for each direction. As is easy to see, in the x-direction the attractor can be described by a Cantor set.

Suppose that among n iterations the map has chosen r times the second alternative offered by the dynamical equation. In this case the stretching rates for n iterations in the direction of the y-axis are

$$\mu_y = a^{-(n-r)} b^{-r} ; \tag{4.77}$$

in the direction of the x-axis the corresponding stretching rates are

$$\mu_x = \xi_a^{(n-r)} \xi_b^r . \tag{4.78}$$

For the direction given by the y-axis we obtain the partial partition function

$$GZ_y(q, \beta, n) = (a^{\beta+q} + b^{\beta+q})^n , \tag{4.79}$$

and the partial free energy

$$GF_y(q, \beta) = \log(a^{\beta+q} + b^{\beta+q}) . \tag{4.80}$$

The associated Lyapunov exponent λ_1 is seen to have the value $\lambda_1 = -a \log a - b \log b$, whereas the generalized "partial" dimensions $\beta_0(q)$ or $-\beta_0(q)/(q-1)$ can be calculated from the equation

$$a^{\beta_{y0}+q} + b^{\beta_{y0}+q} = 1 \tag{4.81}$$

[trivial solution $\beta_{y0} + q = 1$, leading to $\beta_{y0} = -(q-1)$]. For the x-direction we have to reverse the order of iteration. We obtain

$$GZ_x(q, \beta, n) = (\xi_a^\beta a^q + \xi_b^\beta b^q)^{-n} \tag{4.82}$$

and

$$GF_x(q, \beta) = -\log(\xi_a^\beta a^q + \xi_b^\beta b^q) . \tag{4.83}$$

The associated Lyapunov exponent λ_2 turns out to be $\lambda_2 = a \log \xi_a + b \log \xi_b$, whereas the generalized partial dimensions can be calculated by solving the equation

$$\xi_a^{\beta_{x0}} a^q + \xi_b^{\beta_{x0}} b^q = 1 . \tag{4.84}$$

Let us point out for further reference, however, that the two directions x and y are not completely independent. Consider to this end the equation $GF_y(\beta) = \log(a^\beta + b^\beta)$. If for β the value

$$\beta = q + \beta_{x0}(q)\left(\frac{\log(\xi_a/\xi_b)}{\log(a/b)}\right) \tag{4.85}$$

is chosen then $GF_y(\beta)$ is given by

$$GF_y(\beta) = \log\left(a^{q+\beta_{x0}(q)\,[\log(\xi_a/\xi_b)/\log(a/b)]} + b^{q+\beta_{x0}(q)\,[\log(\xi_a/\xi_b)/\log(a/b)]}\right) . \tag{4.86}$$

After some manipulations it turns out that the two functions $GF_y(\beta)$ and $\beta_{x0}(q)$ satisfy the equation

$$GF_y(\beta) = \beta_{x0}(q)\left(\frac{\log \xi_a \log b - \log \xi_b \log a}{\log(a/b)}\right) . \tag{4.87}$$

Note that the second factor on the right-hand side is a constant. Moreover, as is easy to see, the relations $\beta_{x0}(q)/(q-1)|_{q=1} = \lambda_1/\lambda_2$ and $GS_x(\langle\alpha\rangle_0, \langle\varepsilon\rangle_0)/\langle\alpha\rangle_0 = GS_y(\langle\varepsilon\rangle_0)$ are satisfied. For an approach to the same example from a different point of view we refer, among others, to [4.36].

For the sake of completeness, as the last point the relationship between the entropies for the scaling of the measure, the entropies for the scaling of the support and the information-theoretical entropies $K(q)$ should be discussed.

Since starting from (4.27) we know that

$$\sum_{S_n} P^q(S_n) \sim e^{-n(q-1)K_q} , \tag{4.88}$$

it follows that the entropies $K(q)$ can be obtained from the equation

$$(q - 1)K(q) = \sum_i^{j^+} GF_i(q, \beta = 0) , \tag{4.89}$$

where the sum extends only over the expanding directions. In the case of hyperbolic attractors, the procedure of the baker map can always be applied in a generalized sense. The attractor can be factorized at any point into a direct product of different space directions and the partition function GZ can always be obtained as a sum of the products of the Lebesque measure and the SRB measure. Moreover, with the assignment

$$P(S_n(x_0)) \sim e^{-nk(x_0, n)} , \tag{4.90}$$

with each symbolic sequence $S_n(x_0)$ a local or pointwise information-theoretical entropy $k(x_0, n)$ can be associated. Accordingly, $k(x_0, n)$ can be expressed as a sum of the products of partial local dimensions with Lyapunov exponents, see (4.41):

$$k(x_0, n) = \sum_{i=1}^{j^+} \lambda_i(x_0, n)\alpha_i(x_0) . \tag{4.91}$$

Here the sum extends over all positive Lyapunov exponents and $\alpha_k(x_0)$ is the local pointwise dimension in the kth direction. Letting $n \to \infty$, for periodic orbits the expression is zero; for hyperbolic attractors, however, the expression assumes the simple form

$$K(1) = \sum_{i=1}^{j^+} \lambda_i , \tag{4.92}$$

(Pesin's formula) since in this case $\alpha_i = 1$. For repellers the latter statement is no longer true; in this case, the expanding direction has a fractal structure ($\alpha_i \neq 1$). In addition, for nonhyperbolic systems singularities may appear in the measure of the expanding manifold. Correlations due to homoclinic tangency points can become important in the limit $n \to \infty$, so that Pesin's formula can be violated. For a further extended discussion let us refer to Sect. 4.3.

Therefore, as a summary (extension of [4.30]), we can draw the following conclusion. The scaling behavior of a dynamical system can most completely be characterized by means of a generalized free energy and a generalized entropy function. For higher than one-dimensional, hyperbolic systems, partial free energies and entropies can be derived. In the most general case the supports of these functions are given by two-dimensional areas of finite extension, situated in the q–β plane and in the α–ε plane, respectively. Specific functions using different notions of entropy and dimension were obtained in the following way:

- An entropy $GS(\varepsilon)$ for the scaling of the support was given by fixing $q = 0$ in the generalized entropy.

- An entropy for the scaling of the measure was given by $GS(\alpha)$. $GS(\alpha)$ was obtained from the generalized entropy function using the condition $\alpha q + \beta = 0$.
- A generalization of the Hausdorff dimension was obtained from the zeros of the generalized free energy.
- The generalized information-theoretical entropies $K(q)$ were obtained from $GF(q, \beta)$ derived from (4.39) for $\beta = 0$.

From $GF(q, \beta)$ and $GS(\alpha, \varepsilon)$ these functions can be obtained as different cuts along generally curvilinear lines in the q–β plane and in the α–ε plane, respectively. As an example, the generalized entropy of the support is obtained as a vertical section through GF, whereas the generalized information-theoretical entropies are obtained from the restriction on a horizontal section of GF, where q labels the direction of the abscissa. The meaningful range of GF is best characterized by the corresponding values of α and ε of GS, for which GS is not zero.

It may be asked how necessary it is to choose the partition as given in (4.39). The scaling ansatz given in (4.39) gives the most complete information about the scaling behavior of the fractal considered. Two independent "filtering exponents" q, β are used in order to be able to treat cases for which the scaling of the measure depends in a more complicated way than exponentially on the scaling of the support. But, depending on the point of view one is interested in, other partitions can also be chosen. For example, if there is no interest in probabilistic aspects, any generating partition will do. If, however, there is an interest in the probabilistic aspects, a probability measure has to be considered on the attractor. Finally, if the dynamical properties should be reflected, one has to consider a partition generated by the dynamical map. Hence, the partition chosen in (4.39) is the most refined one: it reflects both probabilistic and dynamical information. For experimental systems, of course, the question arises of how the different lengths l_i and probabilities p_i can be chosen or be determined, respectively. For the investigation of a generic system first the number of symbols has to be found (to this end the return map of the system is considered). Then, using an appropriate partition of the phase space (hoping to have chosen a generating one), the nature of the tree has to be investigated. In most cases the structure of the tree will not be of the particularly simple form of the model cases considered. Once the structure of the tree is known, the analogous treatment can be applied to the data originating from the experimental system as before to the models. Using the information of lower branches of the tree (corresponding to short symbolic sequences), however, cannot completely determine the upper branches, since the system may have a high degree of complexity. Furthermore, the values of the set of Lyapunov exponents of given periodic cycles do not therefore specify the partial local dimensions of longer periodic cycles but only give approximations of increasing accuracy to the latter.

In the two following sections we raise the question of how the different scaling exponents and the associated entropy functions can be calculated for experimental systems. Rather than making use of symbolic dynamics our starting point will be the concept of random sampling.

4.3   Generalized Dimensions, Lyapunov Exponents, Entropies

4.3.1   Definition of the Scaling Functions for Sampling Processes

Efforts in evaluating GF and GS for theoretical models as well as for experimental data have been focused essentially upon three different lines: the spectrum of singularities of the measure (or spectrum of generalized dimensions, as it is also sometimes called), the spectrum of generalized Lyapunov exponents and the spectrum of generalized information-theoretical entropies defined in Sect. 4.2.2. We will see how these scaling functions arise as special cases from the entropy function $GS(\alpha, \varepsilon)$ and how the relevant concepts can be realized in the case of random sampling.

First let us consider for this purpose the line corresponding to $\beta_0(q)$ in the graph of $GS(\alpha, \varepsilon)$. In view of (4.46) it is seen that

$$0 = GS(\langle\alpha\rangle_0, \langle\varepsilon\rangle_0) - \langle\varepsilon\rangle_0[\langle\alpha\rangle_0 q + \beta_0(q)] \tag{4.93}$$

and that a function $f(\alpha)$ [4.29] can be introduced through the relation

$$GS(\langle\alpha\rangle_0, \langle\varepsilon\rangle_0) = \langle\varepsilon\rangle_0 f(\langle\alpha\rangle_0) . \tag{4.94}$$

It is verified that the relations

$$\left.\frac{df(\alpha)}{d\alpha}\right|_{\alpha = \langle\alpha\rangle_0} = q \tag{4.95}$$

and

$$\langle\alpha\rangle_0 = -\frac{d\beta_0(q)}{dq} \tag{4.96}$$

are satisfied, so that $f(\langle\alpha\rangle_0)$ is again the Legendre transform of $-\beta_0(q)$: $-\beta_0(q) = \langle\alpha\rangle_0 q - f(\langle\alpha\rangle_0)$. It is a common usage again to write $f(\langle\alpha\rangle_0)$ as $f(\langle\alpha\rangle)$ or simply as $f(\alpha)$. In Fig. 4.23 $f(\alpha)$ is shown for the three-scale Cantor set with measure of Fig. 4.18. Note that $f(\alpha)$ can be interpreted as the partial derivative $\partial GS/\partial\varepsilon$ at $\alpha = \langle\alpha\rangle_0$, $\varepsilon = \langle\varepsilon\rangle_0$; this geometrical meaning could easily be visualized in Fig. 4.19.

In the case of random sampling, if one is interested in the probabilistic aspects of scaling behavior, one has to be aware of the fact that the definitions of $GS(\alpha)$ and $\beta_0(q)$ require the distribution of l or ε. Already in the previous section we have seen that, in general, $GS(\alpha)$ and the entropy function corresponding to $\beta_0(q)$ are different functions. Moreover, $\beta_0(q)$ is not identical with the original function $\tau(q)$ defined in [4.34], which cannot be represented in an obvious way in the graph of $GF(q, \beta)$ in the general case. If the distribution of l and ε is not

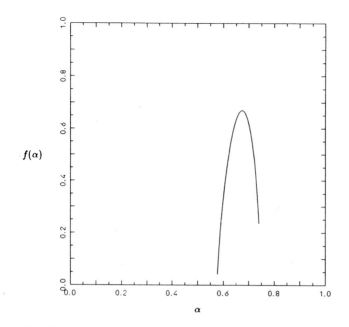

Fig. 4.23. $f(\alpha)$ for the three-scale Cantor set described in Fig. 4.18

known or if the distribution is trivial, a function $\beta_0^*(q)$ may be looked for, such that for $n \to \infty$

$$GZ(q, \beta, n) \sim \sum_j l^{\beta_0^*} p_j^q \tag{4.97}$$

remains finite. Assume [4.37–39], therefore, that there exists a common length scale ε' such that

$$\varepsilon^{-\beta_0^*} \sim GZ(q, 0, n) = e^{nGF(q, 0)} , \tag{4.98}$$

where $\varepsilon = e^{-n\varepsilon'}$. Then

$$-\beta_0^* \sim \frac{nGF(q, 0)}{\log \varepsilon} \sim \frac{\log[GZ(q, 0)]}{\log \varepsilon} =: \tau(q) =: (q - 1)D(q) . \tag{4.99}$$

$D(q)$ are called the *generalized Renyi dimensions* [4.38]. It has been conjectured in [4.29] that $D(q)$ coincides "in most cases" with $D(q)$ calculated using the distribution of length scales ε'. From the definition it is seen that $D(q)$ can be calculated from a randomly sampled attractor according to

$$(q - 1)D(q) = \frac{\log\langle P(B(x, \varepsilon))^{q-1}\rangle}{\log \varepsilon} , \tag{4.100}$$

$\varepsilon \to 0$, where $P(B(x, \varepsilon))$ is the probability of falling into a ball of radius ε around point x, ("fixed-size method" [4.39]). Alternatively, it has been proposed in

[4.40] to consider the implicit relation

$$q - 1 = - \frac{\log \langle \varepsilon(x, p)^{-\tau(q)} \rangle}{\log p}, \tag{4.101}$$

$p \to 0$, where $\varepsilon(x, p)$ is the radius of a ball around point x which contains a portion p of all points of the attractor ("fixed-mass method"). It can be seen that $(q - 1)$ is equal to $D(1)$ if $\tau(q) = 0$, whereas for other values of $q, D(q)$ is more difficult to obtain. Note that for $q \to 1$, $\log \langle P^{q-1} \rangle /(q - 1) \to \langle \log P \rangle$.

In Fig. 4.24 we show the log–log plots used for the evaluation of fractal dimensions. For the present example the chaotic attractor of Hénon's map at standard parameters has been reconstructed from a scalar measurement using an embedding process (the concept of the embedding of experimental signals is discussed in Sect. 4.4). A time series of 10 000 data points in the range from 0 to 10 000 was analyzed. The dimension is then (theoretically) obtained as the slope of the curves for asymptotic embedding dimension in a linear region. In practice, this limit cannot be achieved due to the finite length of the time series. Fixing the parameter q at $q = 1$, the different curves are obtained for different embedding dimensions, increasing from embedding dimension $d_E = 2$ in steps of one. In Fig. 4.24a the uppermost curve corresponds to the largest embedding dimension; in Fig. 4.24b the bottom curve corresponds to the largest embedding dimension.

For the fixed-mass method, the distance of the seventh next neighbor has been considered in order to improve the linear behavior of the curves. For both methods the logarithm of base 1.18 has been used for both axes to allow convenient units. The algorithms have been implemented such that for each integer value on the x-axis a calculated point on every curve is obtained. These points have then been connected by a line to guide the eye, without any further smoothing process.

Both methods yield comparable results (dimension $1.26 + 0.03$ for (a), $1.26 + 0.05$ for (b) when evaluated for the x-intervals [4.35, 55] (a) and [4.45, 55] (b), for embedding dimensions $d_E = 3, 4, 5, 6$. For the calculation of the dimensions in both cases a least-squares procedure was applied. It can be seen that the linear behavior of the curves in (a) extends over a far more extended region as compared to (b). However, at large distances, the applicability of both methods becomes questionable, as will be explained in Sect. 4.4.

The global scaling assumption leads then to the definition of the substitute for $f(\alpha)$ [or $GS(\alpha)$] [4.41, 29]:

$$f^*(\alpha) = \alpha q - \tau(q), \tag{4.102}$$

where it is assumed again that the integral corresponding to (4.25) can be evaluated with the help of the saddle point approximation. It is common practice not to distinguish between f and f^*. For this reason we will drop from now on the asterisk. For $f(\alpha)$ the term *"spectrum of singularities of the measure"* has been coined [4.29]; sometimes the expression "spectrum of dimensions" is

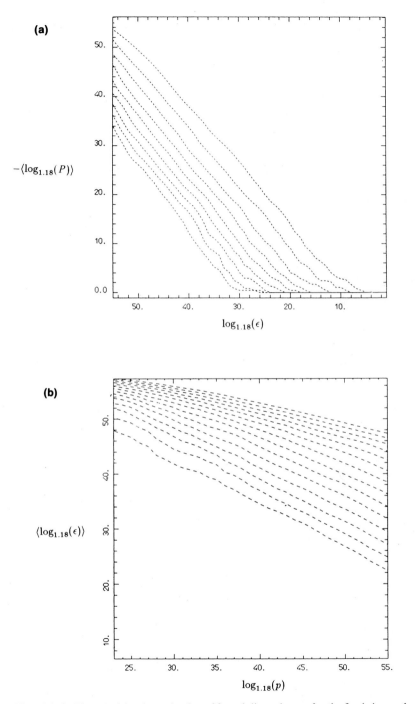

Fig. 4.24a, b. Plots used for the evaluation of fractal dimensions; **a** for the fixed-size method, **b** for the fixed-mass method (see text)

used. For the sake of clarity in the present context, however, we will allow ourselves to call $f(\alpha)$ also the *scaling function* for the *scaling of the measure*.

From a more formal point of view than the one used for the introduction of the scaling functions $GS(\alpha)$ and $GS(\varepsilon)$ in Sect. 4.2.4, the function $f(\alpha)$ instead of $GS(\alpha)$ can be interpreted as the relevant function for the description of the scaling of the measure. This observation is now made evident by a direct comparison of $f(\alpha)$ with the function which corresponds to $GS(\varepsilon)$ in the case of random sampling.

From the dynamical point of view, where we are concerned with the distribution of scaling exponents of the support, we choose $q = 0$ to obtain the scaling function for the scaling of the support, $GS(\varepsilon)$. In this case, the relation $GF(\beta) = GS(\varepsilon) - \varepsilon\beta$ follows immediately and, by adopting the notation $\breve{G}F(\beta) = -GF(\beta)$, the analog Legendre transformation relationship as compared to $\tau(q)$ and $f(\alpha)$ can be obtained between $\breve{G}F(\beta)$ and $GS(\varepsilon)$: $GS(\varepsilon) = \beta\varepsilon - \breve{G}F(\beta)$.

If we want to relate this formalism to the case of random sampling we remember that the dynamical scaling properties are generated by the expanding manifolds. Let us consider a hyperbolic map. If during the application of the dynamical map we can calculate the local stretching exponents belonging to the Lyapunov exponents $\lambda_i > 0$, then we may sample in time to sample the natural or SRB measure, in complete analogy with the probabilistic point of view. In this way the relationship

$$\breve{G}F(\beta, 0) = \lim_{n \to \infty} \frac{\log \langle (DFa^{n,+})^{-(\beta-1)} \rangle}{\log \mathsf{T}} =: \Lambda(\beta) =: (\beta - 1)K(\beta) , \qquad (4.103)$$

where $\mathsf{T} := e^{-n}$, is obtained. The latter relation was also obtained from different starting points in [4.17, 42–44]. The *scaling function* for the *scaling of the support* or spectrum of Lyapunov exponents is then obtained by a Legendre transformation of $\Lambda(\beta)$. In order to be able to treat nonhyperbolic systems without abuse of language, in the following the notation is changed. In place of the scaling function for the scaling of the support $GS(\varepsilon)$, the scaling function of Lyapunov exponents $\phi(\lambda)$ is introduced. In this way the following relation for $\phi(\lambda)$ is obtained:

$$\phi(\lambda(\beta)) = \beta\lambda(\beta) - \Lambda(\beta) . \qquad (4.104)$$

$\Lambda(\beta)$ are sometimes called the generalized Lyapunov exponents, $K(\beta)$, the generalized (dynamical) entropies. It should be noticed that different notations are currently in use to denote different scaling functions of the Lyapunov exponents. The present notation has been chosen in order to underline the analogy with the scaling of the measure. For the sake of clarity, we point out that $\phi(\lambda) \cong h(\gamma)$ of [4.42] and $\phi(\lambda) - \lambda \cong g(\lambda) \cong -\Psi(\Lambda)$ of [4.43, 45], respectively. The algorithms for the evaluation of these quantities from experimental data are outlined in Sect. 4.4.

A short summary of the functions and relations used for the description of the scaling behavior of a generic system can be given as follows [4.44]:

Point of view of scaling: Measure

$$\tau(q) = \lim_{\varepsilon \to 0} \frac{\log \langle P(B(x, \varepsilon))^{q-1} \rangle}{\log \varepsilon} , \tag{4.105}$$

$$P(B(x, \varepsilon)) \sim \varepsilon^{\alpha(x)} , \tag{4.106}$$

$$f(\alpha) = \alpha q - \tau(q) , \tag{4.107}$$

$$\alpha(q) = \frac{d\tau(q)}{dq} , \tag{4.108}$$

$$q = \frac{df(\alpha)}{d\alpha} . \tag{4.109}$$

For the fixed-mass approach it is, furthermore, found that

$$P(\alpha, \varepsilon) \, d\alpha \sim \varepsilon^{\alpha - f(\alpha)} \, d\alpha . \tag{4.110}$$

Point of view of scaling: Support

$$\Lambda(\beta) = \lim_{n \to \infty} \frac{\log \langle (DFa^{n,+})^{-(\beta-1)} \rangle}{\log \mathsf{T}} , \tag{4.111}$$

where $\mathsf{T} := e^{-n}$,

$$P(x, \mathsf{T}) \sim \mathsf{T}^{1/n \, \log[DFa^{n,+}(x)]} , \tag{4.112}$$

$$\phi(\lambda) = \beta\lambda - \Lambda(\beta) , \tag{4.113}$$

$$\lambda(\beta) = \frac{d\Lambda(\beta)}{d\beta} , \tag{4.114}$$

$$\beta = \frac{d\phi(\lambda)}{d\lambda} . \tag{4.115}$$

Furthermore, it is found that

$$P(\lambda, k) \, d\lambda \sim e^{-k(-\phi(\lambda) + \lambda)} \, d\lambda . \tag{4.116}$$

Because of this analogy, the schematic behavior of $\Lambda(\beta)$ and $\tau(q)$ is identical, and likewise $f(\alpha)$ and $\phi(\lambda)$, and $K(\beta)$ and $D(q)$. Observe, for example, that $q' > q$ implies $[1/(q' - 1)]\tau(q') \leq [1/(q - 1)]\tau(q)$, and likewise for $\Lambda(\beta)$. This property can be derived directly or via the convexity of the generalized entropy functions. It should be pointed out that relations (4.113–115) hold only for differentiable functions $\phi(\lambda), \Lambda(\beta)$ ($f(\alpha), \tau(q)$). However, a more general form of Legendre transformation relationships will be introduced in the next section.

Historically, the interest in $D(q)$ focused on three specific values of q. For obvious reasons, for $q = 0$ $D(q)$ is called the *Renyi–Hausdorff* dimension. Since for $q = 1$, as can easily be seen, $D(q)$ is closely related to Shannon's information entropy, $D(q = 1)$ is called the *information* dimension. Finally, $D(q = 2)$, is called the *correlation* dimension [4.46]. The first two quantities can be characterized

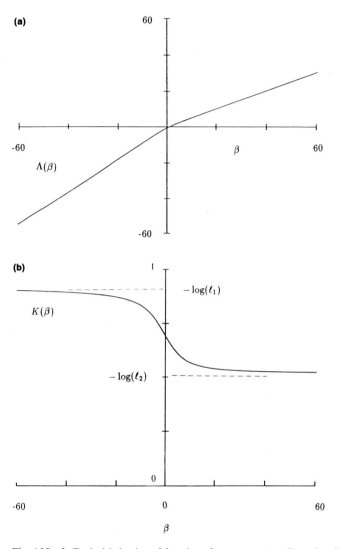

Fig. 4.25a–d. Typical behavior of functions for one- or two-dimensional hyperbolic systems. As a simple example the tent map is taken ($l_1 = 2/5$, $l_2 = 3/5$). Characteristic behavior of **a** $\Lambda(\beta)$ [note that for hyperbolic maps $\Lambda(\beta) = -GF(\beta)$], **b** $K(\beta)$, **c** $\phi(\lambda) - \lambda$, and **d** $\phi(\lambda)$

by means of $f(\alpha)$ in the following way: $\max_\alpha f(\alpha) = D(0)$ and $\alpha = f(\alpha)$ for $\alpha = D(1)$. Notwithstanding the common terminology of "dimensions" for $D(q)$, we recall that we have already pointed out in Sect. 4.2.5 that $D(q)$ is not the only possible generalization of the Hausdorff dimension (and possibly not the most natural one).

Analogous statements can be made for the dynamical entropies of hyperbolic attractors. The maximum of $\phi(\lambda)$ is characterized by $\max_\lambda \phi(\lambda) = K(0)$, where $K(0)$ is the *topological* entropy $[= (1/n)\log \sum_{x \in Fa^{-n}(y)} 1$; see, e.g., (4.27)].

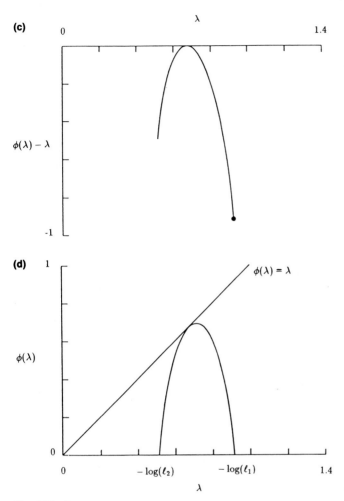

Fig. 4.25c, d.

For $\lambda = \sum \lambda^+$, the sum of positive Lyapunov exponents, $\phi(\lambda)$ is equal to λ. It can be easily seen that in this case $\phi(\lambda)$ is equal to the *Kolmogorov–Sinai* entropy, which can be obtained from $K(\beta)$ for $\beta = 1$.

   Note, further, that since $P(\lambda, k) \, d\lambda \sim \mathrm{e}^{-k(-\phi(\lambda) + \lambda)} \, d\lambda$ and $P(\alpha, \varepsilon) \, d\alpha \sim \varepsilon^{\alpha - f(\alpha)} \, d\alpha$, the observability of an invariant measure is related to the requirements

$$-\phi(\lambda) + \lambda > 0, \quad \text{which implies} \quad \lambda = \phi(\lambda) \quad \text{for } n \to \infty \,,$$

$$-f(\alpha) + \alpha > 0, \quad \text{which implies} \quad \alpha = f(\alpha) \quad \text{for } \varepsilon \to 0 \,,$$

respectively. In [4.42], $-\phi(\lambda) + \lambda$ has been interpreted as the generalized

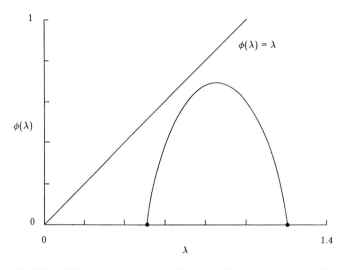

$\phi(\lambda) = \lambda$

$\phi(\lambda)$

λ

Fig. 4.26. $\phi(\lambda)$ for a strange repeller (Tent map, the characteristic condition $l_1 + l_2 < 1$ is satisfied)

escape rate κ from an invisible invariant measure μ. Through the conjecture that the information dimension $D(1)_\mu$ can be expressed as $D(1)_{\mu(\beta)} = \phi(\lambda(\beta))/\lambda(\beta)$, the characteristic relation $D(1)_{\mu(\beta)} = 1 - \kappa/\lambda$ is obtained, which could easily be visualized in the graph of $\phi(\lambda)$. For the discussion of several different invariant measures we refer to [4.3]. For convenience, in Figs. 4.25, 26 the typical behavior of the functions Λ, K, ϕ, τ, D and f of one- or two-dimensional hyperbolic systems is represented. Higher-dimensional or nonhyperbolic systems can give rise to deviations from these characteristic graphs which will be discussed in Sect. 4.3.4. For simplicity, only the functions related to the scaling behavior of the support are indicated. Note, however, that for nonhyperbolic maps and attractors the dynamical entropies $K(\beta)$ and the information-theoretical entropies $K(q)$ may be different functions.

4.3.2 Relation Between the Scaling Functions and a Thermodynamical Formalism

Let us now consider again the description of the scaling behavior in the symbolic dynamics approach for a two-dimensional hyperbolic map. In the mapping onto a thermodynamic formalism [4.43, 47–49] time is considered as a one-dimensional lattice. Negative lattice sites correspond to backward, and positive lattice sites to forward iteration in time. The partition function corresponding to (4.103) can then be rewritten as

$$Z_{c,s}(V, T) = e^{-F_s(V, T)/k_B T} = \sum_{S_n} e^{-n(\beta - 1)\lambda(n, x_0)} \sim e^{-n(\beta - 1)K(\beta)} . \tag{4.117}$$

The subscript c, s is to point out that this partition can be identified with

a canonical ensemble and that the scaling of the support is considered. Since the canonical ensemble consists of all microstates S_n, with fixed volume and different energies, the following identification can then be made (though not in a unique way):

$$E = (-n)\lambda = \text{energy} \leq 0 \,,$$

$$1/k_B T = -(\beta - 1) \,,$$

$$V = n > 0 \,,$$

$$F_s(V, T) = (-n)K(\beta) \,.$$

The microcanonical description looks at the ensembles which consist of all sequences with energies in the interval $[E, E + dE]$. Its partition function

$$Z_{m,s}(E, V) = e^{S(E,V)/k_B} \sim e^{(-n)[\lambda - \phi(\lambda)]} \tag{4.118}$$

relates the following quantities via the saddle point approach, where now the case of nondifferentiable functions ϕ, Λ is considered:

$$\Lambda(\beta) = \inf_\lambda (\beta\lambda - \phi(\lambda)) \,, \tag{4.119}$$

$$\phi(\lambda) = \inf_\beta (\beta\lambda - \Lambda(\beta)) \,, \tag{4.120}$$

from which it follows that

$$K(\beta) = \lambda(\beta) - \frac{[\phi(\lambda) - \lambda]}{\beta - 1} \,. \tag{4.121}$$

The last equation can be interpreted as the classical thermodynamical relation

$$F_{s,v}(T) = E_v - TS_v \,, \tag{4.122}$$

where the subscript v is used to denote the original quantities divided by the volume. As can be checked, all thermodynamical relations follow. In particular, $(\partial S/\partial E)_V = 1/T$ and $(\partial S/\partial V)_E = p/T$, where p denotes the pressure $[p = K(\beta)]$, ensure that this set can be taken as an axiomatic thermodynamical system. By using the analogy between the scaling of the support and the scaling of the measure, a corresponding interpretation can be given for the scaling of the measure (with $-n$ replaced by $\log \varepsilon$). In both cases the corresponding entropy S is a convex function of E and V, E is convex in S and V, etc. From this property of S it follows that a point on the graph of S is either extreme or an inner point of a linear part of it. The former can be identified with pure phases and the latter, with mixtures since points of nonanalytical behavior of $\Lambda(\beta)$ $[\tau(q)$, respectively] will lead to such parts of S [see (4.114) and (4.115)]. They are responsible for a phase-transition-like behavior of the system.

In an earlier approach an analogous thermodynamical formalism has been used to establish an explicit relation between the dynamical system and an Ising model on the basis of the scaling functions for the scaling of the measure. In

[4.50–52] *constant-mass partitions* of mass p were considered, where the balls furnishing a covering of the attractor are chosen at random with respect to the natural measure $\rho(x)$. Using the notation introduced in Sect. 4.2.2, it was shown that the condition $GZ(q, 0) \sim 1$ can be written as $GZ(q) = \sum_{S_n} e^{-[\tau/\alpha(S_n)] \log p} \sim e^{[1 - q(\tau)] \log p}$, where $q(\tau)$ is the inverse function to $\tau(q)$ [which exists due to the monotonic behavior of $\tau(q)$]. As before, the local crowding index $\alpha(S_n)$ is given by $\alpha(S_n) = \log P(S_n)/\log \varepsilon(S_n)$. Again, the sum $\sum_{S_n}$ can be identified with a canonical partition function $Z_c(V, T)$ with the help of the identification

$$\log \varepsilon(S_n) = \text{energy} = E \,,$$

$$- \log P(S_n) = - \log p = \text{volume} \,,$$

$$\tau(q) = \text{inverse temperature} = 1/k_B T \,,$$

and it can be deduced that the associated free energy is $F(V, T) = \log p (q - 1)/\tau(q)$. The microcanonical ensemble consists of balls of equal size $\varepsilon(S_n) = p^{1/\alpha(S_n)}$. The associated partition function has the form $Z_m(E, V) = e^{S(E, V)} \sim \sqrt{|\log p|} p^{[\alpha - f(\alpha)]/\alpha}$. In the limit $p \to 0$ it is obtained that $S(E, V) = k_B [\alpha - f(\alpha)/\alpha] \log p$. Furthermore, again all thermodynamical relations are satisfied [the pressure turns out to be equal to $(q - 1)/\tau(q)$].

The connection with a suitable Ising model can be made explicit by associating with each lattice site i of the symbolic sequence $S_n = \{\ldots, s_{i-1}, s_i, s_{i+1}, \ldots\}$, a symbol $s_i \in [0, M - 1]$, where M denotes the number of symbols needed for the symbolic description. To the symbol s_i at site i, a spin σ_i is attributed by putting $\sigma_i = s_i - (M - 1)/2$. In this way, $\sigma_i \in (- (M - 1)/2, (M - 1)/2)$. To give an explicit example consider the baker map (see Fig. 4.15, Sect. 4.2.5). Take an interval in the x-direction which is generated from the unit interval by n-fold iteration of the map, where j times the first alternative and $n - j$ times the second alternative of the dynamical map has been applied. The size and the measure of this interval can be described using the spin variables σ_i through the expressions [4.60]

$$\log[\varepsilon(S_n)] = (n/2) \log(\xi_a \xi_b) + \sum_i \sigma_i \log(\xi_a/\xi_b)$$

and

$$\log[P(S_n)] = (n/2) \log(ab) + \sum_i \sigma_i \log(a/b) \,.$$

The energy $E = \log[\varepsilon(S_n)]$ can thus be interpreted as the Hamiltonian of an Ising model without exchange interaction but with an external field. The absence of spin–spin interactions is a consequence of the lack of correlation between successive symbols. The local crowding index $\alpha_x(S_n)$, which is obtained from $\log[P(S_n)]$ and $\log[\varepsilon(S_n)]$ as above, converges for $n \to \infty$ towards $D_x(1) = (a \log a + b \log b)/(a \log \xi_a + b \log \xi_b)$. $D_x(1)$ will be interpreted in Sect. 4.3.3 as partial information dimension.

Furthermore, in order to reduce the effect of small-n deviations in the approximation of the scaling functions, the use of the "daughter/mother" scaling function $\breve{\sigma} = \varepsilon_{n+1}(S_{n+1})/\varepsilon_n(S_n)$ can be favorable. Together with $\breve{\sigma}$, in [4.52] a transfer matrix T was considered. T was introduced by writing

$$\langle s_{-n+1}, \ldots, s_0 | T | s_{-n}, s'_{-n+1}, \ldots, s'_{-1} \rangle$$
$$= \breve{\sigma}^{-\tau(q)}(s_{-n}, \ldots, s_0) \delta_{s_{-n+1} s'_{-n+1}} * \cdots * \delta_{s_{-1} s'_{-1}} ; \qquad (4.123)$$

it was shown that the free energy $F(V, T)$ can then be determined numerically from the largest eigenvalue of the transfer matrix. For self-similar systems the scaling function $\breve{\sigma}$ depends only on the last symbol s_{-1} [$\breve{\sigma}(s_{-n}, \ldots, s_0)$ $= \breve{\sigma}(s_{-1}, s_0)$], which corresponds to the behavior at the last point considered itself, while otherwise the whole sequence is relevant. For non-self-similar systems the evaluation of $\breve{\sigma}$ is especially simple if $\breve{\sigma}$ depends more strongly on the head or the tail of $\{s_{-n}, \ldots, s_0\}$. In [4.52] it has been pointed out that for a particular system an adequate covering may furnish this property. As has been mentioned before, the correlation between different symbols in the symbol sequence determines the range of interaction in the Hamiltonian. It is known that in one-dimensional systems for short-range interactions there is no possibility of phase transitions [4.53].

4.3.3 Relation Between the Scaling of the Support and the Scaling of the Measure (Generic Case)

In the first parts of this section we only deal with one-dimensional descriptions of scaling functions for the scaling of the measure [$D(q), f(\alpha)$]. For the example of the baker map the knowledge of $D(q)$ furnishes complete knowledge of the scaling properties of the scaling of the measure since in any point the baker attractor is the product of a one-dimensional Cantor set and a continuum. In order to be able to describe higher-dimensional systems of more involved structures than the baker map, however, some useful local concepts have to be reintroduced, in accordance with Sect. 4.2.

Suppose a nontransient motion is given on an m-dimensional attractor which has a covering of ellipsoids. These ellipsoids are then mapped by Fa into new ellipsoids, situated in a different place with new axes. The direction of the axis which undergoes the strongest expansion determines the direction which is associated with the largest Lyapunov exponent; the next largest, perpendicular to the first, is associated with the second, and so on. The local contribution to the ith Lyapunov exponent is called the ith local Lyapunov exponent λ_i.

The local Lyapunov directions provide each point of a hyperbolic attractor with a *local basis*. In the direction of this basis, local scaling exponents of the measure can be defined as [4.3, 18, 32]

$$\alpha_i(x_0) = \lim_{\varepsilon \to 0} \frac{\log[P_i(x_0, \varepsilon)]}{\log \varepsilon} \qquad (4.124)$$

[compare with (4.23)], where P_i is the projection of the probability P onto the ith axis. These local *partial* scaling exponents α_i can be averaged to yield partial Renyi dimensions according to

$$D_i(q) = \lim_{\varepsilon \to 0} \frac{1}{q-1} \frac{\log \langle P_i(\varepsilon)^{q-1} \rangle}{\log \varepsilon} . \tag{4.125}$$

For SRB measures we have $D_i(q) = 1$, while for fractal directions $D_i(q) \in (0, 1)$. The Renyi dimensions $D(q)$ can be recovered from the partial Renyi dimensions through $D(q) = \sum D_i(q)$, and all corresponding quantities can be factorized in this way:

$$\tau_i(q) = (q - 1)D_i(q) , \tag{4.126}$$

$$\langle \alpha_i(q) \rangle = \frac{d\tau_i(q)}{dq} , \tag{4.127}$$

$$f_i(\langle \alpha_i \rangle) = \langle \alpha_i \rangle q - \tau_i(q) , \tag{4.128}$$

$$f = \sum f_i, \quad \alpha = \sum \alpha_i, \quad \tau = \sum \tau_i . \tag{4.129}$$

Notice that, for systems with $\alpha(x)$ constant over the attractor, α coincides with $D(q = 1)$.

Let us now derive the expressions for the evaluation of the information-theoretical entropies $K(q)$ in the case of random sampling. This then will shed light on one aspect of the relation between the scaling of the support and the scaling of the measure. In the previous section it was argued that

$$P(S_n(x_0)) \sim e^{-nk(x_0, n)} , \tag{4.130}$$

with

$$k(x_0, n) = \sum_{i=1}^{j^+} \lambda_i(x_0, n)\alpha_i(x_0) , \tag{4.131}$$

which led us to Pesin's formula. As remarked earlier for periodic orbits, the analogous sum vanishes (consider the stable and the unstable directions separately):

$$0 = \sum_{i=1}^{m} \lambda_i(x_0, n)\alpha_i(x_0) . \tag{4.132}$$

From definition the family of generalized information-theoretical entropies can be written as averages

$$(q - 1)K(q) = \lim_{n \to \infty} \frac{\log \langle P(S_n(x_0))^{(q-1)} \rangle}{\log \mathsf{T}} , \tag{4.133}$$

where, again, $\mathsf{T} := e^{-n}$. For hyperbolic attractors this expression is, therefore, equivalent to the equation

$$(q - 1)K(q) = \lim_{n \to \infty} \frac{\log \langle \prod_{i=1}^{j^+} e^{(-n)(q-1)\lambda_i(x_0, n)} \rangle}{\log \mathsf{T}} . \tag{4.134}$$

In passing we note that for nonhyperbolic attractors, in addition to partial local dimensions smaller than one, orbits on the expanding manifold may at any time reenter the vicinity of the iterated point x_0. Therefore, the former index j^+ should in this case be chosen depending on the variable q as the index $j^+(q)$ which leads to the maximum of the expression

$$- \lim_{n \to \infty} \frac{1}{n(q-1)} \log \left\langle \prod_{i=1}^{j^+(q)} e^{-n(q-1)\lambda_i(x_0, n)\alpha_i(x_0)} \right\rangle . \tag{4.135}$$

Otherwise an incorrect estimate of $K(q)$ would be obtained.

If no generating partition has been found we must consider the limiting case for the diameters ε of the sets B_j going to zero. In this case it can be seen that the expression (4.130) has to be replaced by the more general form

$$P(S_{n, \varepsilon}(x_0)) \sim e^{-nk(x_0, n)} \prod_{i=1}^{m} \varepsilon^{\alpha_i(x_0)} . \tag{4.136}$$

Analogous to (4.134) the family of entropies $K(q)$ can now be evaluated from

$$(q-1)K(q) = - \lim_{n \to \infty} \frac{1}{n} \lim_{\varepsilon \to 0} \log \left\langle \frac{P(S_{n, \varepsilon}(x_0))^{(q-1)}}{\prod_{i=1}^{m} \varepsilon^{(q-1)D_i(q)}} \right\rangle , \tag{4.137}$$

where m denotes the dimension of the system. The latter relation is the reason why with little extra effort the information-theoretical entropies $K(q)$ can be obtained along with the evaluation of the fractal dimensions $D(q)$. For more details on how this is done the interested reader is referred to Sect. 4.4.1.

Due to the fact that the motion on the attractor is nontransient, *conservation properties* [4.54] relating local Lyapunov exponents and partial local dimensions are to be expected. Since a well-defined local basis is needed in order to have properly defined local quantities, our forthcoming line of argumentation will apply in a strict sense only to hyperbolic systems. The argumentation is, however, also expected to be more or less quantitatively valid for typical attractors, where the influence of homoclinic tangency points (which is where this factorization fails) is not dominant.

Let us start from our basic assumption (4.98) used for the evaluation of the Renyi dimensions $D(q)$

$$\varepsilon^{\tau(q)} \sim GZ(q, 0, n), \quad n \to \infty . \tag{4.138}$$

Equivalently, this assumption may be written in the form

$$\frac{GZ(q, 0, n)}{\varepsilon^{\tau(q)}} \sim 1 , \tag{4.139}$$

or, more explicitly, as

$$\left\langle \frac{P(B(x, \varepsilon))^{q-1}}{\varepsilon^{\tau(q)}} \right\rangle \sim 1 . \tag{4.140}$$

Let us now iterate Fa for k steps and denote again by $\lambda(x_0, k)$ the k-step average of the (local) Lyapunov exponent ("effective Lyapunov exponent of k steps"). An ellipsoid with axes ε is then mapped into an ellipsoid with axes $\varepsilon e^{k\lambda(x_0,k)}$, but with the same probability measure as before. Using the invariance property of the attractor,

$$\left\langle \frac{P(B(x,\varepsilon))^{q-1}}{\varepsilon^{\tau(q)} e^{k\lambda(x_0,k)\tau(q)}} \right\rangle \sim 1 \qquad (4.141)$$

is obtained. Whenever this expression can be factorized into

$$\left\langle \frac{P(B(x,\varepsilon))^{q-1}}{\varepsilon^{\tau(q)}} \right\rangle \left\langle e^{-k\lambda(x_0,k)\tau(q)} \right\rangle \sim 1 , \qquad (4.142)$$

it follows that

$$\left\langle e^{-k\lambda(x_0,k)\tau(q)} \right\rangle = 1 , \qquad (4.143)$$

with $\lambda(x_0, k) \to \lambda$ due to ergodicity in the limit $k \to \infty$. This is the expected conservation property. The factorization assumption can, however, only be justified for hyperbolic attracting systems with short-time correlations between spatial and temporal terms (which is not implicit in hyperbolicity). From (4.143), a fundamental relation between dynamical and probabilistic scaling exponents can be derived (we follow [4.3, 55]).

Suppose that $q < 1$ (an analogous reasoning applies to the case $q > 1$). If all partial Renyi dimensions $D_i(q)$ assumed the maximal value $D_i(q) = 1$ then for $k \to \infty$ the left-hand side of (4.143) would go to zero. Observe, however, that if the exponent is made larger and larger by gradually decreasing D_i, and starting with the highest index i, divergence can finally be obtained. Define now $j(q)$ as the maximal integer, which may depend on q, such that

$$\left\langle \prod_{i=1}^{j(q)} e^{-k\lambda_i(x_0,k)(q-1)} \right\rangle^{-1/k(q-1)} \geq 1 \qquad (4.144)$$

is still true. The aim is now to give an estimate of $D(q = 1)$ in terms of the Lyapunov exponents. To this end consider (4.143) in the limit $q \to 1$, which yields $\sum_{i=1}^{m} D_i(1)\lambda_i = 0$ [note that this is the global version of (4.132)!]. The upper bound for $D(1)$ in (4.143) can then be obtained by making the choice $D_i(1) = 1$ for $i \leq j(1)$, $D_i(1)$ for $i = j(1) + 1$ such that finally the average in (4.144) is made equal to 1, and $D_i(1) = 0$ for $i \geq j(1) + 2$.

In this way, the inequality

$$D(1) \leq D_{KY} := j(1) + \frac{\sum_{i=1}^{j(1)} \lambda_i}{|\lambda_{j(1)+1}|} , \qquad (4.145)$$

which relates the Lyapunov exponents λ_i and the information dimension $D(1)$, is obtained. $j(1)$ can be characterized as the largest index such that the sum of Lyapunov exponents is still positive. Assuming that the heuristic arguments used above hold in a strictly mathematical sense, we refer to (4.145) as the "Kaplan–Yorke relation". D_{KY} is called the Lyapunov dimension. When the equality $D(1) = D_{KY}$ is assumed, we speak of the "*Kaplan–Yorke conjecture*" because this property was originally conjectured by *Kaplan* and *Yorke* [4.56, 57]. Note that for hyperbolic systems the dimension D_{KY} so defined corresponds to the assumption that the attractor is locally the product of $j(1)$ continua with only one Cantor direction along the $(j(1) + 1)$th axis. For ·hyperbolic attractors with more than one contracting direction, the Kaplan–Yorke conjecture will, therefore, not hold generally. For general non-hyperbolic systems the situation is even more complicated: not only does the mathematical derivation become questionable, but also the SRB assumption $D_i(1) = 1$ will be hard to justify.

However, for $F \in C^2$, ρ ergodic with compact support, it has been possible nevertheless to show that $D(1) \le D_{KY}$; for two-dimensional twice differentiable diffeomorphisms of a two-dimensional manifold, and ρ ergodic with compact support, it has even been proven that the local pointwise dimension exists almost everywhere, and $D(1) = 1 + \lambda_1/|\lambda_2|$ [4.58]. Attempts to prove higher-dimensional versions of this theorem have failed so far.

A third possible relation between the scaling of the measure and the scaling of the support is obtained for two-dimensional maps with an SRB measure and constant Jacobian J [4.59]. In this case the spectra f and ϕ are related in a way similar to (4.87). Denoting again the effective Lyapunov exponents of k steps by $\lambda(x_0, k)$, it is evident that

$$\lambda_1(x_0, k) + \lambda_2(x_0, k) = \log|J| . \tag{4.146}$$

From the SRB property it follows that we obtain the simple equations $\tau_1(q) = q - 1$, $\langle \alpha_1 \rangle = f_1(\langle \alpha \rangle) = 1$ and $\Lambda(q) = -\lim_{k \to \infty}(1/k)\log\langle e^{-k\lambda_1(x, k)(q-1)}\rangle$. Together with (4.143) this leads to the equation

$$\Lambda(q - \tau_2(q)) = -\tau_2(q)\log|J| , \tag{4.147}$$

which is one way of expressing the close relationship between the scaling functions for the scaling of the support and for the scaling of the measure. More explicitly, using a Legendre transformation, this relation can be written in the form

$$\frac{f_2(\alpha_2)}{\alpha_2} = \frac{\phi(-\alpha_2 \log|J|/(1 - \alpha_2))}{-\alpha_2 \log|J|/(1 - \alpha_2)} . \tag{4.148}$$

However, in Sect. 4.3.4, where nonanalyticity properties of scaling functions are discussed, we will refer to this example as an exceptional case.

4.3.4 Discussion of Nonanalyticities

As has been motivated in the last section, a discussion of the nonanalyticities of the entropy-like functions is essential. To this end two remarks are necessary. Firstly, assume that $F_s(V, T)$ is continuous, but not differentiable at a point P (the same then holds for the associated Gibbs free energy at point P^*). In this case $\lambda(\beta)$ changes discontinuously; in view of (4.119) and (4.120) this leads to a linear part in $\phi(\lambda)$. The analog situation holds for $\alpha(q)$ and $f(\alpha)$.

Secondly, since the entropy functions of individual "simple systems" are strictly convex, the thermodynamic formalism developed will assign a convex (although no longer strictly convex) function to a "mixture" consisting of a linear combination of two or more simple systems. This function turns out to be the convex hull of the individual convex entropy functions: More precisely, it can be seen that

$$\tilde{GS}(\mu, \xi) = \sup_{n} \sup_{Kn} \sum p_i GS(\mu_i, \xi_i) , \tag{4.149}$$

$Kn = \{p_i, \mu_i, \xi_i / \sum_{i=1}^{n} p_i = 1, \sum_{i=1}^{n} \mu_i p_i = \mu, \sum_{i=1}^{n} p_i \xi_i = \xi\}$, is convex in both variables for any function GS. For uniformly scaling systems this then leads to first-order phase transitions. Higher-order phase transitions may be obtained using nonuniformly scaling systems and imposing appropriate coexistence conditions.

The diagram obtained for the mixture of systems then has the form depicted in Fig. 4.27. The two-humped curve is obtained from a calculation based on a histogram [see (4.116)], while the nonconvex part of this curve is substituted by the segment AB when using the thermodynamical formalism. The endpoints (A, B) of the solid line correspond to "pure phases". The change from A to B is made at fixed "temperature" $\beta(\lambda)$ and $q(\alpha)$, see (4.115) and (4.109), respectively. The points with $\lambda(\alpha)$ below the solid line correspond to mixed states: states with different entropies can coexist at the same temperature.

A simple model system with such behavior has been reported in [4.60]. Although various attempts have been made to find such behavior in experimental systems (notably in the p-Ge experiment, where we expected it), the results obtained so far have not been completely conclusive. An experimental finding which comes nearest to this situation will be discussed later on in connection with Fig. 4.32.

A slightly different type of nonanalytical behavior, however, has been found in different model maps (circle map [4.61], logistic map [4.62], Hénon's map [4.63]), which for higher-dimensional systems might even be generic. The easiest

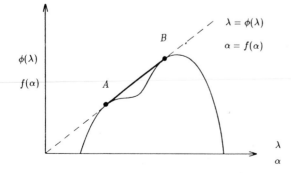

Fig. 4.27. Schematic plot of coexistence between two systems (see text)

insight into the mechanisms which account for such effects is given by the logistic map $x_{k+1} = ax_k(1 - x_k)$ at fully developed chaos, $a = 4$ [4.7, 29, 46]. If the density $\rho(x)$ is investigated on $[0, 1]$, it is easily seen [4.46], e.g., that the range of the scaling exponent $\alpha(x)$ consists of only two values: $\alpha(x) = 1$, $\forall x$ except the endpoints $\{0, 1\}$ and, therefore, $f(1) = 1$; $\alpha(x) = 1/2$ at the two endpoints of the interval and $f(1/2) = 0$ (Fig. 4.28). In [4.50] this nonanalytical behavior was interpreted as a phase transition of a spin system with antiferromagnetic two-spin interactions at a negative temperature.

An infinite numerical resolution would yield two separated systems consisting of an attractor A and a repeller B, respectively, where the straight line between A and B would be removed [4.64], in agreement with the condition for observability (Sect. 4.3.1). A similar effect can also be observed in the scaling function for the scaling of the support. The critical point of Fa, which generates the singularity of the measure, is the repeller, which would not be observable for infinite resolution or $n \to \infty$. The information-theoretical entropies are $K(1) = K(0) = K(q) = \log 2$, $\forall q$, because the map is conjugated to the map $z_{n+1} = 1 - 2|z_n|$, with $x = \sin(\pi z/2)$, and the conjugacy leaves $K(q)$ invariant.

The spectrum of the dynamical entropy (entropy of the support) is easily evaluated. It cannot be expected to coincide with the above-considered spectrum of entropies since the map is not hyperbolic. The range of the definition of $\phi(\lambda)$ is of course $[-\infty, \log 4]$. The behavior of the scaling function could be

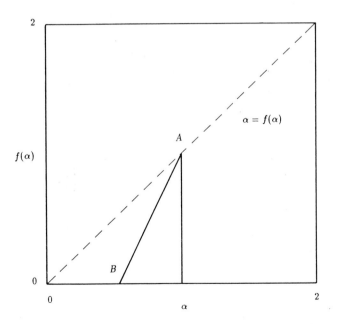

Fig. 4.28. Schematic scaling function of the measure for the logistic map. For infinite resolution, separated systems consisting of an attractor A and a repeller B, respectively, would be obtained. In numerical computations, however, infinite resolution can never be achieved

evaluated from the conjugacy map mentioned above. Here, however, a more general approach [4.45, 65, 66] is followed. Consider a one-dimensional one-humped map of the form $Fa(x) = 1 - a|x|^z$, $z > 1$ and denote the (one-dimensional) derivative of the map by Fa'. The effective Lyapunov exponent of points at a small horizontal distance δ from the maximum can then assume an arbitrarily large negative value. If all orbits of length k are taken, which visit this neighborhood exactly once at the k_0th iterate and all have the same $\lambda(x_0, k) = \lambda(k)$, then

$$\lambda(\delta, k) = \frac{k - 1}{k} \lambda(\delta, k_0, k) + \frac{1}{k} \log |Fa'(\delta)| . \tag{4.150}$$

Compared with $|Fa'(\delta)|$, $\lambda(\delta, k_0, k)$ can be regarded as independent of δ and, therefore, we write

$$\delta(\lambda, k_0)^{(z-1)} \sim e^{k\lambda(k) - (k-1)\lambda(k_0, k)} . \tag{4.151}$$

The invariant density is nearly flat around $x = 0$. The probability for the appearance of orbits with $\lambda(k) < \lambda_b$ is then given by

$$\mu(\lambda(k) < \lambda_b) \sim \sum_{k_0} e^{[1/(z-1)][k\lambda_b - (k-1)\lambda(k_0, k)]} . \tag{4.152}$$

Since $P(\lambda, k) \geq d\mu/d\lambda_b$, the inequality

$$\phi(\lambda) - \lambda \geq \frac{1}{(z-1)} (\lambda - \check{\lambda}) , \tag{4.153}$$

$\lambda \leq \lambda_c$, is obtained, where $\check{\lambda} = \lim_{k \to \infty} \min_{k_0} [\lambda(k_0, k)]$ and λ_c is the largest admissible $\lambda(k)$ such that orbits of the kind considered can exist. The equality is achieved if the contribution of other types of orbits can be neglected. This is the case at least for $\lambda \leq 0$, and for $\phi(\lambda) - \lambda$ a linear part has to be expected with slope $s = 1/(z - 1)$; in our example of the logistic map this slope is therefore equal to one.

On the other hand, the existence of fixed points at the ends of the interval also leads to an effect similar to a phase transition. These fixed points have a maximal expansion rate and lead to a singularity α of the measure, as pointed out before ($\alpha = 1/2$ for the logistic map). As above, consider

$$\lambda(\delta, k) = \frac{k - \tau(\delta)}{k} \lambda(\delta, k_0, k) + \frac{\tau(\delta)}{k} \log |Fa'(0)| , \tag{4.154}$$

where this time $\tau(\delta)$ measures the characteristic time needed to escape from the neighborhood of the unstable fixed point (here we assume, for convenience, that this point is situated at $x = 0$).

Using

$$\tau(\delta) \sim \frac{-\log |\delta|}{\log |Fa'(0)|} , \tag{4.155}$$

$$\delta(\lambda, k_0) \sim e^{-k[\lambda(k) - \lambda(k_0, k)]\lambda_{max}/[\lambda_{max} - \lambda(k_0, k)]} \tag{4.156}$$

is obtained, where $\lambda_{max} = \log|Fa'(0)|$. In this case the probability for $\lambda \geq \lambda_b$ is proportional to $|\delta(\lambda_b, k_0)|^\alpha$ and

$$\mu(\lambda(k) \geq \lambda_b) \sim \sum_{k_0} e^{-k\alpha[\lambda_b - \lambda(k_0, k)]\lambda_{max}/[\lambda_{max} - \lambda(k, k_0)]} . \tag{4.157}$$

Since $P(\lambda, k) \geq d\mu/d\lambda_b$, the inequality

$$\phi(\lambda) - \lambda \geq \alpha(\lambda - \overset{\smallsmile}{\lambda})\lambda_{max}/[\lambda_{max} - \lambda(k_0, k)] , \tag{4.158}$$

$\lambda_c \leq \lambda \leq \lambda_{max}$, is obtained, where this time $\overset{\smallsmile}{\lambda} = \lim_{k \to \infty} \max_{k_0} [\lambda(k, k_0)]$. If the equality occurs for some range of λ, then $\phi(\lambda) - \lambda$ has a linear part with slope $s = -\alpha\lambda_{max}/(\lambda_{max} - \overset{\smallsmile}{\lambda})$, and we may assume that $\overset{\smallsmile}{\lambda}$ is equal to the Lyapunov exponent. In our case, finally, the slope is seen to be equal to -1.

Analogous remarks apply also for the two-dimensional analog of the logistic map, Hénon's map (Fig. 4.29a). Because of the constant Jacobian, only one independent Lyapunov exponent is present; the role of the critical point is taken by the homoclinic tangency points [4.45]. Note that in Fig. 4.29a the scaling function extends also to negative values of λ and that the left part of the scaling function displays a straight line behavior, contrary to the behavior of the hyperbolic analog in Fig. 4.29b. This effect corresponds to a phase-transition-like behavior which is due to the influence of homoclinic tangency points, as explained in the text. It cannot be removed if k is chosen to be larger.

For $a = 1.4$, $b = 0.3$, the Hénon attractor lies at the closure of the unstable manifold of the unstable fixed point $X^*(x^*, y^*)$, $x^* = y^* = \{-1 + b + [4a + (1 - b)^2]^{1/2}\}/2a$ and the unstable manifold lies tangentially to the stable manifold of X^* at the homoclinic tangency points X_i. There, the effective first Lyapunov exponents have negative values, comparable to the situation for the area around the maximum of a nonhyperbolic one-dimensional map. The basin boundary is given by the stable manifold of a saddle point S. Upon increasing the parameter a the attractor is finally destroyed by a heteroclinic crisis involving the manifolds of the points S and X^* (this then leads to a singularity of the measure of the expanding manifold of S). For two-dimensional maps with constant Jacobian, the phase-transition-like effect due to the existence of homoclinic tangency points, which is believed to be generic, can be demonstrated as follows. Suppose the map has a tangency of quadratic order. Around the tangency point, the dynamical map can be linearized. For a point with distance vector $(\varepsilon, \varepsilon^2)$ from the tangency point X_0, situated on the unstable manifold, the first effective Lyapunov exponent λ can be evaluated approximately:

$$\lambda(\varepsilon, k) = \frac{1}{k} \log \{[\varepsilon e^{k\lambda(x_0, k)}]^2 + (\varepsilon^2 e^{k[\log |J| - \lambda(x_0, k)]})^2]^{1/2}\} . \tag{4.159}$$

Thus, it is seen that $\lambda(\varepsilon, k)$ can be written as

$$\lambda(\varepsilon, k) = \frac{1}{2k} \log[\varepsilon^2 e^{2k\lambda(x_0, k)}(1 + \varepsilon^2 e^{2k(\log |J| - 2\lambda(x_0, k))})] , \tag{4.160}$$

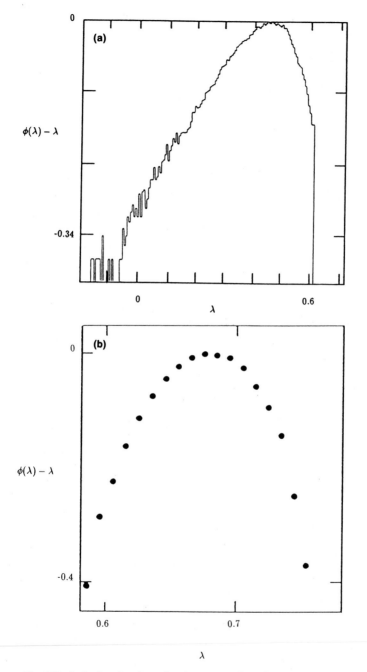

Fig. 4.29a, b. Scaling function of **a** the support for Hénon's map and, for comparison, **b** of the (hyperbolic) Lozi map (from the dynamical equations, $k = 20$)

such that we obtain

$$\lambda(\varepsilon, k) = \frac{\log \varepsilon}{k} + \lambda(x_0, k) + \frac{1}{2k} \log(1 + \varepsilon^2 e^{2k(\log |J| - 2\lambda(x_0, k))}) . \tag{4.161}$$

Therefore, the dependence on ε is given by

$$e^{2k[\lambda(\varepsilon, k) - \lambda(x_0, k)]} \sim 1 + \varepsilon^2 |J|^{2k} e^{-4k\lambda(x_0, k)} . \tag{4.162}$$

A short calculation yields

$$\frac{d\varepsilon}{d\lambda(\varepsilon, k)} \sim k e^{k[\lambda(\varepsilon, k) + \lambda(x_0, k)]} |J|^{-k} . \tag{4.163}$$

Since the point is assumed to be situated on the unstable manifold, which has the shape of a quadratic parabola, the dependence of $P(\varepsilon)$ on ε can easily be found as

$$P(\varepsilon) d\varepsilon \sim \cdot \left(1 + \frac{\varepsilon^2}{2}\right) d\varepsilon . \tag{4.164}$$

Putting this together, it is found that

$$P(\lambda(\varepsilon, k)) = P(\varepsilon) \frac{d\varepsilon}{d\lambda(\varepsilon, k)} \sim k e^{k[\lambda(\varepsilon, k) + \lambda(x_0, k)]} |J|^{-k} \left(1 + \frac{\varepsilon^2}{2}\right) . \tag{4.165}$$

From this it is concluded that

$$\phi(\lambda(\varepsilon, k)) - \lambda(\varepsilon, k) = \frac{1}{k} \log P(\lambda(\varepsilon, k)) \sim + \lambda(\varepsilon, k) + \lambda(x_0, k) - \log |J| + C , \tag{4.166}$$

where C is a constant. In this way a linear part with slope $s = 1$ is obtained for the left-hand side of the scaling function. A linear part on the other tail of $\phi(\lambda)$ occurs only for the special situation corresponding to the case of the logistic map.

For values a in the vicinity of the parameter value a_0 for which the attractor is destroyed due to the collision with the saddle point S, $a < a_0$, a new maximal λ is created, leading to a linear part of $\phi(\lambda)$ also in this region. An analogous effect can be observed for the Lorenz system in situations where the repellent effect of the unstable fixed point becomes dominant (Fig. 4.30). Finally, at another value of the parameter a, an attractor-merging crisis can be observed; as expected from our outline, near this crisis the situation depicted in Fig. 4.27 is found.

As has been observed in [4.65] the partition function for the scaling of the support can be approximated by

$$GZ(\beta, k) \sim \sum_{N \to \infty} |\mu^N_{1, p}|^{-\alpha_1(x^N_p)} e^{-(\beta - 1)k\lambda(x^N_p, k)} , \tag{4.167}$$

where $k < N$. $\mu^N_{1, p} = e^{N\lambda(x^N_p, N)}$ denotes the expanding multiplier of the periodic orbit $Fa^k(x^N_p)$, $k = 1, \ldots, N$, and the sum extends over all periodic points x^N_p of

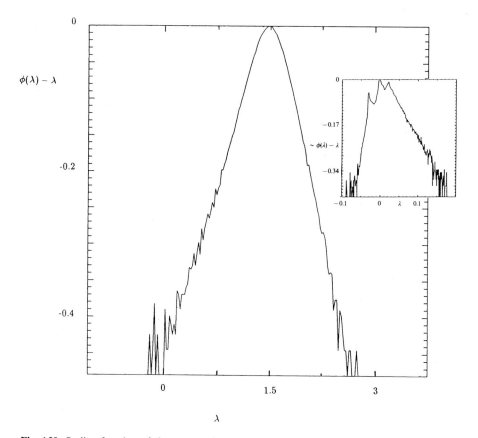

Fig. 4.30. Scaling function of the support for the Lorenz system at parameter values where the repelling influence of one unstable fixed point becomes dominant. A Poincaré section has been done in order to be able to consider a map ($k = 20$). Note the straight line behavior also on the right-hand side of the scaling function; it corresponds to a phase-transition-like behavior which is due to the influence of the repelling fixed point. In the inset the probability distribution of the first effective Lyapunov exponent is shown when no Poincaré section is done ($k = 50$, sampling time $t_s = 0.01$). Although k is not large enough in order to yield the asymptotic scaling function, the graph shows clearly the origin of the linear part on the right-hand side of the scaling function

period N, including all divisors of N. In this way, the spectrum can be estimated from periodic points since $\lambda(x_p^N, k)$ can become negative for pieces of certain periodic orbits with length $k < N$ (this difficulty, however, cannot occur for hyperbolic systems; furthermore, in this case $\alpha_1 = 1$, as pointed out before).

For Hénon's map, in view of (4.146) and (4.147), it can be concluded that if a singularity occurs for $\Lambda(\beta)$, this is inherited by $\tau_1(q)$ and (or) by $\tau_2(q)$. The dynamical spectrum (which in this case can be evaluated directly from the dynamical equations) is believed to give the most reliable scaling function of the measure $f(\alpha)$. α_{max} has, therefore, been estimated for this system in [4.67] with the help of the local Lyapunov dimension of the unstable fixed point of the map, leading to $\alpha_{max} = 1.352$. Bounds for α_{min} are more difficult to obtain [4.68]. Qualitatively, the same effects are expected for higher-dimensional chaotic systems.

In the general case, however, we are faced with the fact that the scaling function of the support and the scaling function of the measure are, in principle, independent; they do not necessarily share the analyticity properties. If, for example, the family of maps [4.29, 69]

$$F(x) = 1 - |x^r - (1 - x)^r|^{1/r}, \quad r > 1 , \tag{4.168}$$

is considered, it is found that the density is $\rho(x) = r(1 - x)^{1/r}$. The probabilistic spectrum $D(q)$ has the values $D(q) = rq/(q - 1)$ for $q \leq q_c = 1/(1 - r)$, while $D(q) = 1$ is found for $q > q_c$. For all values of r, however, there is a dynamical phase transition at $\beta = 1$. In situations where a phase-transition-like effect occurs, the whole range of scaling behavior can be separated into regions of qualitatively different behavior, due to distinct dominant dynamical processes or topological phenomena. The different regions can, for example, be characterized by qualitatively different eigenfunctions of the generalized Frobenius–Perron equation. In this way a better qualitative picture of the different competing scaling processes can be obtained. For the example of the logistic map it has been shown that the associated eigenfunctions have different symmetry properties [4.13]: for $\beta < \beta_c$ the eigenfunctions are symmetric ("disordered phase") while for $\beta > \beta_c$ the symmetry is broken ("ordered phase"), a fact which is very reminiscent of what happens in most real phase transitions.

In one- and two-dimensional *hyperbolic* maps, phase-transition-like effects are not possible, as can be shown in analogy with one-dimensional spin systems, provided that the generating partition has not infinitely many elements and if there are no long-time correlations between the symbols (a situation which corresponds to infinitely many spin states or long-range interactions). For higher-dimensional systems, however, this is no longer true; Lyapunov exponents associated with different Cantor directions can fluctuate in accordance with (4.135) and (4.144), where the index j depends on q.

4.3.5 Evidence of Phase-Transition-Like Behavior in Experimental Observations

In Sect. 4.3.4 different kinds of phase-transition-like behavior have been analyzed for model maps. It has been seen that these effects are due to the long-time memory effects of the dynamical behavior. Let us turn now to scaling functions obtained from experimental time series. We show that this feature can also be observed from experimental data. Note that in order to get the right correspondence with the theory developed for maps, a Poincare section should be used. It can be seen, however, that the essential features of the form of the scaling functions can be preserved if no Poincare section is made.

a) Effect of Homoclinic Tangency Points

As stated before, the presence of homoclinic tangency points leads to a phase-transition-like effect which is best observed if other competing effects are absent. For both the p-Ge semiconductor experiment and the NMR laser experiment we found working conditions for which there is evidence of such a behavior (Fig. 4.31). Note that the two systems are discussed further in Sect. 4.4.3.

b) Effect due to the Influence of a Crisis

Sudden qualitative changes in the dynamical behavior upon variation of the external parameter appear in many diverse systems. They are triggered off either by collisions between the chaotic attractor and some unstable periodic orbit of high order ("crisis") or by tangent bifurcations ("intermittency"). Near the working conditions where the system undergoes a crisis a characteristic dynamical behavior is shown. The trajectory remains for a long time in the phase space region where the periodic orbit is found. This phase is then interrupted by a spatially extended chaotic transient. Note, however, that this behavior is not always convincingly detected in the two-dimensional projections of high-dimensional data. The length of this transient can be studied as a function of the external parameter; it is easily seen that a critical slowing down sets in. This phenomenon leads to the behavior discussed at the beginning of the last section. It can be observed for both experimental and model systems. The scaling functions shown in Fig. 4.32 are consistently interpreted in this way.

In the language of thermodynamics, a set represented by a graph of the type shown in Fig. 4.32a and b would be called thermodynamically unstable because the convexity condition is violated. By choosing k in the definition of effective Lyapunov exponents much larger than above, the convexity can be reestablished. It is clear that for the evaluation of the Lyapunov exponents of a system such a situation is characterized by large oscillatory fluctuations of the time averages of the exponents; this problem, however, can in principle be removed by going to averages over long enough time scales.

Beyond the above-commented properties, the comparison of Fig. 4.32a with Fig. 4.32b reveals a further striking resemblance. Both the range of scaling exponents and the form of the tails of the scaling functions are almost identical. In [4.66] the two-parameter family of circle maps is investigated near (K_c, Ω_c), the values of the order parameters which lead to an attractor-merging crisis of two phase-locked attractors into one phase-unlocked attractor. Figure 4.32b is then interpreted as representing the intermittent hopping between the two repellers with rotation numbers 0/1 and 1/2. Note that the first hump in Fig. 4.32a corresponds to an effective Lyapunov exponent slightly greater than zero. The observed effect might here be related either to a near attractor-merging crisis, such that a hopping between two precursory attractors is observed, or to a slow drift of the experimental system, as will be argued in Sect. 4.4.4. The scaling function for the NMR laser far from crisis was shown in Fig. 4.31b.

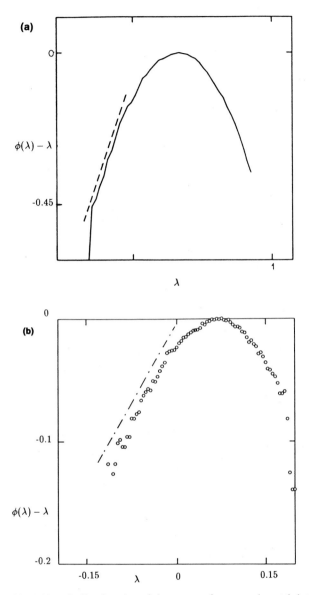

Fig. 4.31. a Scaling function of the support from experimental data (p-Ge semiconductor experiment, hyperchaotic state). The broken line shows again the region where a linear part manifests the influence of the homoclinic tangency points, leading to a phase-transition-like phenomenon. The scaling function was calculated from a time series of 100 000 points at an embedding dimension $d_E = 10$ with $k = 50$. **b** Scaling function of the support from experimental data (NMR laser experiment far from crisis). Note the linear part due to the homoclinic tangency points. A time series of 100 000 integers has been used; $d_E = 8$, $k = 50$

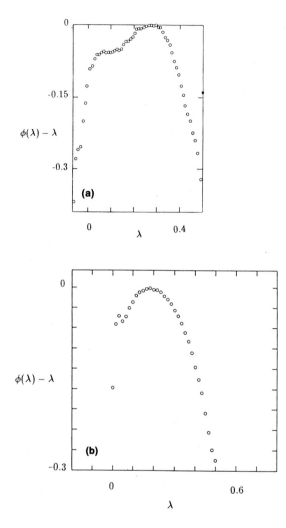

Fig. 4.32. a Scaling function of the support for an experimental NMR laser file. A time series of 100 000 integers has been analyzed using $d_E = 8$ and $k = 50$. The time series considered corresponds to the third file in Fig. 4.47. **b** Scaling function of the rotation number for the circle map calculated directly from the dynamical map for comparison (from [4.66], see text). Both graphs are two-humped, a feature which is discussed in the text

4.4   Evaluation of Experimental Systems

Having already shown various scaling functions calculated either from experimental data or from the dynamical equations, not much has been said about how this is done and how much we can rely on the results obtained. In this chapter we outline both the theoretical principles and the numerical procedures which are involved in the evaluation of the scaling exponents from experimental time series. They are first applied to simulated data generated from well-known model maps or systems and their reliability and intrinsic limitations are discussed. Finally, two experimental systems are investigated with emphasis on the systems as a testing ground of the numerical procedures rather than as an analysis of the systems itself. The results obtained from different algorithms are compared and the origin of deviations explained.

4.4.1 Embedding of a Time Series·

When a physical system is observed and a one-dimensional measurement is made, a smooth function

$$e : M \to \mathbb{R}$$

is generated. In almost all cases when a dynamical system is experimentally observed only little exact information about it is a priori available; not even its algebraic dimension is known. Referring to Sect. 4.3, it can be said that the Hausdorff dimension will be less than or equal to the dimension of the underlying system. This lack of information introduced by the observation process can partly be removed with the help of an application of the *embedding theorem* of *Whitney* [4.70] in the version of *Takens* [4.71, 72]. If M is a compact differentiable manifold of dimension $m \in \mathbb{N}$, and $t : M \to M$ is the (discrete version of the) dynamical map, at least a C^2-diffeomorphism, then it is a generic property that

$$e : M \to \mathbb{R} : x \to (e(x), e(t(x)), \ldots, e(t^{2m}(x)))$$

is an embedding.

In other words, it is possible with the help of the embedding process to get a diffeomorphic picture of a compact manifold of dimension m, consisting of a closed submanifold in a space of dimension less than $2m + 2$. The embedding procedure starts for the simplest case with a scalar measurement, equally spaced in time

$$\{s_1, \ldots, s_n\} \, ,$$

called the *time series*. From these scalars higher-dimensional points $\tilde{x}_i$ are composed by writing

$$\tilde{x}_i = (s_i, \ldots, s_{i+2m}) \, .$$

Remark: There exist examples of m-dimensional manifolds which cannot be embedded in E^{2m} (projective spaces of appropriate dimension). Spaces with certain topological restrictions (which can be formulated in terms of the vanishing of certain homotopy groups) can be embedded in spaces of dimension less than $2m$. If the mapping can be considered as a mapping between two Euclidean spaces, then the diffeomorphism is even an isometry [4.12, 73].

It is true that the embedding theorem states that no matter which (nontrivial) component is measured and no matter what the dimension of the embedding space is, as long as it is larger than $2m + 1$, an embedding is obtained. Implicitly assumed, however, is an infinite amount of noise-free data, a requirement that can never be met in applications. To get the best possible reconstruction of the original system, essentially two parameters, the delay time and the embedding dimension, can be varied, if constant time intervals between two measurements (called delay time) are used. As is well known, choosing smaller delay times leads to more linearly dependent measurements, and the effect of noise becomes dominant. If the delay time is chosen too large, the correlation within the time series is lost and a measurement is obtained which is indistinguishable from white noise. The question of how the delay time should be chosen has been discussed extensively in literature [4.74]. In a new contribution which stresses the connection with information theory, a piecewise linear approximation to the embedding diffeomorphism Φ is considered [4.75]. If the trajectories $s(t)$ of the original system were known, the optimal reconstruction Φ would correspond to the minimum of $\langle (s(t) - L(\Phi(s(t))))^2 \rangle$, where L is the best piecewise linear inverse of Φ. Since the trajectories $s(t)$ are not known, however, this condition has to be reformulated as a condition between random variables, which finally leads to visual criteria for the selection of the optimal delay time. In the present work we assume that an adequate delay time has been chosen by the experimenter if real experiments are considered. In an additional step it is then checked whether with respect to that variable a scaling region exists. The investigations within the context of the stability of results, at the end of this section, indicate that this has been more or less the case and that this issue may not be as critical as suspected by many authors. The second parameter at our disposal is the embedding dimension itself. It should be observed first that the *numerical precision* connected with almost all operations with reconstructed points will decrease if the dimension of the embedding space is enlarged. Therefore, a region of sufficiently large but not too large embedding dimensions is of interest. This can also be seen in accordance with the scaling assumptions of Sect. 4.3: a *scaling region* for the support and the measure, as well as for the embedding process, is required. Note that the embedding dimension for the dynamical approach is limited by a finite "folding time".

For the evaluation of the spectrum of generalized dimensions $D(q)$, definitions (4.100) or (4.101) in Sect. 4.3 can be applied straightforwardly to the points of the embedding space, regardless of the fact that these points are now embedded in a space of a higher dimension as compared with the original system. The spectrum $K(q)$ of generalized information-theoretical entropies can be evaluated from (4.137). Whenever the embedding dimension is chosen larger than $2m + 1$, where m is the dimension of the underlying system, an increase in the embedding dimension by one corresponds to a shift by one in the rest of the coordinates. This effect, in turn, can be associated with the action of the dynamical map. The vertical distance between neighboring log–log curves used for the evaluation of the fractal dimensions in (4.100) or (4.101) are, therefore, a measure of $K(q)$, where it is assumed that the embedding dimension is increased in steps of one. For a more detailed discussion we refer to [4.54]. An evaluation of partial dimensions is very difficult; in any case it is not possible without a procedure which is able to calculate local Lyapunov exponents.

4.4.2 Lyapunov Exponents from the Dynamical Equations and from Time Series

In this section we follow essentially the outline given in [4.76, 77]. Again let $DFa(x_i)$ denote the Jacobian of the dynamical map Fa, i.e., the $N \times N$ matrix of partial derivatives of Fa evaluated at the point x_i. The basic idea is to define an average exponential growth rate of separation in the tangent bundle of the attractor. For a one-dimensional system this average is then given by

$$\lim_{k \to \infty} \frac{1}{k} \log \left| \prod_{i=0}^{k-1} Fa'(x_i) \right| = \lim_{k \to \infty} \frac{1}{k} \sum_{i=0}^{k-1} \log |Fa'(x_i)| . \tag{4.169}$$

In the general case the occurrence of products of noncommuting matrices in (4.156) renders the existence and evaluation of the limits nontrivial. It has been shown, however, by *Oseledec* [4.78] that analogous quantities can be defined and evaluated according to the following theorem ([4.3], for a more extended reference see [4.12]).

Let ρ be a probability measure on a space S and $Fa:S \to S$ measure-preserving such that ρ is ergodic and $DFa:S \to M_{\text{at}}$, the space of $m \times m$ matrices, measurable. Then the following limits exist for almost all $x_0 \in S$ and they are independent of x_0:

$$L = \lim_{k \to \infty} [DFa^k(x_0)^* DFa^k(x_0)]^{1/2k} , \tag{4.170}$$

where the asterisk denotes the adjoint.

Let the distinct eigenvalues of this matrix be denoted by d_i. The logarithms λ_i of the numbers d_i can be ordered $\lambda_1 > \lambda_2 > \dots$ and the subspaces of $\mathbb{R}^N$ corresponding to all eigenvalues $\leq \lambda_i$ can be labelled by E_i. Then, for ρ-almost everywhere,

$$\lambda_i = \lim_{k \to \infty} \frac{1}{k} \| DFa^k(x_0)v \| \quad \text{for } v \in E_i \backslash E_{i+1} . \tag{4.171}$$

Based on this theorem an efficient algorithm for the evaluation of these Lyapunov exponents has been proposed by *Benettin* et al. [4.79]. In the tangent space $\mathbb{R}^N$ of Fa at x_0 an arbitrary complete set of orthonormal basis vectors $\{v_1^0, \dots, v_N^0\}$ is chosen. For $j = 1, \dots, k$ subsequent steps the following calculations are then performed. One determines the vectors

$$w_n^j = DFa(x_{j-1})v_n^{j-1} , \tag{4.172}$$

where $n = 1, \dots, N$. By orthogonalization these vectors can be restricted to the subspaces E_i and stretching factors d_n^j are obtained by

$$d_1^j = \| w_1^j \| ,$$
$$d_m^j = \| \breve{w}_m^j \| , \tag{4.173}$$

with

$$\breve{w}_m^j = w_m^j - \sum_{l=1}^{m-1} (v_l^j, w_m^j)v_l^j , \tag{4.174}$$

where $m = 2, \ldots, N$ and

$$v_n^j = \overset{\cup}{w}_n^j / d_n^j .\tag{4.175}$$

The Lyapunov exponents are then finally obtained from

$$\lambda_n = \lim_{k \to \infty} \frac{1}{k} \sum_{j=1}^{k} \log d_n^j .\tag{4.176}$$

Figure 4.33 shows a schematic illustration of this procedure for a two-dimensional system. Each vector v_n tends to relax into that direction which allows the largest stretching. After some transient steps the first vector v_1^j will point in the direction of fastest separation. This procedure allows a convenient calculation of all Lyapunov exponents of a system. It avoids numerical problems of overflow and computational difficulties of large time and memory consumption. Moreover, the fact that the procedure involves vectors from the tangent space ensures that the motion is restricted to the attractor under consideration, avoiding complications due to the possible existence of different basins of attraction. For another similar procedure see [4.80]; a completely different one is used in [4.81].

For *experimental data* the attractor is given by a number of points obtained with the help of an embedding process. The embedding process can be seen as an on-line creation of new points; therefore, the dynamical map Fa induces in the embedding space of dimension d_E a dynamical map $\tilde{F}a$ with

$$\tilde{F}a(\tilde{x}_k) = (s_{k+l}, \ldots, s_{k+d_E+l-1}) ,\tag{4.177}$$

where

$$\tilde{x}_k = (s_k, \ldots, s_{k+d_E-1}) .\tag{4.178}$$

l denotes a possible delay step. For convenience, we omit the tilde in the

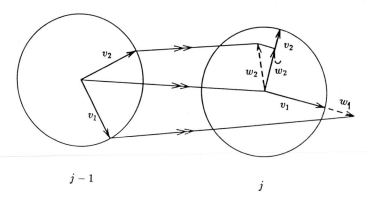

Fig. 4.33. Illustration of the procedure to calculate Lyapunov exponents from the dynamical equations. Vector v_2 is mapped on w_2, orthogonalized with respect to w_1 to give $\overset{\cup}{w}_2$ and rescaled to a new v_2 of unit length

following and identify Fa with $\tilde{F}a$ and x with $\tilde{x}$. To calculate the Lyapunov exponents, $DFa(x_k)$ should be known at every point x_k. Using ideas put forward by *Ruelle* and *Eckmann* [4.3] and by *Sano* and *Sawada* [4.82], the Lyapunov exponents can now be approximately calculated from the information contained in the dynamical map if we replace the linearized flow DFa by a suitable (linear) approximation $\tilde{D}Fa$. To construct this approximation at x_k, the images of various points in the neighborhood of x_k are investigated. Let y_k be a point within a ball of radius r around x_k. Then denote

$$v^{(k)} = y_k - x_k \tag{4.179}$$

and

$$w^{(k+1)} = Fa(y_k) - Fa(x_k) . \tag{4.180}$$

Now the relations

$$w_n^{(k+1)} = \tilde{D}Fa(x_k)v_n^{(k)} \tag{4.181}$$

determine the matrix $\tilde{D}Fa(x_k)$ if d_E linearly independent vectors w_n are given. If the corresponding points y_k can be found within a neighborhood of x_k where a linear approximation can be justified, the constructed $\tilde{D}Fa(x_k)$ will be a good approximation to $DFa(x_k)$. In that case $\tilde{D}Fa(x_k)$ can be used for the evaluation of Lyapunov exponents as outlined in the previous subsection. Figure 4.34 shows a schematic illustration of the situation described.

Implementation of the Algorithm

Two problems arise in practical applications. Usually only a limited set of more or less noisy data points is available. As pointed out in the first section, the distribution of points will be rather inhomogeneous since a smooth distribution of points along the unstable direction and a singular one along the stable (fractal) direction of the manifold is to be expected. To get more precise information about the stable directions, one has to investigate the mapping of M_0 neighboring points to x_k, where M_0 is usually considerably larger than d_E.

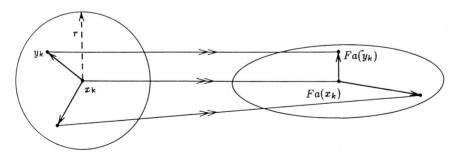

Fig. 4.34. Illustration of the method to determine the linearized flow from time series. An approximation $\tilde{D}Fa(x_k)$ is constructed from the map which sends $y_k - x_k$ to $Fa(y_k) - Fa(x_k)$

The matrix $\tilde{D}Fa(x_k)$ is then determined from a least-squares fit in the following way:

$$\tilde{D}Fa(x_k) = A(x_k)B(x_k)^{-1} , \tag{4.182}$$

where

$$A(x_k)_{\alpha\beta} = \sum_{m=1}^{M_0} w_{m\alpha}^{(k+1)} v_{m\beta}^{(k)} \tag{4.183}$$

and

$$B(x_k)_{\alpha\beta} = \sum_{m=1}^{M_0} v_{m\alpha}^{(k)} v_{m\beta}^{(k)} . \tag{4.184}$$

Due to the limited number of data points available, a compromise has always to be made between choosing a sufficient number of data points within a small ball of radius r to reflect the geometrical structure of the attractor in all directions and selecting only the nearest neighbors such that a linear approximation F can be justified.

As pointed out in [4.82], the average radius r containing M_0 points can be determined roughly by using the fractal dimension of the attractor. But different complications, depending on the embedding dimension, can arise. Generally, points nearer to the origin of the ball than the amplitude of the noise inherent in the system make a more-than-average contribution to the error (the amplitude of noise can be estimated, for example, with the help of the fixed-size method (4.100) for the evaluation of the fractal dimension). In low-dimensional systems the collection of the very nearest points often leads to singular matrices $\tilde{D}F$ since the points on the attractor are most densely distributed in the direction of the unstable manifolds. Even if more points than required by the embedding dimension are taken into account, this can lead to linear dependency. The finite precision of measured data also contributes to this effect. Therefore, the exclusion of points at too small a distance from the center of the ball is favorable. The points to be used are then chosen from a shell of outer radius r_1 and inner radius r_2, where the numerical values of r_1, r_2 should be determined for each particular system along the lines outlined above. An automatic choice of the radii which depend on the point considered can be implemented as follows. The outer radius is fixed large enough, the inner according to the noise and geometrical structure of the attractor. We search then for a fixed number of points (say, 200) within that shell, randomly with respect to the time-ordering of the time series. In this way, the artificial preference of parts of the time-ordered time series can be suppressed. Otherwise, this effect could lead to large oscillations of the local Lyapunov exponents at certain points. Sometimes one does not have to go through the whole data file to find these points; occasionally, not as many as desired can be found. For the least-squares fit, however, we use only a small number (say, 30) of the nearest neighbors (this number depends mainly on the embedding dimension). If the matrix turns out to be singular with that choice of points, we replace some of the innermost ones by the same number of so far unused nearest-neighbor points. If, in spite of all precautions, the matrix $A(x_k)$ becomes singular and cannot be inverted, or more likely, if not as many neighbors as wanted can be found, the directions of the stable and unstable manifolds are no longer determined accurately. Then, again, for some transient steps the algorithm should proceed without taking the contributions to the Lyapunov exponents into account until the vectors v_n have found their directions along the flow again.

In high-dimensional embedding spaces we expect the matrix representing the lower-dimensional dynamical system to become singular. But noise and the finite resolution of data aquisition lead, in general, to nonsingular matrices, introducing a set of what could be called "noise-scaled" additional Lyapunov exponents. Particularly in experimental applications they are often the source of difficulties in the interpretation of the measured set of exponents. To get rid of these it is possible to reduce the dimension of the matrix representing the dynamical system by the following procedure proposed in [4.83]. Instead of

$$(s_k, s_{k+1}, \ldots, s_{k+d_E-1}) \,, \tag{4.185}$$

the l-dimensional vector

$$(s_k, s_{k+p}, s_{k+2p}, \ldots, s_{k+(l-1)p}) \tag{4.186}$$

is taken, where $l = \mathrm{int}((d_E - 1)/p)$, and the same dimension is inherited by the matrix $\widetilde{D}Fa$.

In our experience, the applicability of this procedure, as well as other procedures for saving computation time, depends strongly on the geometrical structure and, especially for experimental systems, on the sampling rate. The difficulties connected with these problems led us to the new algorithm proposed in Sect. 4.5.

Finally, it might be worthwhile to summarize the different sources of intrinsic limitations for the evaluation of Lyapunov exponents from time series. First of all, long-time oscillations of the system itself, which are of the same nature as those for which Lyapunov exponents are calculated from the dynamical equations, can be found. Although with longer calculations convergence can be expected, computation time and the length of the time series set stronger limits compared to computations starting from the dynamical equations. For hyperbolic systems the local crowding index displays complicated oscillatory behavior in accordance with the fact that the dependence of $P(\varepsilon)$ on ε in the direction of the stable manifold resembles a devil's staircase (or Cantor function [4.7]). Since typical hyperbolic systems have no long-range correlations, it is expected that the imprecision introduced by this effect is averaged out with an increasing number of iterations. For generic, i.e., nonhyperbolic systems, however, in the vicinity of homoclinic tangency points long-range correlations are generated. They lead to large fluctuations during the evaluation of Lyapunov exponents. It is worth noting that in this respect we always find similar intrinsic limitations for the evaluation of Lyapunov exponents as encountered in the evaluation of fractal dimensions.

4.4.3   Results and Stability of Results

In this section results obtained from the application of the algorithm described in Sect. 4.4.2 are discussed. First the behavior of the algorithm with respect to simulated data is investigated. Then the influence of noise and sampling rate is briefly worked out. Finally, the algorithm is applied to experimental data.

a) Numerical Examples: Simulated Data

The algorithm has been tested in the following way. For several systems time series $\{s_k\}$ were generated by selecting one coordinate of the map and scaling it to integer values in the range from 0 to 10 000. Then the algorithm was applied and the results were compared to those calculated from the dynamical equations.

The Hénon Map. In Fig. 4.35 the Lyapunov exponents for the Hénon map from time series are shown as a function of the embedding dimension. The directly calculated Lyapunov exponents are shown as dashed lines. In the analysis of the data the radii of the shells in which neighboring points were determined were kept constant and the number M_0 of neighbors was between 20 and 40. The deviation of the largest exponent at $d_E = 8$ can be traced to the fact that for this dimension the linear approximation fails. Repeating the calculation with a smaller ball and more points from a longer time series readjusts the value of λ_1 also at $d_E = 8$. The Hénon map is also considered in Sects. 4.4.4 and 4.5.

The Lorenz System. With respect to the experimental data that are investigated in this section later on, the Lorenz system provides, due to its smoothly evolving flow, a better comparison than the discrete Hénon map. Results obtained from time series generated from the Lorenz model are shown in Fig. 4.36. Again, the convergence of the three Lyapunov exponents to the known values indicated by

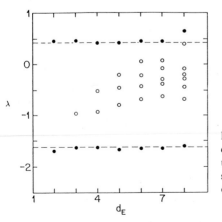

Fig. 4.35. Lyapunov exponents from a time series consisting of 40 000 points of the Hénon map at usual parameters for various embedding dimensions d_E. The dashed lines indicate the values obtained from a direct calculation

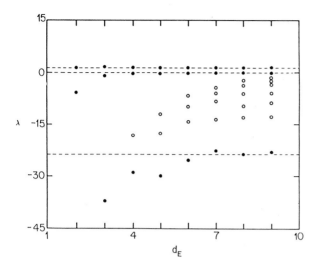

Fig. 4.36. Lyapunov exponents from a time series of 50 000 integers from the Lorenz system at parameter values $\sigma = 16$, $b = 4$, $r = 45.92$, for different embedding dimensions d_E. The dashed lines indicate the true values of the exponents

dashed lines with increasing embedding dimensions is very good. Note that the Lorenz system will further be investigated in connection with the response to noise and the behavior of the algorithm with respect to the sampling time. Results from applications of algorithms based on other approaches will be reported in Sects. 4.4.4 and 4.5.

The NMR Laser Model. As the last example based on simulated time series, results obtained for the NMR laser model [4.4] are discussed. It has been shown that a model of two generalized Bloch equations describes the basic experimental observations reasonably well. If the dimension of the phase space is enlarged to three by modulation of one of the system's parameters, a rich variety of nonlinear phenomena is found. In particular, a Feigenbaum route to chaos is obtained by an increasing modulation amplitude p of the pump (Fig. 4.37). For two values of the pump parameter p, time series consisting of 40 000 integers have been generated and the Lyapunov exponents calculated. Figure 4.38 shows the results obtained from a period-64 oscillation; Fig. 4.39 shows the results obtained just beyond the onset of chaotic behavior. From Fig. 4.29 it can be seen that the most severe limitation for a precise evaluation of the Lyapunov exponents from time series is given in the direction of the stable manifolds of points on the trajectory. The fact that for the most negative exponent not completely arbitrary values have been obtained is essentially due to the high periodicity of the motion and to the fact that the system is low-dimensional.

In all of these examples the algorithm proposed for the calculation of Lyapunov exponents from time series gives reliable estimates for both positive and nonpositive exponents. The results are of the same quality as obtained in [4.83], where a different diagonalization procedure has been used. Compared to the imprecision resulting from the intrinsic structure of the attractor, the use of

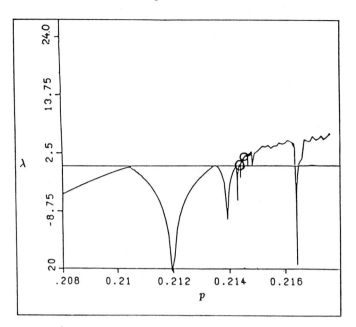

Fig. 4.37. Largest nonzero Lyapunov exponent vs. pump amplitude p for the NMR laser model. A Feigenbaum route to chaos is shown. The circles indicate the p-values for which the time series evaluated in the next two figures were generated

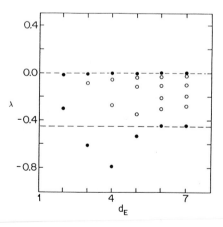

Fig. 4.38. Lyapunov exponents calculated from a time series of the NMR laser model (40 000 points) for a period of 64 for various embedding dimensions. As must be for a periodic motion, the first exponent is zero. Note that the most negative exponent converges slowly towards the correct value

different diagonalization procedures has no influence at all. (The validity of both approaches for the diagonalization of products of random matrices seems to be widely accepted [4.84].)

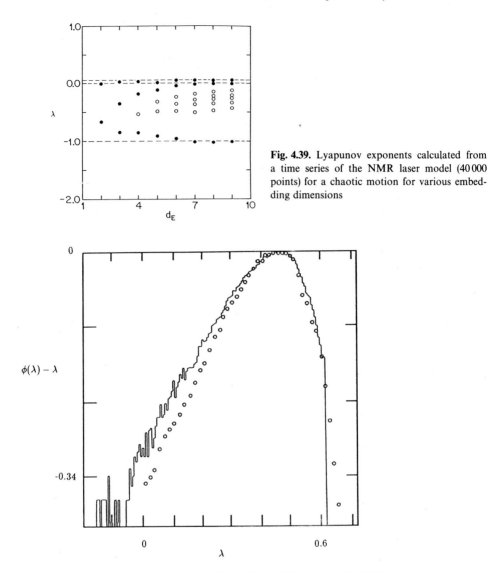

Fig. 4.39. Lyapunov exponents calculated from a time series of the NMR laser model (40 000 points) for a chaotic motion for various embedding dimensions

Fig. 4.40. Approximated scaling function $\phi(\lambda) - \lambda$ for the Hénon map. The *full line* shows results obtained from simulations using the dynamical equation; the *circles* indicate results obtained from the calculation of Lyapunov exponents from a time series of 45 000 points and an embedding dimension of 6. k was chosen equal to 20

Scaling Functions from Simulated Data. A direct comparison of the dynamical scaling functions calculated from the dynamical equations and from time series also gave satisfactory results, as can be seen in Fig. 4.40. The accuracy with which a scaling function could be recovered from a histogram was checked separately for the tent map; the result is shown in Fig. 4.41.

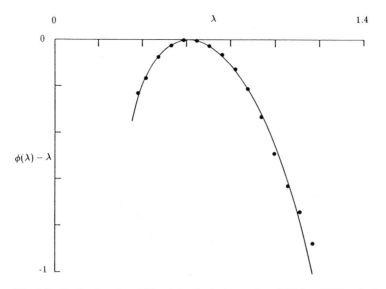

Fig. 4.41. Scaling function $\phi(\lambda) - \lambda$ for the tent map ($l_1 = 3/10, l_2 = 7/10$), calculated analytically (*full line*) and with the help of a histogram (*dots*). k was chosen equal to 25

b) Noise and Transients

An intrinsic property of the concept of Lyapunov exponents is the fact that stability properties are measured, even if the motion considered is not yet asymptotic. When from the power spectrum one can hardly tell whether a motion of high periodicity or a chaotic motion is being observed (Fig. 4.42), a calculation of the Lyapunov exponents gives an immediate answer. This answer is most easily obtained if the dynamical equations of the system are known.

It has been observed [4.85] that even for simple dynamical systems very complex boundary structures separating different coexisting attractors are found, although only the most robust ones can survive in real systems. Upon alteration of the external parameter, one is often led to believe that Feigenbaum bifurcations towards chaos are observed, but the concept of Lyapunov exponents in the form of Floquet multipliers can immediately prove this to be erroneous, as in the case of the NMR laser model. In this way solutions are often lost due to very small basins of attraction, thus leading to poor approximations of the Feigenbaum exponents (whose exact values are found in principle only in the limit of the number of bifurcations going to infinity).

When an embedding of a time series is considered in the presence of noise, it is observed that the local structures of the stable and the unstable manifolds are partially wiped out. Therefore, what is to be expected for the spectrum of Lyapunov exponents is an increase of the positive Lyapunov exponent (due to a larger degree of freedom) and a decrease in absolute magnitude for the

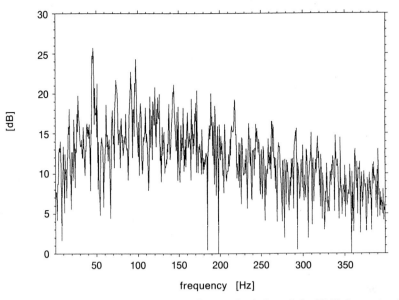

Fig. 4.42. (Real) Fourier spectrum, taken from a simulation of the NMR laser equations with delayed feedback. The corresponding motion is not chaotic, as can be shown by calculation of the Lyapunov exponents, which yield nonpositive values

negative Lyapunov exponents. This feature can be illustrated by the investigation of the change of the exponents of the Lorenz system upon the addition of noise (Fig. 4.43).

In accordance with what could be expected, the first and the second Lyapunov exponents change considerably when the region where a linear approximation makes sense is left. For a time series with rescaled values in $0\ldots10\,000$, this range has a radius of about 200 units. At a noise amplitude of 200, the first exponent becomes twice as large as the true value; at a noise amplitude of 300, four times as large. Note also that the effect on the negative exponent is less drastic.

The essential parameters that can be controlled in the algorithm are the number of points used for the least-squares fit, the minimal and maximal radius of the sphere where the map is expected to be well approximated by its linearization, the sampling time and the delay time. The sampling time can be chosen almost as short (but not as long) as wanted if one is interested only in the value of the positive Lyapunov exponent. The value of the negative exponent, however, depends strongly on an optimal choice. This fact can be advantageous in comparison with fractal dimension calculations (which depend crucially on the "right" sampling time) if, for example, only the value of the first exponent or the distribution of Lyapunov exponents is of interest (compare with Fig. 4.44).

The number of points used for the least-squares fit hardly affects the value of the Lyapunov exponents in the region where it is chosen for typical cases of

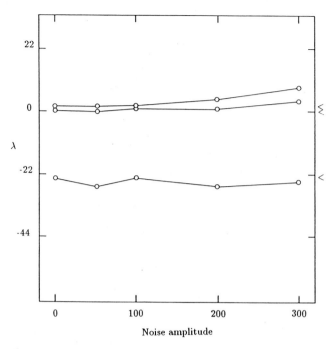

Fig. 4.43. A noisy time series of 60 000 points of the Lorenz system was generated. The abscissa shows the amplitude of the additive, uniformly distributed noise in absolute values, where the noise-free time series has values between 0 and 10 000. All three Lyapunov exponents of the Lorenz system are shown. The sampling time was $t_s = 0.03$, the embedding dimension was chosen equal to 8. The arrows indicate the values of the exponents when calculated directly from the dynamical equations

low-dimensional chaos (between 20 and 35). Since the least-squares fit is the most time-consuming part of the algorithm, k should be chosen nearer to the lower bound. The most critical choice is the radius and the procedure by which points within the ball are chosen. If only the innermost points are considered, for example, the first exponent is overestimated and the most negative exponent is underestimated, and analogously for the other extreme case. Both ways the difference from the true or most reliable value is within 10%. As has already been mentioned, for a good approximation to the linear approximation of the dynamical map from the data, it is important to have a sufficiently small radius of the ball $B(x, r)$ considered. If the linearity region is passed, points behave according to a higher-order Taylor expansion than just the linear approximation. A perturbation analysis shows that the contribution of higher than linear terms to the Lyapunov exponent is of almost the same magnitude as the true exponent; therefore, it is expected that the value of the largest exponent is almost doubled when the range of validity for the linear approximation to the dynamical map is left. A comparison with Fig. 4.43 shows full accordance (see also Fig. 4.56).

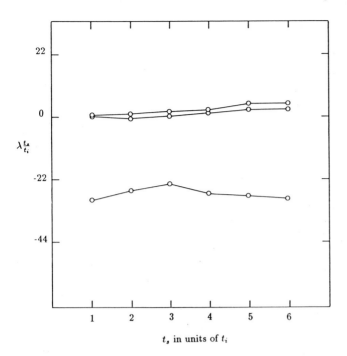

Fig. 4.44. Dependence of the three Lyapunov exponents of the Lorenz system on the sampling time. The abscissa shows the sampling time t_s in units of the integration step ($t_i = 0.01$) and the ordinate, the Lyapunov exponents in units of t_s/t_i. The calculation of Lyapunov exponents is based on 500 steps. Time series of 60 000 points were used; the embedding dimension was equal to 8. It is concluded that the scaling region for the sampling time is between 1 and 3 (or 4). Observe, however, that the most negative exponent does not scale, at variance with the different algorithm used for Figs. 4.56 and 4.57

Within the limits of reasonable manipulations, it can be said that the value of the positive Lyapunov exponent is found within 10% of the mean value with regard to different choices of parameters. This can be concluded from the results obtained by applying the algorithm to "computer-generated" examples, where a direct comparison with the Lyapunov exponents evaluated from the dynamical equations can be made.

The application of the algorithm to the investigation of the chaotic behavior of two experimental systems (NMR laser and p-Ge semiconductor system) is discussed in the following part.

c) Numerical Examples: Experimental Data

For the evaluation of Lyapunov exponents, long data strings have to be recorded under conditions as stable as possible, which is a nontrivial task for almost any system. Only with the help of sophisticated data acquisition systems and control over the system can data strings be obtained which are long enough for the analysis of Lyapunov exponents and generalized dimensions. If these quantities of interest can be evaluated in a satisfactory way, it is due to the efforts made in the experimental as well as in the numerical research field.

Analysis of NMR Laser Time Series. As the first application we discuss the calculation of the Lyapunov exponents from time series which have been directly recorded from a parametrically modulated NMR laser [4.4] at six different values of the modulation amplitude A. Each time series $\{s_i\}$ consisted of 250 000 integer data points scaled to the range from 0 to 4095. The signal-to-noise ratio in the data is estimated to be below 1%, when a sampling rate of about 28 points per main oscillation period was used (the modulation frequency was 110 Hz). The qualitative structure of the attractors for six different files with increasing amplitudes A is shown in a two-dimensional representation in Fig. 4.45. The points are plotted using coordinates (s_{i+1}, s_i). It is seen that the attractor changes its shape and becomes less uniform for increasing modulation amplitude.

From these data the Lyapunov exponents have been calculated. In Fig. 4.46 the results obtained for three characteristic experiments are shown (the delay time was chosen equal to one). The modulation amplitude was increased from (a) to (c). Cases (b) and (c) indicate the appearance of hyperchaos (for discussion of a hyperchaotic experimental system see [4.44]).

The relevant Lyapunov exponents (marked by black dots) become rather independent of the embedding dimension for d_E larger than 7. From Fig. 4.46a, which shows the results of the second file in Fig. 4.45, corresponding to $A = 0.415$, the following values for the three Lyapunov exponents are deduced: $\lambda_1 = 0.09$, $\lambda_2 = 0$ and $\lambda_3 = -0.7$ (in units of the sampling rate $v = 2996 \text{ s}^{-1}$). These values are estimates based on the rate of convergence of the exponents for a fixed value of d_E and on their dependence on the embedding dimension. They lead to a Lyapunov dimension of 2.13.

The analysis of the experiment at the second highest modulation amplitude ($A = 0.665$) clearly gives two positive exponents, as is seen in Fig. 4.46b. This indicates a qualitative change of the system when the modulation is increased. The exponents found for the data at the highest value of $A = 0.785$ are shown in Fig. 4.46c. From these results alone the existence of a second (small) positive Lyapunov exponent might be questioned. These results for the Lyapunov exponents seem, however, to be corroborated by the values of the information dimension which has also been determined for the same set of time series. In Fig. 4.47 both the Lyapunov dimension and the information dimension are plotted against the modulation amplitude. The agreement between the two dimensions, which have been determined by completely different methods, is remarkable. For a more detailed critical discussion of the errors possible in the evaluation of fractal dimensions we refer to the end of this chapter.

These results which are based on the experimental resolution presently available indicate a drastic change of the chaotic behavior of the NMR laser

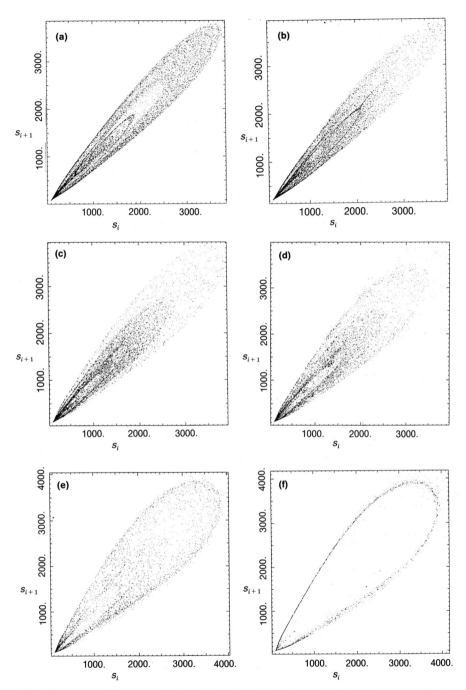

Fig. 4.45a–f. Six files of NMR laser time series, taken at increasing modulation amplitudes A. Points with coordinates (s_{i+1}, s_i) are plotted for parts of the time series

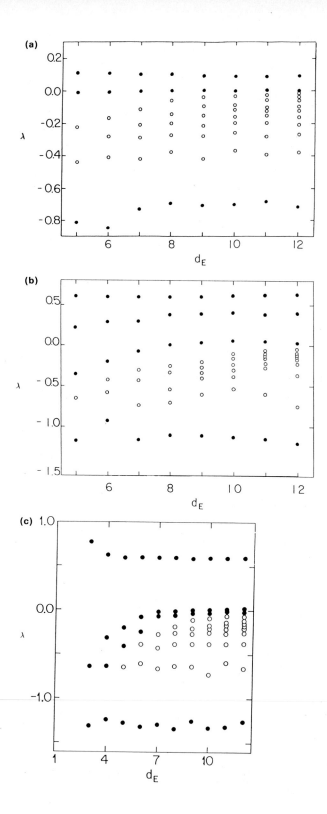

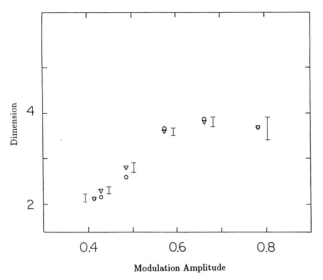

Fig. 4.47. The calculated Lyapunov dimensions plotted against the modulation amplitude of the NMR laser experiment. The values calculated for the information dimension of the same files are marked by triangles; error bars were estimated from the dependence on the embedding dimension with fixed channel range only

as the modulation amplitude is increased above a certain threshold. Further investigations are necessary to clarify the nature of this transition. It should be mentioned that an analogous property for a dye laser has been observed and explained in detail in [4.86], where, however, only fractal dimensions and no Lyapunov exponents have been calculated. A growth of the dimensions above three implies that for a description of the system at high A-values the set of model equations which has been used in [4.4] is no longer sufficient for these amplitudes and must be enlarged. A comparison between the positive Lyapunov exponents and the metric entropy gives a less satisfactory agreement, as is seen from Table 4.1.

Table 4.1. Comparison of Lyapunov exponents and metric entropies. The numerical
values are given in units of the sampling rate
$2996\ \mathrm{s}^{-1}$.

File	λ_1, λ_2	Metric entropies
1	0.09	0.15–0.35
2	0.12	0.20–0.40
3	0.25	0.20–0.40
4	0.45, 0.2	0.30–0.50
5	0.63, 0.4	0.30–0.50
6	0.73, 0.1	0.30–0.60

Fig. 4.46a–c. Lyapunov exponents from three time series of the NMR laser experiment. See text for further discussion

Analysis of the p-Ge Semiconductor System. As the second experiment a *p-Ge semiconductor system* [4.5, 44] has been investigated (see also Chaps. 2 and 3). Here the nonlinear effects are due to the autocatalytic nature of the avalanche breakthrough at low temperatures, whereas the NMR laser was externally modulated. In comparison with the latter the p-Ge semiconductor system evolves on a faster time scale and the evolution can be visually followed on-line. Among others, two different, apparently chaotic, working conditions could be separated; the first obvious question was whether these two cases were characterized by distinct positive Lyapunov exponents and fractal dimensions. The second question concerned the spatio-temporal properties of this spatially extended system. To this end, for both working conditions, from differently localized contacts on the p-Ge sample, time series of 100 000 integers were recorded at a sampling rate of $100\,000$ s^{-1}. The analysis showed that, although the places of the avalanche breakthrough can be localized very precisely, the different localization of the contacts had no influence on the results. The different working conditions, however, could be shown to be characterized by significantly different values of fractal dimensions and Lyapunov exponents. Table 4.2 shows the results obtained from two time series, each one characteristic of one of the two working conditions (subsequently called the chaotic and the hyperchaotic state).

To investigate the two states in a more thorough way, the scaling function of the measure and the scaling function of the support were calculated for both states, with emphasis on whether signs of coexistence of attractors could be found or not (therefore, only the top region of the scaling functions was of interest). Parts of the time series and the different scaling functions are displayed in Figs. 4.48, 49, and 50. Although no evidence of coexisting attractors could be given, the scaling function of the support shows clearly a phase-transition-like behavior for the hyperchaotic state. With a higher resolution for the scaling function of the support, also the phase-transition-like effect due to the presence of homoclinic tangency points is clearly visible (the corresponding scaling function was shown in Fig. 4.31a). For a physical interpretation of the results see [4.44].

Let us now turn to the comparison between the Lyapunov dimension and the fractal dimensions. Although also for this system the comparison again turns out to be quite satisfactory, a word of caution concerning the precision must be

Table 4.2. Lyapunov exponents, dimensions and metric entropies from time series of the p-Ge semiconductor experiment. Again, the numerical values are given in units of the sampling rate $100\,000$ s^{-1}.

File	$\lambda_i, i = 1, \ldots, 3(4)$				$D(1), D(0)$		$K(1), K(0)$	
1	0.095	0.003	−0.72		2.5	2.6	0.09	0.09
Errors	±0.005	±0.005	±0.02		±0.1	±0.1	±0.01	±0.01
2	0.159	0.076	−0.021	−0.77	3.5	3.6	0.15	0.15
Errors	±0.005	±0.005	±0.005	±0.03	±0.1	±0.1	±0.01	±0.01

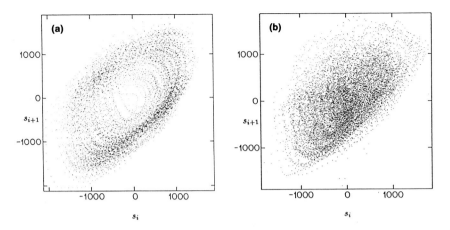

Fig. 4.48a, b. Time series of **a** the chaotic and **b** the hyperchaotic state of the p-Ge semiconductor. Points with coordinates (s_{i+1}, s_i) are plotted for parts of two characteristic time series

added. As pointed out before, an embedding process gives only a local diffeomorphic "copy" of the original system. The requirement $d_E > 2m + 1$, where m is the dimension of the original system, assures that the system stays within the smooth submanifold. If the radius of the ball in which the neighboring points are investigated is chosen too large, not only can the map no longer be approximated with sufficient precision by its linearization, but also parts in the embedding space are considered which do not belong to the local "copy" of the original system. This then is expected to lead to an overestimate of the fractal dimensions.

Roughly, if the fixed-mass procedure is applied and a data basis of 200 000 points is given and the 100 nearest neighbors are considered, this leads, with a fractal of dimension 2.5, to a ball of a radius of one-eighth of the extent of the typical phase space region, where the attractor is found. Furthermore, it has been pointed out in [4.87] that the number of points necessary to evaluate fractal dimensions scales exponentially with the dimension and the analog situation holds for the evaluation of the Lyapunov exponents. This would mean that results of examples with fractal dimensions above 3 are more inaccurate than were expressed by the error estimates which are usually obtained by taking the maximum difference of the results obtained from reasonably fixed scaling regions in the log–log plot and from different reasonable embedding dimensions (the least-squares errors in the cases considered are so small that they can be neglected). We have taken into account that uncertainty in the error estimates given in Table 4.2. Note that practically identical values for the dimensions were obtained by both methods; for a numerical comparison of the two concepts see [4.88].

As a final remark it should be pointed out that, although an estimation of the numerical accuracy of methods using embedded time series is very difficult,

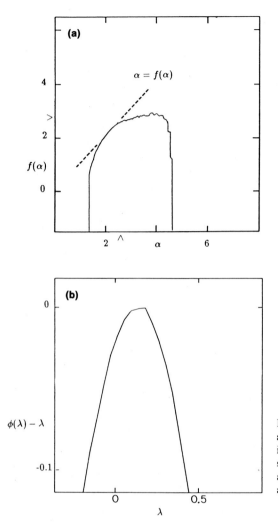

Fig. 4.49a, b. Scaling functions for **a** the probabilistic and **b** the dynamical points of view for the chaotic state of the p-Ge semiconductor. The arrows indicate the values of $D(0)$ and $D(1)$

the consistent pictures obtained for both experimental systems strongly indicate that the methods and the numerical procedures have been used in a sound way. The length of the time series which accounts for the most crucial limitation in calculation accuracy was in all cases checked not to have a major impact on the results as stated. In the case of the p-Ge semiconductor it was seen that the contribution of noise was larger in comparison with the NMR laser. The high sampling rates of the former, however, offered a successful consistency check for the results over an extended range of different sampling times. For the NMR laser this was not possible. When the sampling time was chosen essentially smaller than the sampling time given above, difficulties in the evaluation of Lyapunov exponents arose, which led to inconsistent results. The results of fractal dimension calculations, on the other hand, were affected much less, with

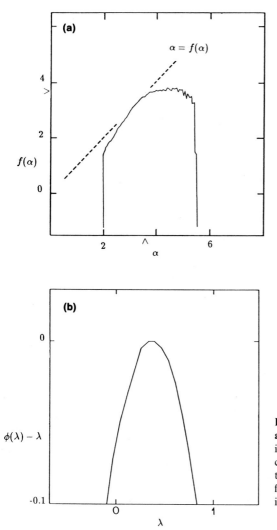

Fig. 4.50a, b. Scaling functions for **a** the probabilistic and **b** the dynamical points of view for the hyperchaotic state of the p-Ge semiconductor. Note the linear part in the scaling function of the measure. The arrows indicate the values of $D(0)$ and $D(1)$

the exception that for "strongly spiking" signals similar problems arose. This indicates that, as a rule, for the evaluation of Lyapunov exponents a sampling rate faster than that for the evaluation of fractal dimensions should be used.

4.4.4 Comparison with Other Methods

A rough approximation for the values of local and global positive Lyapunov exponents can be given by the following formula [4.89]:

$$\lambda \approx \frac{1}{k} \left\langle \log \frac{dist(Fa^k(x), Fa^k(y))}{dist(x, y)} \right\rangle, \tag{4.187}$$

where y is the nearest neighbor of x. Remember that with the method introduced in Sect. 4.4.2 the stretching rates in the tangent bundle of a trajectory were calculated. In the above approximation, instead, an *average* of the *most expanding direction* of the points visited is taken, and solely the exponent corresponding to this direction is taken into account. Note that also generalized dynamical entropies could easily be introduced by raising the quotient of the distances in (4.187) to the $-(\beta - 1)$th power and taking the logarithm of the average of this expression. Note, further, that in this way again the similarity between the scaling of the support and the scaling of the measure [see (4.100) and (4.101)] becomes obvious. Under time-reversal the most contracting exponent can also be estimated. It is reasonable to expect that values of the approximated Lyapunov exponents are overestimated when evaluated over small time scales. Instead, pieces of trajectories corresponding to the largest number of iterations possible without entering the folding region should be considered. The estimate can be improved if we can be sure that the nearest-neighbor point already lies on the expanding manifold when we start to calculate the separation rate. This can be accomplished if we let a few iterations pass, until the motion of the point considered has relaxed sufficiently on the expanding manifold of the reference point. We point out, however, that for times larger than the average folding time, a sharp decrease of the value of the first Lyapunov exponent can occur. (This feature can be easily observed in the following figures.) In this way the above formula is changed into the expression [4.89]

$$\lambda \approx \frac{1}{k} \left\langle \log \frac{dist(Fa^{k+m_0}(x), Fa^{k+m_0}(y))}{dist(Fa^{m_0}(x), Fa^{m_0}(y))} \right\rangle . \tag{4.188}$$

As a consequence of the approach only an approximate estimate of the positive Lyapunov exponent is possible; taking into account the limitations indicated above, an agreement with the correct value of the positive Lyapunov exponent within an order of magnitude can be expected. The implementation of the algorithm proceeds as follows. A set of generic data points is taken. For each of these points the next-neighbor point is sought and the separation from the generic point is calculated, while the map is iterated a fixed number of times, following the above prescription. As pointed out before, this results in a simple program of almost the same structure as the program for the evaluation of fractal dimensions.

The algorithm was applied again to different model systems. The calculations made for these systems corroborated what was expected above. The results obtained for the characteristic examples of the Hénon map and the Lorenz system are shown in Fig. 4.51. As can be seen, the results obtained for the Lorenz system agree over a large time scale with the correct value if the embedding dimension is chosen not too large. For the Hénon map, however, the situation is less favorable.

Both the concept and the obtained results suggest that this method can be very useful for checking purposes; however, other applications seem to be limited by insufficient accuracy.

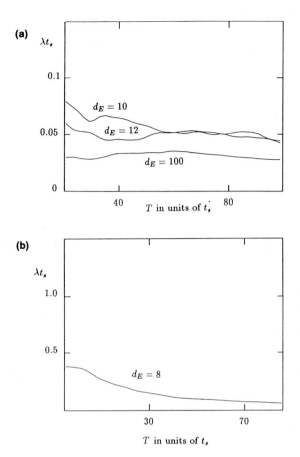

Fig. 4.51a,b. a The largest Lyapunov exponent in units of the sampling time $t_s = 0.03$ calculated from a time series according to Sect. 4.4.2 for the Lorenz system, using various embedding dimensions and $m_0 = 4$. b The largest Lyapunov exponent for the Hénon map at $d_E = 8$ with $m_0 = 2$ and $t_s = 1$. The abscissa shows the time in units of the sampling time t_s during which the orbits were observed

In connection with this approach it was checked whether the numerical results of the algorithm discussed in Sect. 4.4.2 could be improved if special weight was given to the next-neighbor vector when brought into the least-squares fit in (4.182). It was noticed that this procedure led to an additional expanding exponent, leading to the interpretation that the nearest neighbor does not lie with sufficient reliability on the expanding manifold in the tangent bundle.

A different approach to calculate Lyapunov exponents which uses the *filter technique* has been reported in [4.90]. In this method, experimental data is first "filtered", i.e., on the experimental data the effect of a low-pass filter is simulated, for different cut-off frequencies. Then the fractal dimension of the filtered data is calculated. With the help of the Kaplan–Yorke conjecture, the Lyapunov exponents of the original experimental data are then evaluated. Unfortunately, the range of the validity of the Kaplan–Yorke conjecture is not well determined; furthermore, there seem to exist experimental cases in which numerical calculations show a large disagreement between the information dimension and the Lyapunov dimension, although not of order of magnitudes.

An application of this indirect method for the determination of Lyapunov exponents to the NMR laser data showed results that disagreed by orders of magnitude from the results obtained by all other methods [algorithms proposed in Sects. 4.4.2 and 4.4.4, (4.188), information-theoretical entropy $K(q)$ and the algorithm to be proposed in Sect. 4.5] when reasonably small embedding dimensions ($d_E > \sim 2n + 1$) were used. If, however, the embedding dimension was chosen as large as >400 sampling steps (corresponding to about 15 main oscillation periods), the same value for the first Lyapunov exponent was found by the algorithm described in (4.188) and the filter method (the algorithm described in Sect. 4.4.2 cannot reasonably be applied for such high embedding dimensions).

This astonishing fact can be given the following interpretation. Normally, for points with components distributed over so many main oscillation periods, the largest Lyapunov exponent should be zero. This can be immediately checked for the Lorenz system (Fig. 4.52). An analogous calculation has been performed for the NMR laser system. The results presented in Fig. 4.42 were obtained. Possible explanations for this behavior include the presence of a remainder of a previously merged second attractor with a small positive Lyapunov exponent or a slow drift of the experimental situation. Such a drift of the experimental situation during the process of measurement could result in a long-time behavior which can be measured in very high embedding dimensions or in the filtering process (where fast changes of the variable are filtered off). As a physical

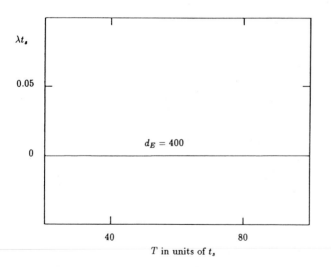

Fig. 4.52. Largest Lyapunov exponent in units of the sampling time $t_s = 0.03$ calculated according to (4.188) for the Lorenz system at the embedding dimension 400. The abscissa shows the time in units of the sampling time during which the average logarithmic separation speed over 20 points has been measured. For points with components sampled over a much larger time scale than the mean time of folding, the measured Lyapunov exponent is zero

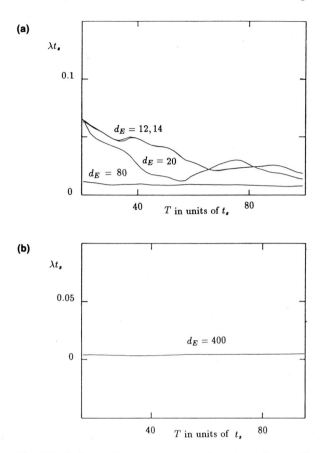

Fig. 4.53a, b. Largest Lyapunov exponent in units of the sampling time calculated according to (4.188) for the NMR laser data at different embedding dimensions. The abscissa shows the time in units of t_s over which the average over 20 points has been taken. **a** Largest exponent calculated at embedding dimensions $d_E = 12, 14, 20, 80$. **b** Largest exponent calculated at an embedding dimension $d_E = 400$. The value obtained in **b** is almost identical to the one obtained by the filter method

reason for this feature, different mechanisms can be imagined, such as a heating of the magnet or related effects.

4.5   High-Dimensional Systems

4.5.1   Singular-Value Decomposition

In this part a new approach to calculate the Lyapunov exponents from experimental time series is discussed. This approach is shown to give more direct information on the dynamics of the underlying system while allowing a faster calculation with a smaller amount of data than the implementation used so far, so that a new class of physical systems can now be treated. The algorithm is applied to simulated data of some well-known nonlinear systems. It is hoped that this algorithm will be of particular help in the evaluation of Lyapunov exponents from small time series or high-dimensional embedding spaces and in modelling experimental systems [4.91, 92].

One of the main reasons why the concept of Lyapunov exponents has found less application to experimental systems than the concept of fractal dimension, besides the fact that its computation is more involved, is the fact that the latter can be applied to a time series of a much smaller size than the former. A modified, and for that purpose improved, algorithm for the calculation of the Lyapunov exponents from experimental time series is now reported. The algorithm proposed deals with a conceptual improvement using *singular-value* decomposition of the algorithm discussed in Sect. 4.4.

In a different context, singular-value decomposition [4.93–96] has already been used to calculate Lyapunov exponents in a number of papers (e.g., [4.95]), where essentially a Gram–Schmitt orthogonalization procedure or a conventional least-squares fit was replaced by singular-value techniques. Here, however, a completely new point of view is taken. It is shown how, with the help of this technique, it is possible to use a subspace of the embedding space of the same dimension as the original dynamical system to calculate Lyapunov exponents. This procedure then removes ambiguities in how to choose the relevant exponents among the rescaled ones and leads to new possibilities for the modelling of experimental systems, since in this submanifold the diffeomorphic linearized dynamical map is known with a precision limited only by inaccuracy resulting from the numerical treatment and finite data. In what follows, the new algorithm is applied to computer-generated time series; the accuracy of the calculation is discussed and advantages in the interpretation of the results as well as the time required for their calculation with reference to previous methods are pointed out. Furthermore, it is shown that the new algorithm can be applied to small data sets in comparison with the commonly used algorithms. Finally, it should be remembered that any algorithm for the computation of Lyapunov exponents from time series can be extended in a natural way to the computation of the dynamical scaling function for an experimental system.

4.5.2   The Modified Approach

First let us recall that the calculation of Lyapunov exponents from time series requires a numerical approximation to the linearized dynamical map $D\tilde{F}a$, which corresponds to DFa, but acts in the d_E-dimensional embedding space M ($d_E > 2\,m$). The motion itself, however, is restricted to an m-dimensional submanifold of the embedding space. Therefore, it makes sense to ask oneself how the approximation to the linearized map $D\tilde{F}a$ can be performed in this subspace. With the help of fractal dimension methods it can be guessed what the dimension m of this submanifold should be.

Now consider the m-dimensional smooth submanifold obtained by the embedding of the time series. A schematic picture of the situation is given in Fig. 4.54. It is well known that the local directions of this submanifold around point x_k can be found with the help of a singular-value decomposition. This decomposition of the "neighboring matrix" $V = (V)_{M_0}^{(k)} = \{v_i^{(k)}\}_{i=1,\ldots,M_0}$ [see Fig. 4.25 and (4.179)] reads as follows:

$$V = UQZ^{(T)} . \tag{4.189}$$

Here U is an orthogonal $M_0 \times M_0$ matrix, Q is a diagonal matrix with elements called "singular values" ordered in descending magnitude and $Z^{(T)}$ is an orthogonal matrix of dimension $d_E \times d_E$; T denotes transposition. The squares of the singular values are eigenvalues of either $VV^{(T)}$ or $V^{(T)}V$:

$$(VV^{(T)})U = U(QQ^{(T)}), \qquad (V^{(T)}V)Z = Z(Q^{(T)}Q) , \tag{4.190}$$

such that the columns of U and Z turn out to be the corresponding eigenvectors. Since Z is orthogonal of dimension d_E, its eigenvectors form an orthonormal basis around x_k in the embedding space. The matrix product VZ can be interpreted as a rotation

$$VZ = V', \qquad v' = vZ . \tag{4.191}$$

Now consider

$$V' = UQ . \tag{4.192}$$

In this rotated space M', those directions e_k which lead to $q_{kk} = 0$ are not spanned by the rotated vectors; on the other hand, those e_k's which correspond to the largest q_{kk}'s give the directions of the "main axes" of the set of vectors V' considered. Each $v_k'^n$ is the component of v'^n in the direction of e_k; the elements

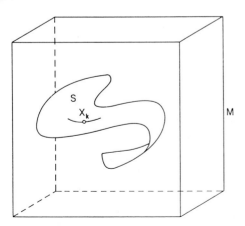

Fig. 4.54. The m-dimensional smooth submanifold S to which the motion of the dynamical system is restricted is shown in the d_E-dimensional embedding space

q_{kk}^2 can be interpreted as the statistical variances of the kth components of the set of neighboring vectors.

Therefore, if we have a true m-dimensional embedded submanifold, we expect

$$q_{ii} = 0, \quad i = m + 1, \ldots, d_E \tag{4.193}$$

(which has been proposed as a means to estimate the dimension of the underlying dynamics). In this case the neighboring matrix V, and therefore the matrix $D\tilde{F}$, would theoretically have rank $< d_E$ and the algorithms for the computation of Lyapunov exponents from time series operating in the full d_E-dimensional space would fail. In reality, however, finite precision and noise "desingularize" the matrix; the calculation can nevertheless be performed. However, a number of spurious exponents are introduced.

Let our intention be to model the dynamics in the m-dimensional space. To obtain an m-dimensional system locally, we set by force

$$q_{ii} = 0, \quad i = m + 1, \ldots, d_E , \tag{4.194}$$

in (4.192). Then one can proceed in two ways:

1. Using (4.192) we can calculate the truncated vectors corresponding to the set of neighboring vectors $v_k''^n$. With the image vectors we proceed analogously. We then project these vectors onto an m-dimensional plane, e.g., $E: x_i = 0, i = m + 1, \ldots, d_E$, perform the least-squares fit and obtain an m-dimensional approximation of $D\tilde{F}a(x_k)$ at every point x_k. With the help of this approximation, we calculate the Lyapunov exponents.

This procedure also takes into account the orientation of the linearized matrix in space. However, by choosing a certain direction of projection, an additional source of expansion is introduced, and the true Lyapunov exponent is slightly overestimated. Therefore, we expect this method to give the best results for maps of the Henon type, with fast-changing coordinates and essential bending of the attractor. In the limiting case when the dimension of the submanifold is equal to the dimension of the embedding space, the new method is equivalent to the old one and the same values are computed. It can be easily seen that the accuracy of the result compared to the true exponents depends crucially on the orientation of the submanifold in space. Therefore, it is recommended to take only a few points more than necessary to determine the m-dimensional submanifold. The number of points used, however, should not be too small to allow a good least-squares fit; so, a sufficiently large embedding dimension is recommended. This procedure will then give the value of the positive exponents and the zero exponent as accurately as possible. However, the information about the contracting direction can be poor. It is then of relevance to include the nearest points on the contracting manifold via back-mapping in time. Since its distance from the center of the linearized region will be relatively large, only its projection onto the submanifold is taken into account for the determination of the approximation to $D\tilde{F}a(x_k)$.

As pointed out before, the results produced by the algorithm depend crucially on the dimension chosen for the submanifold. Particularly, the value of the first exponent can increase considerably with respect to the correct one if the dimension is chosen inappropriately.

2. If we can assume that the trajectories change their orientation rather smoothly on the attractor (which is the case, for example, for the Lorenz attractor), we can perform the least-squares fit for $D\tilde{F}a(x_k)$ for the vectors in the reference frame of M'_{x_k}, provided that M'_{y_k} has about the same orientation. Again, an m-dimensional approximation of $D\tilde{F}a(x_k)$ is obtained. In the present case we need not project on the plane $E: x_i = 0, i = m + 1, \ldots, d_E$.

This approach allows for a much faster computation than the method considered just before; however, the values obtained are only approximate ones, and give good estimates only for a class of attractors with smooth evolution along the flow in comparison to the sampling time.

4.5.3   Results on Simulated Data

To indicate the results that can be expected, this algorithm was applied to a number of "computer-generated examples".

In Fig. 4.55 the Lyapunov exponents for the Hénon map obtained from time series are shown as a function of the embedding dimension. The directly calculated Lyapunov exponents are shown as full lines. Contrary to the old method, only two exponents are calculated. For these results all parameters in the algorithm were kept at fixed values, except the embedding dimension. Two pairs of dashed lines indicate the range of the two relevant Lyapunov exponents obtained by the proposed method for all other reasonable choices of the parameters in the algorithm, at embedding dimensions $d_E = 7$ and $d_E = 6$. The values of the two exponents obtained from a direct calculation are shown as full

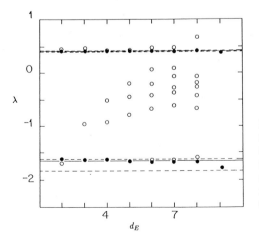

Fig. 4.55. Values of the Lyapunov exponents from time series of the Hénon map at standard parameters obtained by the first method (1) using different embedding dimensions. The results evaluated by the new method are marked by solid circles. For details see text

lines. For a comparison, the results from the old method are shown (compare with Fig. 4.26); they are marked by open circles. The time series used for both algorithms contained 40 000 integers, rescaled to the range between 0 ... 10 000.

Results obtained by the second method (2) from time series generated from the Lorenz model at parameter values $\sigma = 16$, $b = 4$, $r = 45.92$ are shown in Figs. 4.56–58. Again, the convergence of the three Lyapunov exponents to the known values, indicated by dashed lines, with increasing embedding dimension is very good. With the new method only the smallest exponent is less precise than the one obtained by the old method; the other two exponents are obtained with ease and precision. With a more elaborate code the value of the smallest exponent can also be improved. Contrary to the calculation with the old method, the new algorithm did not show the problem of multiples of the first exponent even with as few as 10 000–15 000 data points (Fig. 4.56). Results on the behavior with respect to different sampling rates are shown in Fig. 4.57.

To give an idea of the accuracy of the linear approximation to the dynamical map, in Fig. 4.58 a part of the local Lyapunov exponents as a function of time is shown in comparison with a direct calculation. Here the Lorenz system has been investigated at the usual parameter values, in comparable situations. Note,

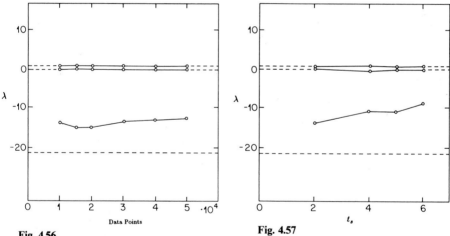

Fig. 4.56   **Fig. 4.57**

Fig. 4.56. Lyapunov exponents from time series as a function of the size of the data basis used (Lorenz system). The values of the directly calculated exponents are shown as dashed lines. If the size of the data basis is decreased, the values of the calculated exponents are unchanged within a large range, in contrast to the behavior of the algorithm described in Sect. 4.4. (compare with Fig. 4.44). The embedding dimension d_E was again equal to 8

Fig. 4.57. Lyapunov exponents from time series as a function of the sampling time t_s. t_s is given in units of the integration time $t_i = 0.01$(Lorenz system). The values of the directly calculated exponents are shown as dashed lines. From $t_s = 2$ to $t_s = 6$ the values obtained for the first two exponents remain almost unchanged.

however, that due to the richness of the temporal behavior, it is difficult to reproduce exactly the same behavior. In particular, the more extended range of the curve in (a) in comparison with the curve in (b) is not due to a defect of the algorithm. The rather "regular" situations have been chosen in order to facilitate a comparison of the two curves. We, point out that another part of the temporal evolution at the same parameter values has been shown in [4.95].

The algorithm was also applied to the hyperchaotic Rössler attractor [4.97]. For this system it is difficult to establish from computer-generated time series the values of the second positive and the negative exponent with satisfactory

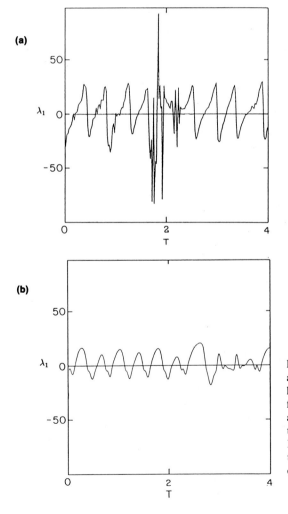

Fig. 4.58a, b. Diagram indicating the accuracy of the approximation to the linearized dynamical map. **a** The first local Lyapunov exponent shown as a function of time, calculated by the proposed algorithm. **b** The first Lyapunov exponent calculated from the dynamical equations (see text), for comparison

precision. While from first attempts with the usual method, on a time series consisting of 100 000 integers, the second positive exponent did not appear, it was obtained with the new method. Since a second positive exponent did not show up when the Lorenz system was considered as a four-dimensional system, it is concluded that this method can be of considerable use in the analysis of experimental hyperchaotic systems.

In the examples considered above the new algorithm proposed for the calculation of Lyapunov exponents from time series gives reliable estimates in all cases for the nonnegative exponents. In high-dimensional embedding spaces and in cases of small data sets the results obtained by the new method are superior to those of the old algorithm. The omnipresent contribution of noise would, in principle, require the use of high-dimensional embedding spaces. While the old algorithm is generally able to calculate Lyapunov exponents from embedding dimensions smaller than $d_E = 16$ from time series of less than 100 000 data points; using the new algorithm, for example, in an embedding dimension of $d_E = 28$ for the Lorenz system the largest exponent obtained is $\lambda_1 = 1.8$ instead of the correct value $\lambda_1 = 1.5$ (time series of 80 000 data points were used). The fact that the old algorithm performs the least-squares fit in the embedding space leads, for high-dimensional embedding spaces, to an increased time consumption of the algorithm. Furthermore, much information is used for the calculation of the rescaled Lyapunov exponents which are not needed. If the information at our disposal is not sufficient to determine the approximation to the linearized dynamical map in this high-dimensional space, orientational problems prevent a successful evaluation of the Lyapunov exponents. For experiments from which only short time series can be extracted the usual method cannot, therefore, be applied. In this case, the proposed method might be the only solution if one tries to get out more than just the first exponent. Furthermore, in all cases investigated, the precision obtained exceeded the precision that could be achieved by methods which only consider the largest exponent. Satisfactory values of the exponents are obtained with less adjustment of the algorithmic parameters and the speed of calculation is improved by a factor of 2 in comparison with the old algorithm. Possibilities of further acceleration in the execution of the algorithm are being tested.

Before ending this discussion let us point out that an application of related ideas for the evaluation of fractal dimensions is likely to yield less satisfactory results. While the scaling of the support is given by the expanding manifolds, the scaling of the measure is built essentially along the directions which have a Cantor structure and belong to contracting manifolds. These, of course, are more difficult to localize by the present method.

As a conclusion, a conceptually modified algorithm to calculate Lyapunov exponents from time series has been proposed and good agreement with already known values of these exponents has been found. For smoothly sampled attractors, a precise estimate of Lyapunov exponents can be given by calculations that can be performed with a smaller data set and shorter computing time than the commonly known algorithms. The proposed method offers extended checking possibilities taking into account the connection between expanding manifolds and the probability measure. When in [4.44] we reported the experimental finding of hyperchaos for the p-Ge semiconductor system, such checks were performed and confirmed our first results. Beyond this, the algorithm can be of special interest because a distinct level of dimensional approximation to the problem can be fixed and direct information on the Jacobian of the dynamical map can be obtained, a feature that could be of considerable use while trying to describe the dynamics of such an experimental system with the help of a model. However, these ideas exceed the scope of this work and are to be pursued in future investigations.

4.6 Lyapunov Exponents, Rotation Numbers and the Degree of Mappings

4.6.1 Characterization of Solutions of Dynamical Systems via the Degree of Mapping

In Sect. 4.4 the algorithm to evaluate Lyapunov exponents was outlined. It could be seen that considerable effort has to be made due to the fact that the basis of the tangent space is rotating along the curve of motion. For the calculation of Lyapunov exponents this rotation is not used further, although additional information about the dynamical system can be obtained (different motions with identical Lyapunov exponents but different rotational properties can, of course, be imagined). In the two-dimensional case the number of rotations of a vector field $v(x)$ can be defined via the mapping $f: x \to v(x)/|v(x)|$ from U', a region of $\mathbb{R}^2$ where the singular points have been excluded, to S^1, the circle. The neighborhood of $f(x)$ can be parametrized by an angle variable φ with a complete differential form $d\varphi = d\arctan(v_2/v_1) = (v_2\,dv_1 - v_1\,dv_2)/(v_1^2 + v_2^2)$. The integral of the form $d\varphi$ along a closed oriented curve is called the *index* of the curve. The generalization of this concept to higher dimensions is the *degree of mapping*.

The degree of mapping was introduced at the beginning of this century by *Kronecker, Poincaré* and *Brower* [4.98–101]. Soon it was realized that this concept could be used as a means of predicting the number of solutions a nonlinear equation might have. The degree can only give a lower bound to this number and was mostly used as a tool for proofs of existence. In a more recent contribution [4.102], making use of the degree of mapping or closely related concepts has been considered as a means of answering a number of interesting questions in connection with the complicated structures of bifurcation diagrams, such as identifications of solutions with respect to a bifurcation tree, prediction of collisions of solutions and prediction of crises. In what follows, we summarize the different mechanisms which are involved with these tools and we sketch the range of possible applications.

Given the dynamical map $\dot{F}a$, denote by A an open, bounded set $A \subset \mathbb{R}^m$, $y \notin F(\partial A)$ and by ∂A the boundary of A. The degree of Fa with respect to a point $y \in A$ is defined as

$$\deg(Fa, A, y) = \sum_{x \in Fa^{-1}(y)} \text{sgn} \det DFa(x) \, . \tag{4.195}$$

As can easily be shown [4.100, 101] this gives a lower bound to the number of solutions of the equation

$$Fa(x) = y \, . \tag{4.196}$$

The nonnegative number $\deg(Fa, A, y)$ has some additional remarkable features:

1. $\deg(I, A, y) = 1$ for $y \in A$; otherwise, $\deg(I, A, y) = 0$.
2. If $\deg(Fa, A, y)$ is not equal to 0 then $Fa^{-1}(y)$ is not empty.
3. Invariance with respect to homotopic maps: Let F, G be maps, $A \subset \mathbb{R}^n$, $B \subset \mathbb{R}^n$ open. F is called a homotopic map with respect to G if there is a continuous map $H : [0, 1] \times A \to B$ such that $H(0, \cdot) = F$ and $H(0, \cdot) = G$. H itself is called a homotopy. If F and G are homotopic in $\mathbb{R}^n \setminus \{y\}$ then they have the same degree deg on A.
4. $\deg(\cdot, A, y)$ is constant on $B(Fa, r) \subset C(\bar{A})$, where $C(\bar{A})$ is the space of continuous functions on the closure of A and $r = \text{dist}(y, Fa(\partial A))$.
5. $\deg(Fa, A, \cdot)$ is constant on each component of $\mathbb{R}^n \setminus Fa(\partial A)$.
6. $\deg(Fa, A, 0)$ can be evaluated as follows:

$$\deg(Fa, A, 0) = \frac{1}{\text{vol}(S^{n-1})} \int_{\partial A} \sum_{i=1}^{n} (-1)^{i+1} \frac{Fa^i}{|Fa|^n} dF^1 \wedge \ldots \wedge dF^i \wedge \ldots , \tag{4.197}$$

where dF means that this element has to be omitted.

The above characteristics indicate that the degree can be used to provide an answer to a number of interesting questions [4.102]. Can two specific solutions be traced back in the space of the outer parameter to a common bifurcation point? Will they collide under a further increase of this parameter? Consider the "relative vector function" linking one orbit to the other (Fig. 4.59). Due to homotopy invariance, the two orbits can interact only if the degree of the relative vector functions with respect to all other existing orbits remains unchanged.

However, the evaluation of all solutions in a dynamical system is normally an endless task, and can only be solved incompletely, even if the system is relatively simple. Therefore, such concepts have so far not been used much for the characterization of chaotic behavior. It should, however, be mentioned that the degree can be successfully applied for the description of a bifurcation process, analogous to the Lyapunov exponents (the degree always remains constant in the interval between two bifurcations and jumps discontinuously at the bifurcation points). Finally, we stress again the fact that the degree is the other piece of information we get from the dynamical map, in addition to the

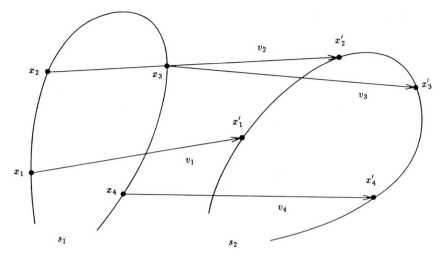

Fig. 4.59. Schematic drawing of the relative vector function v linking two different solutions s_1, s_2

stretching factors or Lyapunov exponents. From this point of view it could be concluded that someday the degree of mapping could play a more important role in the characterization of chaotic systems than at present.

4.6.2 Riemannian Motions on Manifolds of Constant Negative Curvature

Using the fact that for a surface of negative curvature -1 the Lobachevski plane is a universal covering, the geodesics and the tangent bundles of the former and the latter can be identified [4.103]. Then it can easily be shown [4.103] that the geodesic flow on the Lobachevski plane is an Anosov flow [4.10, 103], from which it follows that the motion is chaotic [4.104]. Furthermore, this implies that the geodesic flow on a compact surface of constant negative curvature is structurally stable and has an everywhere-dense set of closed geodesics. For a multidimensional system with negative (not necessarily constant) curvature, the geodesic flow is also an Anosov flow [4.105].

4.7 Conclusions

In this last chapter the characterization of the scaling behavior of an attractor of a dissipative dynamical system has been investigated, for both theoretical models and experimental systems. It has been shown that the whole scaling behavior must be characterized by two essentially independent scaling processes, the scaling behavior of the support and the scaling behavior of the measure. Although for very simple model cases an exact evaluation of the scaling properties can be made analytically, for more difficult cases only a characterization using random sampling can be given. For the scaling of the support, different algorithms for evaluating the relevant scaling quantities have been proposed and discussed. The limitations inherent in the procedures used have been worked out; they are: the amount of data (length of the time series), the quality of the data (noise), the sampling time and the embedding dimension. A sensible application of the methods and concepts developed for theoretical models led to coherent results also for experimental systems, a fact which can open the door to a better understanding of the physical processes involved.

In the first part of the chapter, where a thorough discussion of the scaling behavior of strange attractors and repellers was given from a new point of view, the central role of the Frobenius–Perron equation and the analogy with a thermodynamical formalism was pointed out. It was shown that from there the interrelation between the scaling of the measure and the scaling of the support can be worked out. The discussion of model cases, which contains essential new aspects, showed that the various concepts of entropy describe different scaling properties and that they have to be distinguished. For generic systems the case of random sampling was derived from the concepts developed for model cases.

In the second part the characterization of the scaling behavior of generic systems was examined. Emphasis was laid on the evaluation of the scaling behavior calculated from time series. It was shown that the description of the temporal scaling behavior of the support of a strange attractor can be made in a reliable way with the help of the calculation of effective Lyapunov exponents and their distribution. By use of simulated data, where exact results can be obtained, intrinsic limitations of the methods involved in the calculation of Lyapunov exponents from time series were worked out. As expected, the most negative exponent is usually the most difficult exponent to be obtained within reasonable precision. A precise evaluation of the positive Lyapunov exponents, which is crucial for the evaluation of the scaling functions for the scaling of the support, is less difficult. To estimate the accuracy that can be expected in such a calculation two comparisons are relevant: comparison of a scaling function calculated analytically, and approximately via a histogram, and calculation of a scaling function calculated directly from the dynamical equations, and indirectly via an embedding process from a simulated time series. For both cases good agreement was obtained for our numerical procedures. A short theoretical discussion on phase-transition-like phenomena in dynamical systems was given for model maps and systems. In this work, for the scaling of the support calculated from experimental time series, they have been observed for the first time. For the scaling of the measure a phase-transition-like effect has also been observed. Furthermore, cases of experimental hyperchaos have been found.

In many experimental situations only a small amount of data material is available or high-dimensional embedding spaces should be used, thus leading to severe difficulties in the evaluation of the scaling exponents and scaling functions. For this case a conceptually new algorithm for the evaluation of

Lyapunov exponents from time series has been presented and discussed. It was shown that for these cases this algorithm is superior to all presently known algorithms. Furthermore, it was pointed out that this algorithm, when applied to hyperchaotic files, can independently confirm their hyperchaotic nature. The road to further investigations was indicated. Finally, additional possibilities for the estimation of Lyapunov exponents from time series have been discussed. Disagreement of the results in one case was discussed and resolved.

As the final point other possible tools for the investigation of dynamical systems were discussed and their origin was enlightened. Their connection with the concept of Lyapunov exponents was shown and their limitations were briefly outlined.

As a summary it can be said that a successful numerical characterization of theoretical and experimental systems can be achieved, but only within the limits of accuracy given both by the intrinsic properties of the system and the properties of the numerical procedures, the influence of the former being generally dominant. It cannot, unfortunately, be done in an arbitrarily precise way. A discussion of Lyapunov exponents, dimensions and different concepts of entropy should therefore be done in a very careful way and only in the spirit of a consistent picture. Although the numerical characterization of the scaling behavior of the support is more involved, calling for more complicated algorithms and larger computational time than the characterization via fractal dimensions, it can be done in a reliable way. On the other hand, the dynamical approach can be seen to give more direct information about the mathematical model of the system under investigation, so that increased efforts can be justified.

Beyond these facts, however, it cannot be denied that from a strictly mathematical point of view we are confronted with a rather unsatisfactory situation when we calculate exponents and scaling functions from time series for the characterization of the scaling behavior of a strange attractor (and this is equally true for both characterizations of the scaling behavior!). The application of exact mathematical concepts is generally problematical for generic experimental situations. When exact mathematical concepts can be applied to a situation, the situation, in general, will be nongeneric.

Problems

4.1 Consider Hénon's map. Decompose it into a vertical compression, followed by a vertical shear, followed by a reflection through the line $y = x$. Derive from this decomposition some consequences concerning the fixed points of the map.

4.2 Reduce the Lorenz system to a one-dimensional mapping by means of a Poincaré cross section. (See [4.6].)

4.3   Calculate in a straightforward way the logarithmic separation rate of neighboring points for the case of the Hénon map. Why is the result so poor? Compare it with the results obtained from a method which uses the separation rate in the tangent space, the common Lyapunov exponent algorithm.

4.4   Apply different methods to draw the stable and unstable manifolds of the Hénon map. Note that there exists one particularly clever way: Look for analytic functions $x(t)$, $y(t)$ and a real number l such that $F(x(t), y(t)) = (x(lt), y(lt))$.

4.5   Consider the simple planar map $x' = x + y'$, $y' = y + kx(x - 1)$ with $0 < k < 4$. Of what type are the fixed points? Can you show the homoclinic "tangle"?

4.6   The Baker's transformation B of the torus T_2 into itself is given by $x' = 2x \bmod 1$, $y' = (1/2)(2x - x' + y) \bmod 1$. What happens to rectangles $P_0 = [0, 1/2) \times [0, 1)$ and $P_1 = [1/2, 1) \times [0, 1)$? Show that every point p in T_2 can be obtained as $p = \bigcap \{n: -\inf, \inf\} B^n P_{\sigma_n}$, where σ is a bi-infinite sequence of $\{0, 1\}$. Use this fact to show that $B(p) = \bigcap \{n: -\inf, \inf\} B^n P_{1(\sigma)_n}$, where 1 is the left shift.

4.7   Think for a moment about some people's claim that the geometric world is essentially self-similar. Even if we are not aware of it, the point of view of traditional geometry just represents the lowest-order approximation to this self-similar world. Use an iterated function system to construct a leaf of a tree which disguises its self-similarity. Where could such a self-similar world come from? Take the point of view of a physicist.

4.8   Calculate the topological entropy of a three-scale Cantor set with forbidden sequence ... CCC ... . Different approaches are possible.

4.9   Take advantage of one-dimensional Poincaré cross sections obtained for the Lorenz system, in order to investigate its grammatical rules at the usual parameter values. What is the topological entropy of the system considered? And that of the Rössler system?

4.10 What is the order of the phase transitions that are described in the text of the present chapter? Can you think of transitions other than the ones met generically?

4.11 Consider a stable manifold of saddle type in a two-dimensional space separating the basins of two attracting sets. Along this separatrix, an effect similar to the sensitivity to initial conditions can often be observed. Which numerical procedure for calculating Lyapunov exponents is not suitable in this case?

4.12 Derive rough estimates of the number of points necessary to calculate fractal dimensions and Lyapunov exponents from numerical data.

4.13 Program an algorithm to plot the rotation number for a period-doubling way to chaos. (See [4.106].)

References

Chapter 1

1.1   J. Peinke, J. Parisi, B. Röhricht, O.E. Rössler, W. Metzler: Z. Naturforsch. **43a**, 287 (1988)
1.2   Anaxagoras: On Nature (in Greek), Fragment 12, reprinted and translated in G.S. Kirk, J.E. Raven: *The Presocratic Philosophers. A Critical History with a Selection of Texts* (Cambridge University Press, London 1957) p. 372
1.3   O.E. Rössler: Z. Naturforsch. **38a**, 788 (1983)
1.4   S. Smale: Bull. Am. Math. Soc. **73**, 747 (1967)
1.5   O.E. Rössler: Phys. Lett. **57A**, 397 (1976)
1.6   M.W. Hirsch, S. Smale: *Differential Equations, Dynamical Systems, and Linear Algebra* (Academic, New York 1974)
1.7   O.E. Rössler: In *Structural Stability in Physics*, ed. by W. Güttinger, H. Eikemeier, Springer Ser. Syn., Vol. 4 (Springer, Berlin, Heidelberg 1979) p. 290
1.8   M. Hénon: Commun. Math. Phys. **50**, 69 (1976)
1.9   I. Gumowski, C. Mira: *Dynamique Chaotique, Transformations Ponctuelles, Transition Ordre-Désordre* (Cepadues, Toulouse 1980)
1.10 O.E. Rössler: Ann. N.Y. Acad. Sci. **316**, 379 (1979); in *Iteration Theory and Functional Equations*, Lect. Notes Math., Vol. 1163, ed. by R. Liedl, L. Reich, G. Targonski (Springer, Berlin, Heidelberg 1985) p. 149
1.11 O.E. Rössler: Z. Naturforsch. **31a**, 1664 (1976)

Chapter 2

2.1   I.L. Ivanov, S.M. Ryvkin: Sov. Phys. – Tech. Phys. **3**, 722 (1958)
2.2   R.D. Larrabee, M.C. Steele: J. Appl. Phys. **31**, 1519 (1960)
2.3   Proc. 5th Int'l Conf. on Semiconductor Physics, Prague, 1960 (Academic, New York 1961); Proc. 7th Int'l Conf. on the Physics of Semiconductors, Paris, 1964 (Academic, New York 1965); Proc. 9th Int'l Conf. on the Physics of Semiconductors, Moscow, 1968 (Nauka, Leningrad 1968)
2.4   V.L. Bonch-Bruevich, I.P. Zvyagin, A.G. Mironov: *Domain Electrical Instabilities in Semiconductors* (Consultants Bureau, New York 1975)
2.5   J. Pozhela: *Plasma and Current Instabilities in Semiconductors* (Pergamon, Oxford 1981)
2.6   E. Schöll: *Nonequilibrium Phase Transitions in Semiconductors*, Springer Ser. Syn., Vol. 35 (Springer, Berlin, Heidelberg 1987)
2.7   B.K. Ridley: Proc. Phys. Soc. **82**, 954 (1963)
2.8   J. Peinke, D.B. Schmid, B. Röhricht, J. Parisi: Z. Phys. B **66**, 65 (1987)
2.9   K.M. Mayer, J. Peinke, B. Röhricht, J. Parisi, R.P. Huebener: Phys. Scr. **T19**, 505 (1987)
2.10 M.A. Lampert, R.B. Schilling: In *Semiconductors and Semimetals*, Vol. 6, ed. by R.K. Willardson, A.C. Beer (Academic, New York 1970) p. 1
2.11 A.A. Kastalsky: Phys. Status Solidi (a) **15**, 599 (1973)
2.12 M. Glicksman: In *Solid State Physics*, Vol. 26, ed. by H. Ehrenreich, F. Seitz, D. Turnbull (Academic, New York 1971) p. 275

2.13 K. Yamada, N. Takara, H. Imada, N. Miura, C. Hamaguchi: Solid-State Electron. **31**, 809 (1988);
      K. Yamada, N. Miura, C. Hamaguchi: Appl. Phys. A **48**, 149 (1989)
2.14 D.G. Seiler, C.L. Littler, R.J. Justice, P.W. Milonni: Phys. Lett. **108A**, 462 (1985);
      X.N. Song, D.G. Seiler, M.R. Loloee: Appl. Phys. A **48**, 137 (1989)
2.15 K. Aoki, T. Kobayashi, K. Yamamoto: J. de Phys. **42**, C7–51 (1981); J. Phys. Soc. Jpn. **51**, 2373
      (1982);
      K. Aoki, K. Yamamoto: Phys. Lett. **98A**, 72 (1983);
      K. Aoki, K. Miyamas, T. Kobayashi, K. Yamamoto: Physica **117/118 B + C**, 570 (1983);
      K. Aoki, O. Ikezawa, N. Mugibayashi, K. Yamamoto: Physica **134B**, 288 (1985);
      K. Aoki, K. Yamamoto, N. Mugibayashi: J. Phys. Soc. Jpn. **57**, 26 (1988);
      K. Aoki, K. Yamamoto: Appl. Phys. A **48**, 111 (1989)
2.16 A. Brandl, T. Geisel, W. Prettl: Europhys. Lett. **3**, 401 (1987);
      J. Spangler, A. Brandl, W. Prettl: Appl. Phys. A **48**, 143 (1989);
      U. Frank, A. Brandl, W. Prettl: Solid State Commun. **69**, 891 (1989)
2.17 S.W. Teitsworth, R.M. Westervelt, E.E. Haller: Phys. Rev. Lett. **51**, 825 (1983);
      S.W. Teitsworth, R.M. Westervelt: Phys. Rev. Lett. **56**, 516 (1986);
      E.G. Gwinn, R.M. Westervelt: Phys. Rev. Lett. **57**, 1060 (1986); ibid. **59**, 157, 247 (1987);
      S.W. Teitsworth: Appl. Phys. A **48**, 127 (1989)
2.18 J. Peinke, A. Mühlbach, R.P. Huebener, J. Parisi: Phys. Lett. **108A**, 407 (1985);
      J. Peinke, B. Röhricht, A. Mühlbach, J. Parisi, Ch. Nöldeke, R.P. Huebener, O.E. Rössler:
      Z. Naturforsch. **40a**, 562 (1985);
      J. Peinke, J. Parisi, B. Röhricht, B. Wessely, K.M. Mayer: Z. Naturforsch. **42a**, 841 (1987);
      U. Rau, J. Peinke, J. Parisi, R.P. Huebener, E. Schöll: Phys. Lett. **124A**, 335 (1987);
      R. Stoop, J. Peinke, J. Parisi, B. Röhricht, R.P. Huebener: Physica **35D**, 425 (1989);
      J. Parisi, J. Peinke, R.P. Huebener, R. Stoop, M. Duong-van: Z. Naturforsch. **44a**, 1046 (1989);
      J. Peinke, J. Parisi, R.P. Huebener, M. Duong-van, P. Keller: Europhys. Lett. **12**, 13 (1990)
2.19 J. Spinnewyn, H. Strauven, O.B. Verbeke: Z. Phys. B **75**, 159 (1989)
2.20 K.A. Pyragas, J. Pozhela, A.V. Tamasevicius, J.K. Ulbikas: Sov. Phys. – Semicond. **21**, 335
      (1987);
      J. Pozhela, A.V. Tamasevicius, J.K. Ulbikas: Solid-State Electron. **31**, 805 (1988);
      J. Pozhela, A. Namajunas, A.V. Tamasevicius, J.K. Ulbikas: Appl. Phys. A **48**, 181 (1989)
2.21 G.N. Maracas, W. Porod, D.A. Johnson, D.K. Ferry, H. Goronkin: Physica **134B**, 276 (1985);
      G.N. Maracas, D.A. Johnson, H. Goronkin: Appl. Phys. Lett. **46**, 305 (1985)
2.22 W. Knap, M. Jezewski, J. Lusakowski, W. Kuszko: Solid-State Electron. **31**, 813 (1988)
2.23 G.A. Held, C. Jeffries, E.E. Haller: Phys. Rev. Lett. **52**, 1037 (1984);
      G.A. Held, C. Jeffries: Phys. Rev. Lett. **55**, 887 (1985); ibid. **56**, 1183 (1986)
2.24 S.B. Bumeliene, J. Pozhela, K.A. Pyragas, A.V. Tamasevicius: Physica **134B**, 293 (1985);
      S.B. Bumeliene, J. Pozhela, A.V. Tamasevicius: Phys. Status Solidi (b) **134**, K71 (1986);
      J. Pozhela, Z.N. Tamaseviciene, A.V. Tamasevicius, J.K. Ulbikas, G.V. ·Bandurkina: Phys.
      Status Solidi (a) **110**, 555 (1988)
2.25 P. Cvitanovic (ed.): *Universality in Chaos* (Adam Hilger, Bristol 1984)
2.26 O.E. Rössler: Phys. Lett. **71A**, 155 (1979); Lect. Appl. Math. **17**, 141 (1979); Z. Naturforsch. **38a**,
      788 (1983)
2.27 M. Duong-van (ed.): Proc. Int'l Conf. on the Physics of Chaos and Systems far from Equilib-
      rium, Monterey, 1987 (North-Holland, Amsterdam 1987); special issue Nucl. Phys. B (Proc.
      Suppl.) Vol. 2 (1987)
2.28 A.R. Bishop, G. Gruener, B. Niclaenko (eds.): Proc. Workshop on Spatio-Temporal Coherence
      and Chaos in Physical Systems, Los Alamos, 1986 (North-Holland, Amsterdam 1986); special
      issue Physica **23D** (1986)
2.29 R.P. Huebener, K.M. Mayer, J. Parisi, J. Peinke, B. Röhricht: Nucl. Phys. B (Proc. Suppl.)
      2, 3 (1987);
      J. Peinke, J. Parisi, B. Röhricht, K.M. Mayer, U. Rau, R.P. Huebener: Solid-State Electron. **31**,
      817 (1988)
2.30 K.M. Mayer, J. Parisi, J. Peinke, R.P. Huebener: Physica **32D**, 306 (1988)
2.31 V.L. Bonch-Bruevich: Phys. Scr. **T19**, 491 (1987)

2.32 S.M. Sze: *Physics of Semiconductor Devices* (Wiley Eastern, New Delhi 1979)
2.33 K. Seeger: *Semiconductor Physics*, 5th edn. Springer Ser. Solid-State Sci., Vol. 40 (Springer, Berlin, Heidelberg 1991)
2.34 R.L. Jones, P. Fisher: J. Phys. Chem. Solids **26**, 1125 (1965);
     E.E. Haller, W.L. Hansen: Solid State Commun. **15**, 687 (1974);
     L. S. Darken: J. Appl. Phys. **53**, 3754 (1982)
2.35 R.P. Huebener: In *Advances in Electronics and Electron Physics*, Vol. 70, ed. by P.W. Hawkes (Academic, New York 1988) p. 1
2.36 J. Peinke, J. Parisi, B. Röhricht, K.M. Mayer, U. Rau, W. Clauß, R.P. Huebener, G. Jungwirt, W. Prettl: Appl. Phys. A **48**, 155 (1989)
2.37 J. Parisi, U. Rau, J. Peinke, K.M. Mayer: Z. Phys. B **72**, 225 (1988);
     U. Rau, J. Parisi, K.M. Mayer, W. Clauß, J. Peinke, B. Röhricht, R. P. Huebener: In Proc. 3rd Int'l Conf. on Shallow Impurities in Semiconductors, Linköping, 1988, ed. by B. Monemar (Institute of Physics, Bristol 1989) p. 167
2.38 B. Röhricht, J. Parisi, J. Peinke, R.P. Huebener: Z. Phys. B **66**, 515 (1987);
     E. Schöll, J. Parisi, B. Röhricht, J. Peinke, R.P. Huebener: Phys. Lett. **119A**, 419 (1987);
     E. Schöll, H. Naber, J. Parisi, B. Röhricht, J. Peinke, S. Uba: Z. Naturforsch. **44a**, 1139 (1989)
2.39 W. Metzger, R.P. Huebener, K.-P. Selig: Cryogenics **22**, 387 (1982)
2.40 W. Mönch: In *Festkörperprobleme* (*Advances in Solid State Physics*), Vol. 9, ed. by O. Madelung (Vieweg, Braunschweig 1969) p. 172
2.41 H. Haken: *Advanced Synergetics*, Springer Ser. Syn., Vol. 20 (Springer, Berlin, Heidelberg 1983)
2.42 H. Haken: *Information and Self-Organization*, Springer Ser. Syn., Vol. 40 (Springer, Berlin, Heidelberg 1988)
2.43 E. Schöll: Z. Phys. B **46**, 23 (1982)
2.44 W. Clauß, J. Peinke, J. Parisi, U. Rau, A. Kittel, R.P. Huebener: Solid-State Electron. **32**, 1197 (1989);
     W. Clauß, U. Rau, J. Parisi, J. Peinke, R.P. Huebener, H. Leier, A. Forchel: J. Appl. Phys. **67**, 2980 (1990)
2.45 J. Peinke, J. Parisi, A. Mühlbach, R.P. Huebener: Z. Naturforsch. **42a**, 441 (1987)
2.46 K. Aoki, U. Rau, J. Peinke, J. Parisi, R.P. Huebener: J. Phys. Soc. Jpn. **59**, 420 (1990);
     U. Rau, K. Aoki, J. Peinke, J. Parisi, W. Clauß, R.P. Huebener: Z. Phys. B **81**, 53 (1990)
2.47 U. Rau, W. Clauß, A. Kittel, M. Lehr, M. Bayerbach, J. Parisi, J. Peinke, R.P. Huebener: Phys. Rev. B **43**, 2255 (1991);
     A. Kittel, U. Rau, J. Peinke, W. Clauß, M. Bayerbach, J. Parisi: Phys. Lett. **147A**, 229 (1990)
2.48 D. Armbruster, G. Dangelmayr: Phys. Lett. **138A**, 46 (1989)
2.49 E. Schöll: Physica **134B**, 271 (1985); Phys. Rev. **B34**, 1395 (1986); J. Phys. Chem. Solids **49**, 651 (1988); Phys. Scr. **T29**, 152 (1989); Solid-State Electron. **32**, 1129 (1989)
2.50 J. Peinke, J. Parisi, U. Rau, W. Clauß, M. Weise: Z. Naturforsch. **44a**, 629 (1989)
2.51 B. Röhricht, J. Parisi, J. Peinke, O.E. Rössler: Z. Phys. B **65**, 259 (1986);
     J. Parisi, J. Peinke, B. Röhricht, U. Rau, M. Klein, O.E. Rössler: Z. Naturforsch. **42a**, 655 (1987);
     B. Röhricht, J. Parisi, J. Peinke, O.E. Rössler: Dynamics Stability Syst. **5**, 99 (1990)
2.52 K.M. Mayer, R. Gross, J. Parisi, J. Peinke, R.P. Huebener: Solid State Commun. **63**, 55 (1987);
     U. Rau, J. Peinke, J. Parisi, R.P. Huebener: Z. Phys. B **71**, 305 (1988);
     U. Rau, K.M. Mayer, J. Parisi, J. Peinke, W. Clauß, R.P. Huebener: Solid-State Electron. **32**, 1365 (1989)
2.53 K.M. Mayer, R.P. Huebener, U. Rau: J. Appl. Phys. **67**, 1412 (1990)

Chapter 3

3.1 R.H. Abraham, Ch.D. Shaw: *Dynamics, the Geometry of Behavior* (Aerial, Santa Cruz 1983)
3.2 J. Guckenheimer, P. Holmes: *Nonlinear Oscillations, Dynamical Systems, and Bifurcations of Vector Fields*, Appl. Math. Sci., Vol. 42 (Springer, Berlin, Heidelberg 1983)

3.3   H. Haken: *Advanced Synergetics*, Springer Ser. Syn., Vol. 20 (Springer, Berlin, Heidelberg 1983)

3.4   H.G. Schuster: *Deterministic Chaos* (Physik, Weinheim 1984)

3.5   P. Cvitanovic (ed.): *Universality in Chaos* (Adam Hilger, Bristol 1984)

3.6   B.-L. Hao (ed.): *Chaos* (World Scientific, Singapore 1984); *Directions in Chaos*, Vols. 1, 2 (World Scientific, Singapore 1987, 1988)

3.7   A.V. Holden (ed.): *Chaos* (Manchester University Press, Manchester 1986)

3.8   R.L. Devaney: *An Introduction to Chaotic Dynamical Systems* (Benjamin/Cummings, Menlo Park, CA 1986)

3.9   J.M.T. Thompson, H.B. Stewart: *Nonlinear Dynamics and Chaos* (Wiley, Chichester, UK 1986)

3.10 P. Berge, Y. Pomeau, Ch. Vidal: *Order within Chaos* (Hermann, Paris 1986)

3.11 R.W. Leven, B.-P. Koch, B. Pompe: *Chaos in dissipativen Systemen* (Akademie, Berlin 1989)

3.12 E.N. Lorenz: J. Atmos. Sci. **20**, 130 (1963)

3.13 J.-P. Eckmann: Rev. Mod. Phys. **53**, 643 (1981)

3.14 J.D. Farmer, E. Ott, J.A. Yorke: Physica **7D**, 153 (1983)

3.15 N.H. Packard, J.P. Crutchfield, J.D. Farmer, R.S. Shaw: Phys. Rev. Lett. **45**, 712 (1980);
      F. Takens: In *Dynamical Systems and Turbulence*, *Lect. Notes Math.*, Vol. 898, ed. by D.A. Rand, L.-S. Young (Springer, Berlin, Heidelberg 1981) p. 366

3.16 A.A. Andronov, L. Pontryagin: Dokl. Akad. Nauk. SSSR **14**, 247 (1937)

3.17 C. Jacobi: Poggend. Annal. **33**, 229 (1834)

3.18 H. Poincaré: Acta Math. **7**, 259 (1885)

3.19 R. Thom: *Structural Stability and Morphogenesis* (Benjamin, New York 1975)

3.20 J. Carr: *Applications of Centre Manifold Theory*, Appl. Math. Sci., Vol. 35 (Springer, Berlin, Heidelberg 1981)

3.21 E.C. Zeeman: *Catastrophe Theory* (Addison-Wesley, Reading, MA 1977)

3.22 I. Stewart: Rep. Prog. Phys. **45**, 185 (1982)

3.23 P.T. Saunders: *Catastrophe Theory* (Vieweg, Braunschweig 1986)

3.24 R.P. Khosla, J.R. Fischer, B.C. Burkey: Phys. Rev. B **7**, 2551 (1973)

3.25 E. Schöll: Z. Phys. B **46**, 23 (1982)

3.26 M. Lehr, R.P. Huebener, U. Rau, J. Parisi, W. Clauß, J. Peinke, B. Röhricht: Phys. Rev. B **42**, 9019 (1990)

3.27 E. Hopf: Ber. Math.-Phys. Kgl. Sächs. Akad. Wiss. Leipzig **94**, 1 (1942)

3.28 H.B. Stewart: Z. Naturforsch. **41a**, 1412 (1986)

3.29 E.C. Zeeman: In *Papers in Algebra, Analysis, and Statistics*, ed. by R. Lidl (American Mathematical Society, Providence, RI 1982)

3.30 J. Peinke, U. Rau, W. Clauß, R. Richter, J. Parisi: Europhys. Lett. **9**, 743 (1989)

3.31 F.T. Arecchi, A. Lapucci, R. Meucci, J.A. Roversi, P.H. Coullet: Europhys. Lett. **6**, 677 (1988)

3.32 M.M. Peixoto: Topology **1**, 101 (1962)

3.33 P. Bak, T. Bohr, M.H. Jensen: In *Directions in Chaos*, Vol. 2, ed. by B.-L. Hao (World Scientific, Singapore 1988) p. 16

3.34 M.H. Jensen, P. Bak, T. Bohr: Phys. Rev. A **30**, 1960 (1984)

3.35 T. Bohr, P. Bak, M.H. Jensen: Phys. Rev. A **30**, 1970 (1984)

3.36 K. Kaneko: *Collapse of Tori and Genesis of Chaos in Dissipative Systems* (World Scientific, Singapore 1986)

3.37 V.I. Arnol'd: *Geometric Methods in the Theory of Ordinary Differential Equations*, 2nd edn., Grundlehren math. Wissenschaften, Vol. 250 (Springer, Berlin, Heidelberg 1988)

3.38 H. Haucke, R. Ecke: Physica **25D**, 307 (1987);
      M. Dubois: Nucl. Phys. B (Proc. Suppl.) **2**, 339 (1987)

3.39 J. Parisi, J. Peinke, B. Röhricht, K.M. Mayer: Z. Naturforsch. **42a**, 329 (1987);
      J. Peinke, J. Parisi, B. Röhricht, B. Wessely, K.M. Mayer: Z. Naturforsch. **42a**, 841 (1987)

3.40 U. Rau, J. Peinke, J. Parisi, R.P. Huebener, E. Schöll: Phys. Lett. **124A**, 335 (1987)

3.41 E.N. Lorenz: J. Atmos. Sci. **20**, 130, 448 (1963); Tellus **16**, 1 (1964)

3.42 T.Y. Li, J.A. Yorke: Am. Math. Mon. **82**, 985 (1975)

3.43 O.E. Rössler: Phys. Lett. **71A**, 155 (1979); Lect. Appl. Math. **17**, 141 (1979); Z. Naturforsch. **38a**, 788 (1983)

3.44 O.E. Rössler: Phys. Lett. **57A**, 397 (1976)

3.45 R.S. Shaw: Z. Naturforsch. **36a**, 80 (1981)

3.46 F. Hausdorff: Math. Ann. **79**, 157 (1919)

3.47 M.J. Feigenbaum: J. Stat. Phys. **19**, 25 (1978)

3.48 B.B. Mandelbrot: *The Fractal Geometry of Nature* (Freeman, San Francisco 1982)

3.49 E. Ott, E.D. Yorke, J.A. Yorke: Physica **16D**, 62 (1985)

3.50 P. Grassberger, I. Procaccia: Physica **9D**, 193 (1983)

3.51 A. Renyi: *Probability Theory* (North-Holland, Amsterdam 1970)

3.52 P. Grassberger, I. Procaccia: Physica **13D**, 34 (1984)

3.53 P. Grassberger: Phys. Lett. **107A**, 101 (1985)

3.54 R. Badii, A. Politi: J. Stat. Phys. **40**, 725 (1985)

3.55 R. Stoop, J. Peinke, J. Parisi, B. Röhricht, R.P. Huebener: Physica **35D**, 425 (1989)

3.56 G. Mayer-Kress (ed.): *Dimensions and Entropies in Chaotic Systems*, Springer Ser. Syn., Vol. 32 (Springer, Berlin, Heidelberg 1986)

3.57 J.-P. Eckmann, D. Ruelle: Rev. Mod. Phys. **57**, 617 (1985)

3.58 T.C. Halsey, M.H. Jensen, L.P. Kadanoff, I. Procaccia, B.I. Shraiman: Phys. Rev. A33, 1141 (1986)

3.59 J.L. Kaplan, J.A. Yorke: In *Functional Differential Equations and Approximation of Fixed Points*, Lect. Notes Math., Vol. 730, ed. by H.O. Peitgen, H.-C. Walther (Springer, Berlin, Heidelberg 1979) p. 204;
P. Frederickson, J.L. Kaplan, E.D. Yorke, J.A. Yorke: J. Differ. Equ. **49**, 185 (1983)

3.60 P. Grassberger: In *Chaos*, ed. by A.V. Holden (Manchester University Press, Manchester 1986) p. 291

3.61 K. Pawelzik, H.G. Schuster: Phys. Rev. A **35**, 2207 (1987)

3.62 J.-P. Eckmann, I. Procaccia: Phys. Rev. A **34**, 659 (1986)

3.63 M. Sàno, S. Sato, Y. Sawada: Prog. Theor. Phys. **76**, 945 (1986)

3.64 S. Grossmann, S. Thomae: Z. Naturforsch. **32a**, 1353 (1977)

3.65 Y. Pomeau, P. Manneville: Commun. Math. Phys. **77**, 189 (1980)

3.66 D. Ruelle, F. Takens: Commun. Math. Phys. **20**, 167 (1971)

3.67 S. Newhouse, D. Ruelle, F. Takens: Commun. Math. Phys. **64**, 35 (1978)

3.68 C. Grebogi, E. Ott, J.A. Yorke: Phys. Rev. Lett. **51**, 339 (1983)

3.69 S.W. Teitsworth, R.M. Westervelt, E.E. Haller: Phys. Rev. Lett. **51**, 825 (1983)

3.70 J. Peinke, A. Mühlbach, R.P. Huebener, J. Parisi: Phys. Lett. **108A**, 407 (1985)

3.71 R. Richter: Diploma Thesis, Tübingen (1989)

3.72 J.Y. Huang, J.J. Kim: Phys. Rev. A **36**, 1495 (1987)

3.73 J. Sacher, W. Elsässer, E.O. Göbel: Phys. Rev. Lett. **63**, 2224 (1989)

3.74 M. Dubois, M.A. Rubio, P. Berge: Phys. Rev. Lett. **51**, 1446 (1983)

3.75 G. Baier, K. Wegmann, J.L. Hudson: Phys. Lett. **141A**, 340 (1989)

3.76 R. Richter, J. Peinke, W. Clauß, U. Rau, J. Parisi: Europhys. Lett. **14**, 1 (1991)

3.77 J.E. Hirsch. B.A. Huberman, D.J. Scalapino: Phys. Rev. A **25**, 519 (1982)

3.78 P. Manneville: J. de Phys. **41**, 1235 (1980)

3.79 C. Grebogi, E. Ott, J.A. Yorke: Physica **7D**, 181 (1983)

3.80 M.J. Feigenbaum, L.P. Kadanoff, S.J. Shenker: Physica **5D**, 370 (1982)

3.81 S.J. Shenker: Physica **5D**, 405 (1982)

3.82 A.P. Fein, M.S. Heutmaker, J.P. Gollub: Phys. Scr. **T9**, 79 (1985)

3.83 M.H. Jensen, L.P. Kadanoff, A. Libchaber, I. Procaccia, J. Stavans: Phys. Rev. Lett. **55**, 2798 (1985)

3.84 G.A. Held, C. Jeffries: Phys. Rev. Lett. **56**, 1183 (1986)

3.85 E.G. Gwinn, R.M. Westervelt: Phys. Rev. Lett. **57**, 1060 (1986); ibid. **59**, 157, 247 (1987)

3.86 Y. Kim: Phys. Rev. A **39**, 4801 (1989)

3.87 J. Peinke, J. Parisi, R.P. Huebener, M. Duong-van, P. Keller: Europhys. Lett. **12**, 13 (1990)

3.88 A.R. Bishop, G. Gruener, B. Niclaenko (eds.): Proc. Workshop on Spatio-Temporal Coherence and Chaos in Physical Systems, Los Alamos, 1986 (North-Holland, Amsterdam 1986); special issue Physica **23D** (1986)

3.89 P. Keller, M. Duong-van: Nucl. Phys. B (Proc. Suppl.) **2**, 603 (1987)
3.90 J.P. Crutchfield, K. Kaneko: In *Directions in Chaos*, Vol. 1, ed. by B.-L. Hao (World Scientific, Singapore 1987) p. 271
3.91 K. Kaneko: Physica **34D**, 1 (1989)
3.92 A. Brandl, W. Kröninger, W. Prettl, G. Obermair: Phys. Rev. Lett. **64**, 212 (1990)

Chapter 4

4.1   P. Cvitanovič (ed.): *Universality in Chaos* (Adam Hilger, Bristol 1985);
      B.-L. Hao (ed.): *Chaos* (World Scientific, Singapore 1984)
4.2   B.B. Mandelbrot: *The Fractal Geometry of Nature* (Freeman, New York 1982)
4.3   J.-P. Eckmann, D. Ruelle: Rev. Mod. Phys. **57**, 617 (1985);
      for related topics, see [4.47–49] and
      O.E. Lanford: *Statistical Mechanics*, C.I.M.E. Lectures (Birkhäuser, Boston 1976)
4.4   P. Boesiger, E. Brun, D. Meier: Phys. Rev. Lett. **38**, 602 (1977);
      E. Brun, B. Derighetti, M. Ravani, G. Broggi, P.F. Meier, R. Stoop: Phys. Scr. **T13**, 119 (1986)
4.5   J. Peinke, A. Mühlbach, B. Röhricht, B. Wessely, J. Mannhart, J. Parisi, R. P. Huebener: Physica **23D**, 176 (1986);
      K.M. Mayer, J. Parisi, J. Peinke, R.P. Huebener: Physica **32D**, 306 (1988)
4.6   C. Sparrow: *The Lorenz Equations*, Appl. Math. Sci., Vol. 41 (Springer, Berlin, Heidelberg 1982)
4.7   R.L. Devaney: *An Introduction to Chaotic Dynamical Systems* (Benjamin, Menlo Park, CA 1986)
4.8   M. Hénon: Commun. Math. Phys. **50**, 69 (1976)
4.9   P. Billingsley: *Ergodic Theory and Information* (Wiley, New York 1965);
      V.I. Arnold, A. Avez: *Ergodic Problems of Classical Mechanics* (Benjamin, New York 1968);
      P. Walters: *Ergodic Theory, Introductory Lectures*, Lect. Notes Math., Vol. 458 (Springer, Berlin, Heidelberg 1975);
      M. Denker, C. Grillenberger, K. Sigmund: *Ergodic Theory on Compact Spaces*, Lect. Notes Math., Vol. 527 (Springer, Berlin, Heidelberg 1976);
      M. Reed, B. Simon: *Functional Analysis* (Academic, London 1972)
4.10 Z. Nitecki: *Differentiable Dynamics* (MIT Press, Cambridge, MA 1971)
4.11 J.D. Farmer, E. Ott, J.A. Yorke: Physica **7D**, 153 (1983)
4.12 A. Katok, J.-M. Strelcyn: *Invariant Manifolds, Entropy and Billards; Smooth Maps with Singularities*, Lect. Notes Math., Vol. 1222 (Springer, Berlin, Heidelberg 1986)
4.13 T. Bohr, T. Tél: In *Directions in Chaos*, Vol. 2, ed. by B.-L. Hao (World Scientific, Singapore 1988)
4.14 S. Newhouse: *Dynamical Systems*, C.I.M.E. Lectures on Dynamical Systems (Birkhäuser, Boston 1980)
4.15 J. Guckenheimer, P. Holmes: *Nonlinear Oscillations, Dynamical Systems, and Bifurcations of Vector Fields*, Appl. Math. Sci., Vol. 42 (Springer, Berlin, Heidelberg 1986)
4.16 R. Lozi: J. de Phys. **39**, 9 (1978)
4.17 H. Fujisaka: Prog. Theor. Phys. **70**, 1264 (1983)
4.18 P. Grassberger, I. Procaccia: Physica **13D**, 34 (1984)
4.19 H.E. Stanley: *Introduction to Phase Transitions and Critical Phenomena*, 2nd edn. (Oxford University Press, New York 1987)
4.20 H.E. Stanley: In *Statistical Mechanics and Field Theory*, ed. by R.N. Sen, C. Weil (Halsted, New York 1972)
4.21 P. Collet, J.-P. Eckmann: *Iterated Maps on the Interval as Dynamical Systems* (Birkhäuser, Boston 1980)

4.22 M.J. Feigenbaum: J. Stat. Phys. **21**, 669 (1979)
4.23 P. Grassberger: In *Complex Systems*, 12th Gwatt workshop, Internal Report (1988)
4.24 P. Cvitanovič, G. Gunaratne, I. Procaccia: Phys. Rev. A **38**, 1503 (1988);
      G. Gunaratne, M.H. Jensen, I. Procaccia: Nonlinearity **1**, 157 (1988)
4.25 P. Grassberger, H. Kantz: Phys. Lett. **113A**, 235 (1985)
4.26 A. Cohen, I. Procaccia: Phys. Rev. A **31**, 1872 (1985)
4.27 M.J. Feigenbaum: Commun. Math. Phys. **77**, 65 (1980)
4.28 A.B. Pippard: *The Elements of Classical Thermodynamics* (Cambridge University Press, Cambridge 1957);
      D. Ruelle: *Statistical Mechanics* (Benjamin, Reading, MA 1969);
      A. Münster: *Statistical Thermodynamics*, Vol. 1 (Springer, Berlin, Heidelberg 1969)
4.29 T.C. Halsey, M.H. Jensen, L.P. Kadanoff, I. Procaccia, B. Shraiman: Phys. Rev. A **33**, 1141 (1986)
4.30 M. Kohmoto: Phys. Rev. A **37**, 1345 (1988)
4.31 Y. Oono, Y. Takahashi: Prog. Theor. Phys. **63**, 1804 (1980)
4.32 V.N. Shtern: Dokl. Akad. Nauk. SSSR **270**, 582 (1983)
4.33 F. Hausdorff: Math. Ann. **79**, 157 (1919)
4.34 H.G. Hentschel, I. Procaccia: Physica **8D**, 435 (1983)
4.35 L.P. Kadanoff, C. Tang: Proc. Natl. Acad. Sci. USA **81**, 1276 (1984)
4.36 T. Yoshida, B. Chol So: Prog. Theor. Phys. **79**, 1 (1988)
4.37 A. Renyi: *Probability Theory* (North-Holland, Amsterdam 1970)
4.38 P. Grassberger: Phys. Lett. **97A**, 227 (1983)
4.39 P. Grassberger: Phys. Lett. **107A**, 101 (1985)
4.40 R. Badii, A. Politi: Phys. Rev. Lett. **52**, 1661 (1984)
4.41 G. Parisi: Appendix to "Fully developed turbulence and intermittency", by U. Frisch. In Proc. Int'l School on Turbulence and Predictability in Geophysical Fluid Dynamics and Climate Dynamics, ed. by M. Ghil (North-Holland, Amsterdam 1984);
      U. Frisch: Phys. Scr. **T9**, 137 (1985)
4.42 M. Sano, S. Sato, Y. Sawada: Prog. Theor. Phys. **76**, 945 (1986)
4.43 J.-P. Eckmann, I. Procaccia: Phys. Rev. A **34**, 659 (1986)
4.44 R. Stoop, J. Peinke, J. Parisi, B. Röhricht, R.P. Huebener: Physica **35D**, 425 (1989)
4.45 T. Horita, H. Hata, H. Mori, T. Morita, K. Tomita: Prog. Theor. Phys. **80**, 923 (1988)
4.46 H.G. Schuster: *Deterministic Chaos* (Physik, Weinheim 1984)
4.47 Ya. G. Sinai: Russ. Math. Surv. **27**, 21 (1972)
4.48 R. Bowen: *Equilibrium States and the Ergodic Theory of Anosov Diffeomorphisms*, Lect. Notes Math., Vol. 470 (Springer, Berlin, Heidelberg 1975)
4.49 D. Ruelle: *Thermodynamic Formalism*. Encyclopedia of Mathematics and its Applications, Vol. 5 (Addison-Wesley, Reading, MA 1978)
4.50 D. Katzen, I. Procaccia: Phys. Rev. Lett. **58**, 1169 (1987)
4.51 M.J. Feigenbaum: J. Stat. Phys. **46**, 919, 925 (1987)
4.52 M.H. Jensen, L.P. Kadanoff, I. Procaccia: Phys. Rev. A **36**, 1409 (1987)
4.53 R.S. Ellis: *Entropy, Large Deviations, and Statistical Mechanics*, Grundlehren math. Wissenschaften, Vol. 271 (Springer, Berlin, Heidelberg 1985)
4.54 P. Grassberger: In Proc. Conf. on Chaos in Astrophysics, Palm Coast, FL, 1984, ed. by J. Perdang et al. (Reidl, Dordrecht 1985);
      P. Grassberger: In *Chaos*, ed. by A.V. Holden (Manchester University Press, Manchester 1986)
4.55 R. Badii, A. Politi: Phys. Rev. A **35**, 1288 (1987)
4.56 J.L. Kaplan, J.A. Yorke: In *Functional Differential Equations and Approximation of Fixed Points*, Lect. Notes Math., Vol. 730, ed. by H.O. Peitgen, H.-C. Walther (Springer, Berlin, Heidelberg 1979) p. 204;
4.57 P. Fredrickson, J.L. Kaplan, E.D. Yorke, J.A. Yorke: J. Differ. Equ. **49**, 185 (1983)
4.58 L.S. Young: J. Erg. Theor. Dyn. Syst. **2**, 109 (1982)

4.59 H. Hata, T. Morita, K. Tomita, H. Mori: Prog. Theor. Phys. **78**, 721 (1987)

4.60 R. Badii: Ph.D. Thesis, Zurich (1987)

4.61 P. Cvitanovic: In Proc. Workshop in Condensed Matter, Atomic and Molecular Physics, Trieste, 1986, unpublished

4.62 T. Bohr, D. Rand: Physica **25D**, 387 (1987)

4.63 P. Grassberger: In Proc. Workshop on Dynamics of Fractals and Hierarchies of Critical Exponents, Orsay, 1986, unpublished

4.64 M. Duong-van: Private communication (1988)

4.65 H. Hata, T. Horita, H. Mori, T. Morita, K. Tomita: Prog. Theor. Phys. **80**, 809 (1988)

4.66 T. Horita, H. Hata, H. Mori, T. Morita, S. Kuroki, H. Okamoto: Prog. Theor. Phys. **80**, 793 (1988)

4.67 D. Auerbach, P. Cvitanovič, J.-P. Eckmann, G. Gunaratne, I. Procaccia: Phys. Rev. Lett. **58**, 2387 (1987)

4.68 P. Grassberger: Internal Report, University of Wuppertal (1988)

4.69 P. Szépfalusy, T. Tél, A. Csordás, Z. Kovačs: Phys. Rev. A **36**, 3525 (1987)

4.70 H. Whitney: Ann. Math. **37**, 645 (1936)

4.71 F. Takens: In *Dynamical Systems and Turbulence*, ed. by D.A. Rand, L.-S. Young, Lect. Notes Math., Vol. 898 (Springer, Berlin, Heidelberg 1981) p. 366

4.72 R. Mañé: In *Dynamical Systems and Turbulence*, ed. by D.A. Rand, L.-S. Young. Lect. Notes Math., Vol. 898 (Springer, Berlin, Heidelberg 1981) p. 230

4.73 J. Dieudonne: *Treatise on Analysis*, Vol. 3 (Academic, New York 1972); S. Sternberg: *Lectures on Differential Geometry* (Prentice Hall, Englewood Cliffs, NJ 1964)

4.74 A.M. Fraser, H.L. Swinney: Phys. Rev. A **33**, 1134 (1986)

4.75 A.M. Fraser: Ph.D. Thesis, Austin (1988)

4.76 R. Stoop, P.F. Meier: Nucl. Phys. B (Proc. Suppl.) **2**, 582 (1987)

4.77 R. Stoop, P.F. Meier: J. Opt. Soc. Am. B **5**, 1037 (1988)

4.78 V.I. Oseledec: Moscow Math. Soc. **19**, 197 (1968)

4.79 G. Benettin, L. Galgani, J.M. Strelcyn: Phys. Rev. A **14**, 2338 (1976)

4.80 J. Froyland, K.H. Alfsen: Internal Report, University of Oslo (1984); J.M. Greene, J.-S. Kim: Physica **24D**, 213 (1987)

4.81 A. Wolf, J.A. Vastano: In *Dimensions and Entropies in Chaotic Systems*, ed. by G. Mayer-Kress, Springer Ser. Syn., Vol. 32 (Springer, Berlin, Heidelberg 1986)

4.82 M. Sano, Y. Sawada: Phys. Rev. Lett. **55**, 1082 (1985)

4.83 J.-P. Eckmann, S. Oliffson Kamphorst, D. Ruelle, S. Ciliberto: Phys. Rev. A **34**, 4971 (1986)

4.84 T.M. Hegland: Private communication (1988)

4.85 G. Broggi, P.F. Meier, R. Stoop, R. Badii: Phys. Rev. A **35**, 365 (1987)

4.86 H. Atmanspacher, H. Scheingraber: Phys. Rev. A **34**, 253 (1985)

4.87 J.-P. Eckmann, D. Ruelle: Internal Report, I.H.E.S Bures-sur-Yvette (1989)

4.88 E.J. Kostelich, H.L. Swinney: In *Chaos and Related Nonlinear Phenomena*, ed. by I. Procaccia, M. Shapiro (Plenum, New York 1987)

4.89 S. Sato, M. Sano, Y. Sawada: Prog. Theor. Phys. **77**, 1 (1987)

4.90 R. Badii, G. Broggi, B. Derighetti, M. Ravani, S. Ciliberto, A. Politi, M.A. Rubio: Phys. Rev. Lett. **60**, 979 (1988)

4.91 R. Stoop, J. Parisi: Physica **50D**, 89 (1991)

4.92 R. Stoop, J. Parisi, J. Peinke: In *A Chaotic Hierarchy*, ed. by G. Baier, M. Klein (World Scientific, Singapore 1991) p. 341

4.93 J. Stoer, R. Bulirsch: *Introduction to Numerical Analysis* (Springer, Berlin, Heidelberg 1980)

4.94 W.H. Press, B.P. Flannery, S.A. Teukolsky, W.T. Vetterling: *Numerical Recipes* (Cambridge University Press, Cambridge 1986)

4.95 J.M. Greene, J.-S. Kim: Physica **24D**, 213 (1987)

4.96 D.S. Broomhead, G.P. King: In *Nonlinear Phenomena and Chaos*, ed. by S. Sarkar (Adam Hilger, Bristol 1986)

4.97 O.E. Rössler: Phys. Lett. **71A**, 155 (1979)

4.98   E. Hopf: *Topologie* (Springer, Berlin, Heidelberg 1935)

4.99   P. Alexandroff: *Combinatorial Topology*, Vol. 3 (Graylock, Rochester 1960)

4.100 H. Amann: *Gewöhnliche Differentialgleichungen* (de Gruyter, Berlin 1983)

4.101 K. Deimling: *Nichtlineare Gleichungen und Abbildungsgrade* (Springer, Berlin, Heidelberg 1976)

4.102 H.G. Solari, R. Gilmore: Private communication (1988)

4.103 V.I. Arnol'd: *Geometrical Methods in the Theory of Ordinary Differential Equations*, 2nd edn., Grundlehren math. Wissenschaften, Vol. 250 (Springer, Berlin, Heidelberg 1988)

4.104 H. Brauchli, H. Weber, R. Stoop: Unpublished

4.105 D.V. Anasov: Tr. Steklov **90**, 1 (1967)

4.106 J. Veerman: Physica A **134**, 543 (1986)

Subject Index